GEOPOETICS IN PRACTICE

This breakthrough book examines dynamic intersections of poetics and geography. Gathering the essays of an international cohort whose work converges at the crossroads of poetics and the material world, *Geopoetics in Practice* offers insights into poetry, place, ecology, and writing the world through a critical-creative geographic lens.

This collection approaches geopoetics as a practice by bringing together contemporary geographers, poets, and artists who contribute their research, methodologies, and creative writing. The 24 chapters, divided into the sections "Documenting," "Reading," and "Intervening," poetically engage discourses about space, power, difference, and landscape, as well as about human, non-human, and more-than-human relationships with Earth. Key explorations of this edited volume include how poets engage with geographical phenomena through poetry and how geographers use creativity to explore space, place, and environment.

This book makes a major contribution to the geohumanities and creative geographies by presenting geopoetics as a practice that compels its agents to take action. It will appeal to academics and students in the fields of creative writing, literature, geography, and the environmental and spatial humanities, as well as to readers from outside of the academy interested in where poetry and place overlap.

Eric Magrane is an assistant professor of geography at New Mexico State University. His work takes multiple forms, from scholarly to literary to artistic. He is co-editor of the hybrid field guide/anthology *The Sonoran Desert: A Literary Field Guide*.

Linda Russo, a clinical associate professor at Washington State University, teaches creative writing and literature and directs EcoArts on the Palouse. Her published works include *Meaning to Go to the Origin in Some Way* and *Participant*, both poetry, and the co-edited *Counter-Desecration: A Glossary for Writing Within the Anthropocene*.

Sarah de Leeuw, a professor with the Northern Medical Program of UBC's Faculty of Medicine, is a poet, critical geographer, and anti-colonial feminist researcher whose multidisciplinary work focuses on marginalized peoples and places. She is the author of multiple journal papers, entries, chapters and books (both creative and academic) and a Canada Research Chair in Humanities and Health Inequities.

Craig Santos Perez is an Indigenous Chamorro poet and scholar from the Pacific Island of Guam. He is the author of four collections of poetry and the co-editor of three anthologies. He is an associate professor in the English department at the University of Hawai'i, Mānoa.

Routledge Research in Culture, Space and Identity

Series editor: Dr. Jon Anderson, School of Planning and Geography, Cardiff University, UK

The *Routledge Research in Culture, Space and Identity Series* offers a forum for original and innovative research within cultural geography and connected fields. Titles within the series are empirically and theoretically informed and explore a range of dynamic and captivating topics. This series provides a forum for cutting edge research and new theoretical perspectives that reflect the wealth of research currently being undertaken. This series is aimed at upper-level undergraduates, research students and academics, appealing to geographers as well as the broader social sciences, arts and humanities.

Affected Labour in a Café Culture
The Atmospheres and Economics of 'Hip' Melbourne
Alexia Cameron

Creative Representations of Place
Alison Barnes

Artistic Approaches to Cultural Mapping
Activating Imaginaries and Means of Knowing
Edited by Nancy Duxbury, W.F. Garrett-Petts, Alys Longley

Geopoetics in Practice
Edited by Eric Magrane, Linda Russo, Sarah de Leeuw, and Craig Santos Perez

For more information about this series, please visit: www.routledge.com/ Routledge-Research-in-Culture-Space-and-Identity/book-series/CSI

GEOPOETICS IN PRACTICE

WITHDRAWN

Edited by
Eric Magrane, Linda Russo,
Sarah de Leeuw, and Craig Santos Perez

Routledge
Taylor & Francis Group

LONDON AND NEW YORK

First published 2020
by Routledge
2 Park Square, Milton Park, Abingdon, Oxon OX14 4RN

and by Routledge
52 Vanderbilt Avenue, New York, NY 10017

Routledge is an imprint of the Taylor & Francis Group, an informa business

British Library Cataloguing-in-Publication Data
A catalogue record for this book is available from the British Library

Library of Congress Cataloging-in-Publication Data
Names: Magrane, Eric, editor.
Title: Geopoetics in practice / [edited by] Eric Magrane, Linda Russo, Sarah de Leeuw, and Craig Santos Perez.
Description: Abingdon, Oxon ; New York, NY : Routledge, 2020. | Series: Routledge research in culture, space and identity | Includes bibliographical references and index.
Identifiers: LCCN 2019041425 (print) | LCCN 2019041426 (ebook) | ISBN 9780367145378 (hardback) | ISBN 9780367145385 (paperback) | ISBN 9780429032202 (ebook)
Subjects: LCSH: Poetics. | Geography in literature. | Material culture in literature. | Human ecology in literature. | Nature in literature.
Classification: LCC PN1042 .G38 2020 (print) | LCC PN1042 (ebook) | DDC 808.1—dc23
LC record available at https://lccn.loc.gov/2019041425
LC ebook record available at https://lccn.loc.gov/2019041426

ISBN: 978-0-367-14537-8 (hbk)
ISBN: 978-0-367-14538-5 (pbk)
ISBN: 978-0-429-03220-2 (ebk)

Typeset in Bembo
by Apex CoVantage, LLC

MIX
Paper from
responsible sources
FSC
www.fsc.org   FSC™ C013985

Printed in the United Kingdom
by Henry Ling Limited

CONTENTS

TABLE

FIGURES

PREFACE

Coming to terms with geopoetics

Poetry, to be frank, has always scared me. I enjoy succumbing to the cadence of a poem or puzzling over its symbols, but I never quite feel up to the task of fully engaging with it. In short, poetry unnerves me: I feel like I never quite "get" it as a reader and certainly not as a writer.

That said, the recent rapprochement between geography and poetry—one formulation of geopoetics—has somewhat eased my relations with poetry. Scholar-poets, including many of the authors here, generous with their time, erudite in their writings, and expansive with their patience, have enabled me—and others—to approach the geographies of poetry. By this I mean the ideas of place, space, environment (and so on) found within the poem, as well as the geographies of its production and consumption (from the writing desk to the spaces of the page).

Of course, geopoetics is not new. From the oral traditions of epic poetry with its imaginary journeys, to the geopoetics of von Humboldt's eighteenth-century cosmologies, geopoetics is an expansive concept indeed. Amongst the effects of its more recent expressions has been an attention to words and form that feels increasingly pressing. Given this, it seems appropriate to spend a bit of time with the linguistic imbrication that the compound of "geo" and "poetic" implies.

"Geos" of geopoetics

As a geographer, one of the most powerful questions geopoetics poses for me is what is meant by the "geo." Such "geos" of course are myriad and might take form in the geographies of poetic production and consumption, as well as in the force of poetry as a world-making practice. In such a generous poetics we find not only the imaginative force of words and their co-constitution of peoples and places but also the geographies of the page and practices of reading (alone or collectively) and listening, the spaces of site-specific poetry, and the post-medium practices of immersive poetics. To appreciate a geopoetics—as critic and/as writer—is, I think, always to have these multiple geographies in mind. Yet, I think there is another, expanded, geography of geopoetics that refers to its imbrication within our wider disciplinary landscapes. This expanded geography is a poetics of world and words that exceeds poetry in the sense of the latter as a particular medium or form (e.g., the haiku, the epic, or the sonnet). Instead it embraces a wider poetics, where care with words and a concern with form escape the page of the poetry book to course throughout our disciplinary practices.

Far from exhorting geographers to become poets (perhaps a rather scary thought), it feels as though geopoetic practices have attuned many geographers to the affective force and possibilities of words anew. Resonating with growing attention to embodiment and affect, geopoetics seems to me to be important in a recent reshaping not only of geographic expression but also of how geographic writers approach the space of the page, where composition as well as content becomes a geographic practice. Our writing practices feel refreshed, with the forms of the journal article and the book chapter, as we see here, reinvigorated as mediums in their own right, ripe for reclaiming as sites for experimentation and as such demanding new modes of critique. Such an expanded geopoetics attunes us to words but also to the conceptual promise of style. No longer the gloss on the substance of an article or a turn of phrase to render reading more pleasant, a geopoetic writing practice has become an epistemological tool and, for some, an ontological necessity. It is an orientation toward the form and style of an "output" that recognizes these as an indivisible part of what is said.

Rethinking the "geo" of geopoetics

It has been observed of the eighteenth-century polymath John Ruskin that his geology might as well have been a geopoetry: His was an impassioned yet precise mode of observation rendered in words, pencil, and watercolor. Ruskin was, like many of that era, possessed of an imagination shaped by, and shaping, relations of the arts and sciences. If the Enlightenment myth of a cleaving of the arts and sciences has started to fall away, the advent of the Anthropocene, that era in which humans have left a mark on the geological record, has intensified the practical, linguistic, and intellectual rapprochement between arts and sciences, not least around geology.

The struggles the Anthropocene appears to frame—with the forces, materialities, and intensities of the Earth—have, it seems, led us back to words in new ways. Sometimes these are linguistic stories of loss and failure: We mourn not only words we have lost but also the failure of language, literature, and other creative forms to confront the social and environmental challenges of our times. For others, the linguistic struggles to name and delineate the Anthropocene and its configurations of human–non-human (organic and inorganic alike) offer a point of potential. Linguistic infrastructures (from glossaries to taxonomies) are being retooled and environmental lexicons recomposed in attempts to both save old ways and forge new means to make sense of changing worlds. Querying who tells our stories and how we listen has invigorated discussions of the ways that words matter. This is not all the work of the new; non-western languages and linguistic practices offer insights into compositions of environmental relations, attuning us to poetics of the non-human, from animals and birds to rocks and glaciers. An attentive "geo" poetics for the Anthropocene seems to require that we not just recognize an expanded sense of what our poetics might be, but also question who/what creates poetry for what purpose and how non-human worlds do, and partake in, word-work.

Geopoetics as geopoesis

A third entanglement of "geo" with poetics requires us to confront what it is that geographers do with cultural forms, as critics and as creators of poetry. One answer is that they understand them as making worlds. If a recent form of ecocriticism has injected a valuable tempering into claims for what poetry can do, this does not preclude recognizing the co-constitutive nature of words and worlds; the geographies conjured on pages don't stay on those pages but have a world-making force. If geographer-as-cultural-critic has long recognized this, geographer-as-poet has perhaps been less interested in interpreting in advance the force of their words, and perhaps rightly so. In an oft-quoted phrase, Audre Lorde (1977) extols us not to consider poetry "what we need to dream, to move our spirits most deeply and directly toward and through promise" as "a luxury," but rather as an integral part of how we imagine and so create our futures. This directs us perhaps not to downplay the claims for what poetry can do but to make these claims differently. This is a geopoetics in which "poetics" constitutes the "geo," and vice versa. In which the spaces of the page and the imagination are bound up with those of the street, country, home, city, border, and environment in ways that are material and meaningful.

Musing on three engtanglings of the "geo" and the "poetic" only points to myriad other implications of these ideas and their promise. The linguistic proximities of the geopoetic compound are alchemical ones, most interesting for the admixtures of ideas that illuminate and even shape relations between the "geo" and the "poetic." Maybe such a linguistic disaggregation is a false start, and the real question lies in what analytic propositions and world-making possibilities accrue when we refuse disaggregation. But that, perhaps, is the work of the remainder of this volume.

Harriet Hawkins

INTRODUCTION

Geopoetics as route-finding

Eric Magrane, Linda Russo, Sarah de Leeuw, and Craig Santos Perez

Geopoetics is a field of inquiry that is increasingly inspiring new and innovative modes of practice and conceptualization. This collection launches a multivocal investigation into the forms, theories, and poetics of earth-making—because, at its etymological root, that is precisely what geopoetics is: Earth-making. The 24 contributors gathered here explore geopoetic concerns including place, space, and landscape; power, publics, positionalities, and bodies; relationships in the Anthropocene, including human (race, gender, disability), more-than-human, and other-than-human relationships; and, of course, language. Several contributors consciously situate their practice within a geopoetics framework. Others, while they don't explicitly articulate theoretical precedents, work within the geopoetic field as such, clearly highlighting that this collection is an *opening up* of geopoetics as opposed to a territorialization or laying down of boundaries.

We present geopoetics here as an action, as something to do and to be done, as something *to be worked at*. While geopoetics can be manifest in artistic/literary work or scholarly practice or can take an analytic attitude toward such work, geopoetics can also take the form of route-finding, living/dwelling, or other kinds of making practices. What perhaps unites geopoetics as a field (as a collective of practitioners and practices) is a shared concern not only with the past and present (what has been or is being made) but also with the possibility of shaping new futures. Thus, this collection invites readers to become practitioners, moving to the ground, to the word, through poetics. Perhaps the ultimate goal of geopoetics as a route-finding practice, then, is to acquire perspective, gain empathy, and try to make something that does not fuck things up further but instead reimagines and resets earthly relationships in the hope of creating more informed and more equitable dynamics.

In the spirit of the multivocality of this collection, we (Eric, Linda, Sarah, and Craig, the four editors of the collection) approach this introduction as a conversation. We have organized our conversation around some important overlapping themes and questions: conceptualizing geopoetics, geopoetics methods and methodologies, and organization and uses of the text. Here goes:

Conceptualizing geopoetics

ERIC MAGRANE (EM): As a geographer, I think about geopoetics in the context of current trends in the geohumanities and cultural geography. A growing number of geographers incorporate creative/artistic practices into their geographic research, and a growing number of artists and writers seem to be entering the field of geography. With texts such as *GeoHumanities: Art, History, Text at the Edge of Place*; *Envisioning Landscapes, Making Worlds: Geography and the Humanities* (Daniels et al. 2011; Dear et al. 2011); the advent of AAG's *GeoHumanities* journal (Cresswell and Dixon 2015); Sarah's co-edited issue of the *Geographical Review* on geography and creativity (Marston and de Leeuw 2013); and the (re)flourishing of creative geographies (Hawkins 2013, 2015; de Leeuw and Hawkins 2017), to name just a few, the intersections between the humanities broadly and geography—as a discipline that spans the physical and social sciences as well as the humanities—are currently quite lively.

In the article "Situating Geopoetics" (Magrane 2015), I began to outline different "modes" of geopoetics: as creative geography, in which poetry itself is produced as geography and geography is practiced as poetry/poetics; as literary geography, in which geographers interpret poetic texts; and as geophilosophy, in which the "earth-making" of geopoetics is theorized in forms that are not necessarily recognizable as "poetry." In outlining the modes of geopoetics in that article, I looked to aesthetic and creative traditions within geography to historicize the project of geopoetics, including work on space and representation, as well as on humanistic geography traditions. Here I also would like to acknowledge Kenneth White's work on geopoetics (1992), which has inspired geopoetic works and groups such as the Scottish Centre for Geopoetics. However, in "Situating Geopoetics," I was thinking about some of the different ways to open up the field particularly in the context of contemporary cultural geographic/geohumanities thought and practice.

Concurrently, as a poet, I was finding many overlaps between contemporary discussions on ecopoetics that hadn't yet been fully articulated in relation to geography. In this context I'm thinking about ecopoetics as "house-making," as articulated by Jonathan Skinner in the introduction to his *ecopoetics* journal (2001), and about the more critically and politically infused work of ecopoetics that has clearly digested critiques of "nature" and "wilderness" into its viscera and situated itself as much in an experimental tradition (at least in the US context) as in a lyric nature writing or pastoral tradition (Hume and Osborne 2018).

So the "eco" of ecopoetics has begun to encapsulate any kind of poetics engaging with ecology, with systems thinking, with spatial thinking, with Anthropocene/catastrophe thinking, with "the nature formerly known as nature" thinking, to quote Brenda Hillman (n.d.). However, in our conversation to open this book,

I want to begin by proposing that geography, as an edge discipline (physical/social sciences as well as humanities), may have even more to offer a poetics engaged with the ongoing paradox of the twenty-first century. Many of the concerns being taken up as ecopoetics—such as environmental justice or political ecology, which play out within particular spatial relationships—might be more accurately understood as geopoetics. Here I should also be clear that I'm thinking about ecology as it's practiced as a scientific discipline rather than as it's understood as a movement. Let's try a little table (see Table 0.1).

At a certain point, of course, the distinction between an "eco" poetics and a "geo" poetics does not matter—most categorizations are provisional exercises to find some sense of order in the world. Here I'm also thinking of Linda's introduction to *Counter-Desecration* (Russo, "Entangled" 2018), in which, Linda, you write, "*Anthropocene writing* is our preferred term because it downplays the tendency for categorizing and instead emphasizes the act of writing and the unifying fact of the times we live in" (11–12). Of course, the Anthropocene itself is a contested term.

LINDA RUSSO (LR): I agree: Where the "geo" and "eco" overlap, there—at an intersection of the material and the "insubstantial"—we all are. I hadn't even heard of the term "Anthropocene" (much less any debates surrounding it) when I set out to write an essay (in 2006) identifying mine as a "spatial practice"—a poetic practice rooted in a place *as* an interspecies "environment" (in the basic sense of that term, what environs or surrounds one). I needed to address those relationships that long preceded my presence, that preceded colonial-settlement where I was living at the time. And I've found the distinction Gary Snyder makes, between "the world of culture and nature, which is actual . . . and the insubstantial world of political and rarefied economies" (Snyder 1999, 192), to be helpful. The "actual" world of the former—where all species perform their earth-making—is eclipsed by the latter, which "passes for reality" (to quote Snyder).

To refer to your Table (see Table 0.1), the political world correlates to the "social theory" aspect that geopoesis brings to the surface more readily than ecopoesis—as much as we might wish that weren't so. Ecopoetics as it is variously practiced *is* often rooted in Romantic ideologies, in "nature poetry," though

TABLE 0.1 Geopoetics–Ecopoetics

Geopoetics	*Ecopoetics*
"Earth"	"House"
place + space	organism + species
social theory	biology
climate	weather

many contemporary poets resist and subvert that. In a Euro-American/settler lineage we might trace geopoetics back to the "grounding" of a poet's sense of place within the larger cultural and historical forces that undergird that place. Here I'm thinking of, as examples: Charles Olson's work as "archaeologist"; Muriel Rukeyser's taking "roads . . . into your own country" (as she begins the first poem in *The Book of the Dead*) and the resulting eco-justice documentary poetics; and Joanne Kyger's bioregionalist embrace of a complex present, where her "story" exists within the ancient "story" of the coastal Miwok Territory where she lived. Geopoetics is an attentive remaking, a way of moving forward while locating and digging down into the past. It's all "within" the Anthropocene—a contested term, yes, but the fact of dominance of certain power structures remains: *capitalobscene*, or "disaster capitalism," to quote Naomi Wolf. Isn't the emergence of geopoetics in some way an acknowledgement of that obscenity in our age of the "drastically collective" (to quote my *Counter-Desecration* co-editor Marthe Reed quoting Timothy Morton) confrontation with global climate instabilities? Maybe ecopoetics:weather :: geopoetics:climate.

EM: Linda, I love this idea of ecopoetics:weather :: geopoetics:climate. I've gone ahead and added it to Table 0.1. Now I've got Bob Dylan's "Subterranean Homesick Blues" rolling around in my head. Perhaps geopoetics can help us to "know which way the wind blows" (1965).

Part of knowing which way the wind blows, as you point out, Linda, is historicization. You mentioned Olson's archaeological poetics. One touchstone in tracing the relationship between contemporary poetics and cultural geography is the relationship between Olson and cultural geographer Carl Sauer (Parsons 1996; Magrane 2013), whose work on the morphology of landscape (Sauer 1925) is iconic within an American cultural geography tradition, particularly in its understanding of the interplay between landscape and culture as multidirectional. This compels us to state that we aim not necessarily to reassert that lineage, but to "remake" it. Moving forward from this early-twentieth-century work, critiques of the reification and essentialization of terms such as culture (Mitchell 1995) might then be understood as epistemological (postmodern, poststructuralist) precursors to some of the current critiques of the Anthropocene that take the homogenization of the "human" in the Anthropocene to task. Those critiques point out that it's certain organizations of humans (e.g., capitalism, patriarchy, settler colonialism) that have brought us to this point rather than all humans.

In other words, when we talk about "human," there's so much difference there— maybe this is a little more of the social theory coming in. And the Sauer–Olson lineage/genealogy is just one to trace as part of contextualizing geopoetics. To return to geopoetics as a route-finding practice: Radical and feminist critical geographies are crucial to thinking and doing geopoetics. Here, for example, I'm thinking about Cindi Katz's (1996) cultural geographic readings of Audre Lorde and Adrienne Rich and to your work, Sarah. In particular, I'm thinking about how the introduction to the issue of *Geographical Review* on "Geography and Creativity"

(2013) that you and Sallie Marston wrote called for a close attention to the politics of doing creative geographies. It also pointed to the politics of canons, citation, and genealogies. Poet Brenda Iijima's contention that poetics "recalibrates the social—how we function dynamically in space, in time, with each other" (2010, 276) is also something that continues to stick in my mind and that has resonances with Rukeyser's work, which you mentioned earlier, Linda. Across disciplines I also hear concepts of social reproduction. At a really basic level, each action that one takes either reproduces certain organizations of matter/humans or resists certain organizations of matter/humans.

In my introductory undergraduate World Regional Geography courses, I've started taking a full class period at the beginning of the semester to ask students "what is human nature?" and discuss their responses. How one answers this ontological question has far-reaching effects on how one imagines—or even can imagine—alternative socioecological futures. Maybe it's also a key question for geopoetics.

CRAIG SANTOS PEREZ (CSP): Even though geography has been an important theme in my poetry, I am not trained as a geographer, and I was not familiar with the term "geopoetics" until I read Eric's seminal essay, "Situating Geopoetics" (Magrane 2015). Eric, I was struck by the discussion of geography as "earth writing," and especially by your remarks about how poets and geographers represent our place on the earth through creative and critical geography, literary geographies of poetry, and geographical poetics.

My own developing conception of geopoetics is shaped by my background as an Indigenous Pacific Islander poet with a foundation in decolonial studies. I grew up in the geographical spaces of islands, jungles, beaches, reefs, shores, and tidelands. The constellation of archipelagoes, as well as the vast horizon and unfathomable depths of the ocean, sparked my imagination.

Indigenous Pacific conceptions of geography teach us that the earth is the sacred source of all life, and all beings are interconnected in a complex kinship network. Thus, Pacific environmental ethics revolve around ideas of reverence, respect, and sustainability. Moreover, native islanders believe that geographies are storied places, and Pacific oral literatures have transmitted these stories across generations.

These beliefs were displaced by the history and impacts of imperialism as colonial mappings of the Pacific were imposed, and the islands and ocean were desecrated and viewed only as sites for plantations, incarceration, military bases, nuclear testing, tourist destinations, and extractive logging, mining, and oil drilling. At the same time, colonial forces silenced Indigenous geographical stories and replaced them with settler place names and narratives.

The emergence of Pacific decolonization studies, political movements, and literature in the twentieth century has also shaped how I conceive of geopoetics. Decolonial Pacific geopoetics critiques and protests colonial geographies and stories and reclaims and revitalizes Indigenous beliefs about the lands and water—and the stories associated with these places. Ultimately, Indigenous Pacific geopoetics are symbolic spaces for reconnection, inspiration, empowerment, and healing.

SARAH DE LEEUW (SNDL): Following Craig's observations, my own academic commitment to anti-colonial scholarship, and what many people and communities in Northern British Columbia (Canada) have taught me about the importance of locating oneself (of never neutralizing genealogy or geographic lineage), it feels important here to identify as a colonial settler, as a non-Indigenous white woman with a Dutch last name that reveals a paternal ancestry firmly located in the Netherlands, where my dad was born. This act of locating is central to being a poet, an essayist, and a geographer—it's a crucial part of how I think and write about geopoetics. I was a creative writer before I was a geographer. Before doing a PhD in historical-cultural geography, I attained an undergraduate degree in creative writing. I registered for geography classes as a vague afterthought, undertaken to fill elective slots. I was a poet (or so I thought) in my earliest grade school years, trying desperately (and in an embarrassingly pathos-filled way) to make sense of, and to properly be in, the places I was growing up. Mostly, the places I was trying to properly be in and to make sense of were places on the margins, places guidebooks write off and popular imaginations run roughshod over as redneck backwaters: *Working class resource extraction, northern places full of poor health outcomes, clearcuts and environmental devastation, higher than national average rates of drug and alcohol use, and tensions between loggers, truckers, commercial fishermen (yes, they were men—because it was just that gendered where I grew up), and the Haida or Nisga'a or Tsimshian or Wet'suwet'en or Gitxsan peoples who were, since time immemorial, thriving in the places where my family and I were busily making our very new colonial home.*

My earliest poetry couldn't be categorized as geographic (not really), and I wasn't writing about ecological issues. I was writing to emulate cheap paperback teen romance novels and lyricizing about how I hated my parents and was in love with someone who didn't know I was alive. I had no idea that precedent-setting legal cases regarding Indigenous rights and titles over unceded territory were unfolding outside my back door. These understandings came much later, as I matured into a critical cultural geographer (who wrote poetry). I was a terrific naïve colonial kid 'cause I could be. Before I matured into that critical cultural geographer (who wrote poetry), I would return (usually looking for work) to the places I grew up: I was a bar waitress, a logging camp cook, a tugboat driver, a counselor to women fleeing domestic violence, and a reporter. And I was (still am!) a feminist. A feminist slowly becoming a geographer, still writing poetry, increasingly disenchanted with colonial violence, with white heteronormative (sometimes toxic) masculinity. All these identities (feminist, creative writer, geographer) are tethered to the earth, to place, to ground and soil and territory (to physical geography), in specific ways, through specific tendrils of power.

All of these details are germane to geopoetics. Because, for me, poetry anchored in geography is not *just* poetry anchored in or focused on the soil and ecology and dirt of the earth: It's poetry that writes peopled-places too. And it's poetry that demands a conversation about who is writing it, who is being written out; it's poetry written by people (who as we note at the beginning of this conversation, are folks trying not to fuck things up more than we already have) and it's poetry about people in places (places of the kind that Craig is calling attention to). It's this I try to remember: Geopoetics is not disembodied. To draw further on Haraway, whom Linda discusses later, geopoetics is not a place of god-tricks, written by no one. *We* write geopoetics. Someone writes the *ground geology geography soil topography earth* that is geopoetics. Being attentive to the "someones" is being attentive to the hand of geopoetics—the hands that have crafted all the poetry in this collection.

LR: Sarah, your description of your younger self's attempts at "making sense," at trying to "properly be in place"/figuring this out "on the margins," while simultaneously being unaware of a larger political context that dis/empowered, while *also* being acutely aware of gender ideologies, really resonates with my Eastern US-suburban experience. My margins were the wooded swaths near my home. I realize now I've always been a student of place and a geo-grapher in the sense of being attuned to how relationships are inscribed by and into places. One of my earliest memories of writing is sitting with a notebook beneath the buzzing ConEdison power lines that cut through the woods: Locating myself inescapably in the (material and immaterial) power grid as a way to achieve a socially uninscribed site where I could freely write. Struggling to resist and reinscribe values, right there in the heart of settlement: Infrastructure. Perhaps a geopoetic begins when one sees the irony of writing self and place when these are always already written—to return to the idea of the "someones" Sarah raises earlier: Who is being written in/out when. It begins by recognizing a struggle.

Poetry first presented the possibility of articulating and shaping an *inner* geography; later, poets including Lorine Niedecker and Larry Eigner helped me shape a practice rooted in witness, withness, and betweenness as modes of ecopoethics, of rescaling attention to the rough biotic periphery of interspecies inhabitance: "When habitation moves beyond a passive residence upon a particular ground to ethical relationship via the topophilic sense, being, making, and ethical relation become integrated," in the words of Marthe Reed (2018, 37). Being located *among* to call out the necessity of ground-level connectedness of earth relations. Donna Haraway's *chthulucene* (her preferred term for our present era) bespeaks the condition of earthliness: It joins *khthôn* (beings of earth) and *kainos* (now, fresh, new) to emphasize the ways we can "demonstrate and perform the material meaningfulness of earth processes and critters" (Haraway 2). To my mind, works of geopoetics critically emphasize and engage with *trouble*—and here I'm thinking of Craig's Indigenous geo-ecopoetics of encounter with clashing military, personal, and ecological histories. Haraway's "staying with the trouble" (as she titles the book from which I've just quoted) means attempting to change the narrative. It's no surprise that geopoetics is a product of Late Capitalist flows and processes, that it is emerging as a field at the moment when extractive industries (coal, oil, gas) are facing powerful coalitional resistance: NoDAPL, the Wet'suwet'en blockade, and other efforts to defend treaty rights (and, more broadly, the "rights of nature" movement) are challenges to destructive colonial legacies, attempts to change the narrative.

EM: Linda, I was also raised in the Eastern US, though in a rural rather than suburban environment. I grew up in the Lakes Region of Western Maine, in the foothills of the White Mountains and the Appalachian Trail. My landscape margins were the bogs and marshes, the lakes and forests. I was vaguely aware of contested histories on the landscape but not in any great depth. My margins were also class-based—I was raised by a single mom who worked full time at a regional hospital and also waited tables at night to make ends meet. Although I loved the landscape, I got out of there—I went to college in Vermont and then eventually made my way to the Southwestern US. As a first-generation college student, I had to learn how to think critically about my privilege as a white male in this society and to situate that privilege within settler colonial and imperial legacies.

I had a late landing in geography. Well before I was even aware of geography as a discipline, I completed an MFA in poetry. After that, I spent time on the margins of the academic world: A small stretch of time in Mexico and then about a decade where I worked as a hiking guide around Tucson, Arizona. The University of Arizona has a vibrant geography department, and I started showing up to geography colloquia out of sheer interest and getting to know the variety of conceptual tools that geographers draw upon. I met Sallie Marston and enrolled in one of her cultural geography seminars as a non-degree seeking student. That eventually led to my enrolling in and completing a PhD in geography. This experience exposed me to critical and feminist geographies, as well as to conceptual frameworks such as environmental and climate justice and political ecology. This has informed my critical-creative practice of geopoetics as a marginal or edge practice itself, which leads to our next section.

Geopoetic methods and methodologies

EM: One of the difficulties/tensions in putting together a collection like this is how to find a meeting space for both analytic and creative/expressive impulses. For one thing, at a practical level, different disciplinary fields have different sets of literatures and conventions. For example, texts in literary and cultural studies usually adhere to MLA style formatting, while the social sciences often adhere to APA or Chicago. (In this book we have followed Chicago style, similar to AAG's *GeoHumanities* journal.) And then poetry puts more pressure on close attention to the layout and geographies of the page and of the book object.

Some of the chapters in this book progress fairly straightforwardly in structure; others read more like artist statements; still others like hybrid lyric essays. The chapters don't all draw on the same literatures, with some, for example, drawing more on cultural geography, others on literary or cultural studies. While issues such as citation style and disciplinary differences across this book could be considered somewhat mundane, they also bring us to a question of methodology and epistemology. While methods could be considered tools, techniques, or practices of geopoetics—such as physically placing poems within landscapes for people to encounter (such as in Turnbull's chapter), designing site-based art installations (such as in Farooq and Stanley or Pluecker's chapters) or blending literary criticism with creative writing (such as in Cresswell or Noel's chapters)—methodology refers more to epistemology: How one conceptually makes sense of the world and how that is connected with positionality, politics, and ethics.

CSP: I appreciate the aesthetic diversity of geopoetic methodologies: narrative poems that story certain places, lyric poems that express emotions about certain landscapes, modernist poems about the fragmentation of space, postmodern poems that deconstruct geographical concepts, documentary poems that include archival sources about places, visual poems that embody geographical features, map poems that reimagine cartography, and so much more.

Geopoetic methodologies also change when viewed through an Indigenous Pacific lens. Indigenous methodologies center critical and creative practices within native ecologies, languages, stories, genealogies, customs, values, traditions, aesthetics, and epistemologies. Pacific methodologies center the cultural, linguistic, historical, social, and political themes and contexts of Pacific Islands and islanders. These methodologies also foreground the ongoing impact of colonialism and the struggle for cultural revitalization and political decolonization. Lastly, they also emphasize the importance of community engagement in the sense that the scholarship and poetry Indigenous Pacific people create should contribute to the community. While I am describing a certain kind of geopoetic methodology, you can see that these values can be found in much of the work in this anthology.

SNDL: Extending the lines of inquiry Eric and Craig have raised, I feel invited to consider geopoetics as both a method and a methodology—as an "and/both" as opposed to an "either/or." My reasoning is, primarily, that *writing*—and to a great extent oration, if we're thinking spoken word poetics—always expresses a philosophical and theoretical orientation to the world, to the ecologies and geographies in which we (as poets or geographers) exist. Writing by geographers, if it's possible to generalize, tends not to focus on poetic expression. Conversely, *very few* poets I've ever worked with, or even know of, pick up academic geography journals for poetic inspiration. Geographers and poets, in other words, don't stimulate each other. And, what's more, they don't dialogue very much with each other because, in many respects, we use writing (as a method) differently from each other. This in turn leads to different methodological outcomes. The conversations in this introductory discussion and the conversations that will happen by virtue of a series of chapters being collected together with one spine invite new methods (and methodologies) of writing, of geo-graphing, of being in and representing and making sense of the world. Practicing geopoetics will generate new ways of seeing, understanding, and being in the world. Those new ways of seeing, understanding, and being will hopefully inspire and invite and encourage entirely new modes of geopoetics. We do, with and through geopoetics, have the possibility of making new worlds, geo-graphing anew.

Organization and uses of the text

EM: This book is organized in three sections: Documenting, Reading, and Intervening. Within each of the sections, chapters are organized alphabetically by author. The first section, Documenting, includes chapters that detail and describe geopoetic projects and practices, from gallery installations, to site-based performances, to poems placed within landscapes, to Anishinaabeg picto-poetics. These chapters all in one way or the other describe, document, or archive geopoetic practices by their authors. The chapters in the second section, Reading, foreground geopoetics as a reading practice; the chapters in this section often bring in aspects of literary criticism or critical-creative analysis of texts other than the author's own. The third section, Intervening, highlights geopoetic practices and projects that intervene in environments, places, and/or texts, working—often explicitly—to reorganize power relations.

Many—most—of this collection's chapters could fit in more than one of these sections, and we trust that readers will find many echoes between and among chapters and sections of the book. So, we ask the reader not to take too much time mulling over why a chapter is in one section rather than another. Since we simply had to order the book in some manner, think of these sections as a starting place. Here, I would like to build on that starting place to offer some other possible groupings of the collection's chapters, in the hopes that these groupings may be useful for readers using the text in the undergraduate or graduate classroom. Think of this as an analog version of metadata.

For example, a grouping of chapters that engage Geopoetics and Visual/Performance Art might include Dickson and Clay, Farooq and Stanley, Heit and Kuppers, Jordheim, Pluecker, Rawlings, Ryan, Turnbull, Scappettone, and Ward. A grouping on Theorizing Geopoetics might include Banazek, Boyd, Last, and Scappettone. A grouping on Indigenous Geopoetics might include Blaeser, Pluecker, and Perez. A grouping on Urban Geopoetics might include Clifford and Williams, Skinner, and Ward. Of course, these lists are not exhaustive.

LR: Along with method ("tools") and methodologies ("concepts") that Eric mentioned earlier, I want to add a *field of inquiry*. What's exciting to me about the variety of chapters and especially the hybridity of works included in this collection is how each describes its own field of inquiry through its methods and methodologies. In this collection we can see the overlaps as well as an array of combinatory (interdisciplinary, multigraphical) practices. So, I could see grouping entries by how practice engages field of inquiry, for example, walking or otherwise passing through (Clifford and Williams, Edwards, Heit and Kuppers, Rawlings, and Turnbull); or acknowledging other-than-human agencies that shape (write) place and influence human ways of noticing and documenting (Blaeser, Giscombe, Rawlings, Ryan, and Turnbull).

SNDL: As a writer fascinated by pauses, by spaces, and by breaks in lines, I find these remarks, Linda, compelling as much for what they say as for what they don't say, for what they sagely *imply* as opposed to the space the words take up. When I read poetry, I am often drawn to the silences between the lines, to the space between the words. To what is not said, to what echoes or reverberates in the moments after I've read a poem. It's these silences and pauses that are so often a source of generative inspiration. I applied a similar lens to the organization of this text. I looked for how the text was *not* organized, how the chapters were *not* bound and collated, how we chose to *not* group authors and chapters. What might have happened, in other words, if the text were organized by geographical location of the author? Or by the language the chapter was written in? Or by the levels of postsecondary education achieved by the authors? Or by any given chapter's focus on *place*? Or by the levels of fleshy sexuality contemplated by the authors? If we start to think about it further, what might become apparent is that, if we hadn't organized the text as we did, many of the chapters could easily have fallen into a single grouping. That single grouping would have been: Authors with University-Level Educations from the Global North (myself included), Writing in English to Publish, About Places Often Also in the Global North, from Quite Disembodied Postition/alities. I don't say this in a cranky or complaining way, because I also agree with everything Eric, Linda, and Craig have written. I'm just observing that organizing and categorizing can sometimes, in the ways it is *not* done, highlight that more work in the area needs to be done. I would *love* to see a book of geopoetics edited entirely by Indigenous poets and geographers from the Global South, featuring work only by authors in adjoining geographies. I would also *love* to see a text focused on the geopoetic sensualities, spaces, and sexualities of bodies. This book of course is not one of those books—those books I would love to see. But in what it's *not*, this book *is* a beginning—a recognition, perhaps in the very ways that the chapters are *not* organized, that so much more work must be done.

CSP: Just as there are many possible ways to group the chapters, there are many ways to teach these diverse contributions. There are chapters and poems here that fit within a variety of courses, including Geography, Cultural Studies, Literary Studies, Poetics, Indigenous Studies, Anthropology, Political Ecology, Environmental Humanities, Visual and Performance Arts, and more. As a professor in an English department, I am most excited by the pedagogical possibilities of teaching this anthology in a Creative Writing course. After reading and discussing the chapters with my students, I can ask them to study maps of their own places of origin and to respond to these maps through poetry. We can compose visual poetry based on the geographies of places to which they feel a connection or disconnection. We can walk around our campus, the city, the beaches, valleys, and mountains and thus compose geo-poems that capture the landscapes and experience of walking through space. We can write poems that interrogate and critique politicized geographies, such as bordered, militarized, extracted, polluted, or colonized places. Through critical and creative engagements, students can explore the complex intersections between geography and poetry, as well as the profound cartographies of writing the earth.

EM: Craig, I love the idea that the text can instigate assignments and practices that get creative writing students physically out into landscapes. A number of the pedagogical ideas you noted could also be translated to a cultural geography or geohumanities seminar, or to a qualitative research/critical methodologies class. I'd also love to see the text used in interdisciplinary or co-taught settings. That's just within the academic context. Here's hoping that the chapters within are of use to poets, artists, geographers, activists, and anyone interested in traversing new geopoetic routes.

References

Cresswell, T., and D. Dixon. 2015. "Imagining and Practicing the Geohumanities: Past, Present, Future." *GeoHumanities* 1 (1): 1–3.

Daniels, S., D. DeLyser, J. N. Entrikin, and D. Richardson. 2011. *Envisioning Landscapes: Making Worlds: Geography and the Humanities*. London and New York: Routledge.

Dear, M., J. Ketchum, S. Luria, and D. Richardson, eds. 2011. *GeoHumanities: Art, History, Text at the Edge of Place*. London and New York: Routledge.

De Leeuw, S., and H. Hawkins. 2017. "Critical Geographies and Geography's Creative Re/turn: Poetics and Practices for New Disciplinary Spaces." *Gender, Place & Culture* 24 (3): 303–324.

Dylan, Bob. 1965. "Subterranean Homesick Blues." *Bringing It All Back Home*. Columbia Records.

Haraway, Donna. 2016. *Staying with the Trouble*. Durham: Duke University Press.

Hawkins, H. 2013. *For Creative Geographies: Geography, Visual Arts and the Making of Worlds*. New York: Routledge.

———. 2015. "Creative Geographic Methods: Knowing, Representing, Intervening. On Composing Place and Page." *Cultural Geographies* 22 (2): 247–268.

Hillman, B. (n.d.). "Poetics of Drought: Brenda Hillman." *Omniverse*. Accessed March 6, 2019. http://omniverse.us/poetics-of-drought-brenda-hillman/.

Hume, A., and G. Osborne. 2018. "Ecopoetics as Expanded Critical Practice: An Introduction." In *Ecopoetics: Essays in the Field*, edited by Angela Hume and Gillian Osborne, 1–16. Iowa City: University of Iowa Press.

Iijima, B. 2010. "Metamorphic Morphology (with gushing igenous interlude) Meeting in Language: P as in Poetry, Poetry Rhetorical in Terms of Eco." In *The Eco Language Reader*, edited by B. Iijima, 275–292. Callicoon, NY: Nightboat Books.

Ingold, T. 2007. *Lines: A Brief History*. Abingdon: Routledge.

Katz, C. 1996. "Towards Minor Theory." *Environment and Planning D: Society and Space* 14 (4): 487–499.

Magrane, E. 2013. "Cryptic Species and Compass Points: A Few Gestures from the Conference on Ecopoetics (an experimental review)." *Emotion, Space and Society* 11: 116–118.

———. 2015. "Situating Geopoetics." *GeoHumanities* 1 (1): 86–102.

Marston, S. A., and S. de Leeuw. 2013. "Creativity and Geography: Toward a Politicized Intervention." *Geographical Review* 103 (2): 1–26.

Mitchell, D. 1995. "There's No Such Thing as Culture: Towards a Reconceptualization of the Idea of Culture in Geography." *Transactions of the Institute of British Geographers* 20 (1): 102–116.

Parsons, J. J. 1996. "'Mr. Sauer' and the Writers." *Geographical Review* 86 (1): 22–41.

Reed, M. 2018. "Somewhere Inbetween: Speaking-Through Contiguity." In Russo and Reed (eds.), 77–90.

Rukeyser, M. 2005. "The Book of the Dead." In *The Collected Poems of Muriel Rukeyser*, edited by Janet E. Kaufman, Anne F. Herzog, and Jan Heller Levi, 73–110. Pittsburgh: University of Pittsburgh Press.

Russo, L. 2018. "Entangled & Worldly: Approaches to Anthropocene Writing." In Russo and Reed, 1–15.

Russo, L., and M. Reed, eds. 2018a. *Counter-Desecration: A Glossary for Writing Within the Anthropocene*. Middletown, CT: Wesleyan University Press.

———. 2018b. "Entries." In Russo and Reed (eds.), 17–76.

Sauer, C. 1925. *The Morphology of Landscape*. Berkeley: University of California Press.

Skinner, J. 2001. "Editor's Statement." *Ecopoetics* 1: 5–8.

Snyder, G. 1999. *Gary Snyder Reader*. Washington, DC: Counterpoint.

White, K. 1992. "Elements of Geopoetics." *Edinburgh Review* 88: 163–168.

PART I

Documenting

1

BODIES BELONG TO THE WORLD

On place, visuality, and vulnerability

Kerry Banazek

Late 2010—Missoula, Montana. I have been writing little poems that grammatically collapse my history as an organism and the bodies of landscapes. These poems are tiny, stalling machines. Their forms more than their phrasing suggest they are about limitation; they are my first bid at *talking about* how my embodiment has been held against me. They look something like this:

> *ANDROGYNY, A BORROW PIT. Alchemy in a mechanical culture.*
> *Aerial excursions, mouth-viewed views and all the types of*
> *invasion. Weeds. Glacial silt and-or ordinary runoff. Phototropes.*
> *A body cellar. A sudden flock or orchard or warren. Some*
> *beautiful-moving-slowly that makes a tree mean part of the sky.*

An established writer asks why the poems "do not sound like me." She accuses me of being "very eloquent in person." We work together to create a catalogue of ways punctuation can cut lyricism off mid-flight. I do not tell her that I've been writing this way (in part) because I am afraid of language, or because using incomplete phrases makes the poems feel almost photographic for me, or because this helps me "get them out." I like the poems being little rectangles with some visual intrigue. Form is the thing that binds them to my life, not content or insight. I do not admit to the established writer that for many years I almost never spoke unless I was spoken to directly. I do not yet know how to articulate that I haven't come to Montana to be a poet or a writing teacher *per se*. I am here, instead, because I am carrying around an idea: If I don't learn how to do *something* with the trauma stored in my body, I won't survive.

★ ★ ★

This is an essay about being a particular body. A body that has moved through particular spaces. It is about training a self (my self) to talk about landscape photography in media studies terms, while training a self to work as a poet, and about how these experiences co-inform my ways of being, thinking, writing, and researching. Which is to say, it is about the training that underlies my understanding of methods and methodologies. I mean to "merely" observe and describe some salient features about one way of sensing and engaging the world, a way I think is well described by the adjective "geopoetic." I have access to these descriptions via somewhat accidental coincidences, which is part of the point of this chapter.

When I moved to Montana, the so-called Big Sky state, one of the first things I did was buy a wide-angle lens for my favorite camera. Technically, any lens with a wider field of view than the human eye can be called wide-angle; these lenses have short focal lengths and are good for capturing dramatic landscapes. They enhance perspective (a scene's straight lines appear to converge faster than when viewed with the naked eye; distinctions between grounds are somewhat amplified). These lenses deal in scale-change and distortion. They can produce decidedly awkward portraits or make everyday environments otherworldly. Handled well, they often make it feel like viewers have stumbled into a scene; we might say they trade in immersion more than realism.[1] Or that these lenses deal in groundedness rather than authenticity.[2] Immersion and groundedness are also concepts that have helped me coalesce, name, and engage with the value of geopoetic methods. Immersion, groundedness, and geopoetics are all terms that speak to *styles* of connection, to the existence of relationships that implicate individuals and environments in equal measure. Indeed, as methodological terms, they all assert the ongoing value of acknowledging individual entities while also insisting on each individual's porousness and vulnerability.

Geopoetics are plural. They partake of what we might call wide-angle disciplinary trends, which include, among others: the spatial turn in the humanities and social sciences (e.g., as engaged by Warf and Arias 2008; Pugh 2009; or Withers 2009), invitations to include affective objects and artistic methods in the mainstream of both English and media studies (e.g., Parikka 2015; Kara 2015; Sommer 2014), cultural geography's phenomenological and artistic sides (e.g., Turchi 2004; Harmon 2003; Wood 2010), and interdisciplinary debates surrounding social and technical functions of locative media (e.g., Farman 2011; Gabrys 2016). These disciplinary trends are important, decidedly non-exhaustive affiliations. Like wide-angle optical devices, trends invite participants to experience scholarship in sweeping ways. They distort relationships between researchers, methodologies, objects of study, and disciplines. Where such distortions can be dangerous, they can also be productive, as long we recognize that a distortion is happening. Turns, trends, and other wide-angle images of disciplinary action expose urgencies and exigencies when taken up together and in context. For instance, a simultaneous surge of interest in both poetic articulation and spatial, geographic thought across humanities and social science disciplines can be (non-exhaustively) associated with a number of global phenomena: the rise of new technical regimes (e.g., regimes structured by the increasing affordability of Geographic Information Systems [GIS] devices or the commercialization of drone technology); the expansion of infrastructures of global capitalism; and anthropogenic climate change.

Observation itself is a situated practice performed by individuals *and* a dynamic epistemic category (e.g., in the histories collected by Daston and Lunbeck 2011). Moreover, observing evolving phenomena—including physical or cultural geographic phenomena and the emergence of poetic or academic trends—requires us to evolve our methodologies. In other words, active and imaginative observation of trends has the potential to expose the limits that disciplinary histories impose; whatever our home disciplines, technical and political changes serve as explicit invitations to borrow from neighboring fields. Responding when hailed in this way can be both risky and thrilling.

All research invites a measure of idiosyncrasy. No writer, scholar, or individual can know everything in their own field equally well, and each individual brings an observational history to their work. Phenomena related to researchers' identities and idiosyncrasies are magnified in interdisciplinary research and writing. When a body acts as an interface between fields, it has a filtering effect: Not everything passes through. The more methods one tries to encompass, the more aberrations threaten—sometimes productively, sometimes dangerously. In this way, all inter-disciplines engage with poetics. While the term poetics is most frequently glossed in rhetorical or aesthetic terms, thinking expansively, it is possible to understand poetics as a name for "coming together," for arts of assembly that extend theory and praxis into one another or otherwise interfere with disciplinary bonds. Where poetics borrows from poetries, it forces attention to gaps surrounding work and understanding. If you write poems, Glyn Maxwell argues, the hungry whiteness that surrounds words or precedes them can't be understood as "nothing" (the way it can for people who only write prose); rather, this whiteness must be understood as "half of everything" (2012, 11). Poetic interdisciplinary research, then, is research that works with gaps surrounding existing disciplinary knowledge *and* gaps in researcher experience; it acknowledges the "left out" and asks readers to do the same. Geopoetic research acts on this premise doubly: It draws together long literary and geographic lineages, calling attention to resultant glitches and disjunctures, and it invites the *practice* of poetry itself into places and conversations of interest to geographers.

Here's one more explicit way to think about the importance of defining geopoetic methods as a subset of poetic methods: Geopoetic methods call attention to collisions between locative, non-dominant media philosophies and the politics of form. Geopoetics have the potential to reconfigure expectations surrounding the roles personal experience can play in academic research precisely because—whether or not geopoetics know it—they insist:

Bodies belong to the world.

This is a complicated lesson—one that cannot ever be fully learned, recalled, or applied. Both bodies and the world resist definition. Belonging suggests both the comfort of communion and the violation of being made or becoming property. And despite what we all might like to believe, these are not mutually exclusive suggestions. Encompassing can be a form of domination. Even when we can tell the difference between benevolent and malevolent belongings, the issue of scale sings between bodies and the world.

★ ★ ★

When I decided to move to Montana, I latched onto the idea that living in a new landscape might make living itself easier, and that making images with the wide-angle lens might help me connect with the landscape. Photography—as both a medium that evolves with the technical devices that support it *and* as a disciplined form of artistic practice foregrounding the social and creative potential in human and machinic ways of seeing—has a long-standing relationship with the idea that landscapes always have social dimensions. This relationship is fundamentally complex and conflict-wrought. It is both rhetorically and politically constitutive. The same might be said for the somewhat irrational idea I had that the Bitterroot Mountains and the Clark Fork River and the ponderosa pines might be able to help my young adult self figure out the world in ways that features of my other landscapes (mainly places in Upstate New York and Western Washington) hadn't been able to.

Of course, having had *this kind of idea* wasn't notable. It is the kind of idea that can be understood as cliché, imperialistic, and naïve, especially when it takes shape in relation to a landscape like Southwestern Montana. This is a landscape where a contemporary settler might, with unwitting ease, encounter a diorama celebrating the journeys of Lewis and Clark at a local shopping mall but then experience significant difficulty trying to figure out where to go to learn a few details relevant to picturing the landscape's co-constitutive relationship with Indigenous cultural histories, present-tense philosophies, and experiences. This kind of idea, then, is simultaneously the kind of idea that squeaks with the whiteness of the American transcendental literary tradition and the kind of idea that recalls what trauma and addiction specialists call "the geographic cure" (the name for the often false hope that a change in location might solve one's problems, which appears in both Alcoholics Anonymous lore and psychiatric literature, e.g., in work following Maddux and Desmond [1982]; it is common enough parlance to warrant an urban dictionary entry and to serve as a stand-alone title for a range of think pieces, e.g., Flynn [2012]). But imperialism and avoidance aren't the *only* lineages that support related ideas. The premise that specific landscapes have specific things to teach and offer specific balms (e.g., see Kimmerer 2013) also has something to do with the careful crediting of emplaced, more-than-human knowledges, as well as with the wide richness of eco-literatures, Western and non-Western, colonial and Indigenous. In other words, the limits and distortionary tendencies of ideas like the ones that my young adult self pinned to Southwestern Montana do not negate the ability of related ideas to instigate meaningful, wide-angle engagements with the world. These tendencies *do* ask that we proceed with *care* if we choose to proceed.

Roger Cicala suggests it can be helpful for the optically minded to remember that, compared with standard-range and telephoto lenses, wide-angle lenses are "more prone to distortion and aberration in general" and, consequently, "very wide angle reverse-telephoto lenses are among the most complex lenses made today, usually containing more than a dozen elements in an effort to correct all the aberrations" (2011, n.p.). In this suggestion there is an important metaphorical lesson not only for those who seek to widen personal views by moving through new places but also for scholars who seek to widen their commitments and understanding by moving through multiple disciplines. Academics like to use the metaphor of theory as lens, as if all lenses are composed of a single curve of glass. Thinking of the theoretical kits we borrow as multi-element photographic lenses, however, is perhaps more apt a metaphor for a structure inclined to produce compelling research. This way, one needs to think of collecting a range of elements that work well together—and about carefully aligning those elements.

★ ★ ★

Mid-2012–Pittsburgh, Pennsylvania. I am working on two new writing projects. One is an academic study of aerial images depicting toxic environments. The other is a cycle of poems based on the visceral, aesthetic experiences defining my move from Western Montana to Western Pennsylvania. In Pittsburgh, most features of the landscape function primarily as unwelcome reminders of my Upstate New York childhood. The correspondence between this landscape and the one that first shaped me is not all about me; it owes much to physical geography and the social history of industrial activity. The poems I am writing have become looser, although they still stagger and restart. In these poems,

> *Things don't just pile up: they deluge.*

All the objects and ideas I try to describe feel, overwhelmingly and illogically,

> *cracked with*
> *little latches of similarity.*

These poems are not the only place where my insecurity in the face of the physical world shows up; neither are they the only place where correspondences driven by my experience of different landscapes structure my attention. When, as a scholarly researcher, I am drawn to the work of photographer, mixed media installation artist, and activist David T. Hanson, I cannot say to what extent it is because of the geographic claustrophobia I've been feeling or to what extent it has to do with long-standing intellectual interests in land use, land management, the history of photography, and related rhetorics. I act as if my time in Montana is what connects me to this Montana photographer's work, but it is also true that the personal history I am trying to hold at bay, by working on this research project, is in part populated with snapshots of things such as school trips to "heritage" sites on the shores of one of the most polluted lakes in the United States.

I focus myself by working outward from Hanson's *Waste Land* series. Originally composed in 1985 and 1986, *Waste Land* comprises 67 triptychs. Each triptych depicts a Superfund site and includes the official government text designating the site, a modified USGS map showing the boundaries of the site, and an aerial image of the site composed by Hanson. The featured sites represent "a cross-section of American geography and industrial waste activities," and texts embedded in the project reveal, among other things, "some of the elaborate legal strategies that corporations and individuals have used to avoid responsibility for the contamination and cleanup" of these "highly hazardous" sites (Hanson 1997, 53). *Waste Land* is about the confluence of the constructive and the destructive; it acts poetically (aesthetically) to materialize a specific relationship between viewers, literal wide-angle imageries, and the kind of wide-angle interdisciplinary thinking that I've begun describing here.

Landscapes are sometimes understood, colloquially, as separate from lived-in places, as backdrops against which human life unfolds. Hanson's work, however, promotes a more nuanced understanding of landscape, one that resonates with a broad body of existing geographic and philosophic work. Theory rooted in the field of geography gives us language for the kind of engagement his images ask viewers to enact. It helps us name how "landscape reflects the way humans have cared for, built in and exploited the surroundings" (Palang et al. 2017, 128). It encourages us to more directly engage with the ways "culture represents both problem and possibility, form and process" (128). Landscapes are, among other things, symbolic environments, and "when events or technological innovations challenge the meanings of these landscapes, it is our conceptions of ourselves that change through a process of negotiating new symbols and meanings" (Greider and Garkovitch 1994, 2). In part, what I like about Hanson's work is that the longer I spend with it, the more intensely it disrupts my relationship to lineages of pastoral imagery, American landscape photography, and abstract formalisms. I begin to talk about *Waste Land* as a machine that, like the wide-angle lens or the camera itself, has the potential to instruct viewers in ways of seeing and ways of reacting to the limits of sight.

The aerial photographs included in Hanson's series are filled with rich colors and dominated by geometric patterns. At a glance, they are pleasurable—more formally reminiscent of color field paintings than of other critical documentary photographs set in the American Outdoors.[3] My initial reactions to these images are consistent with Kim Sichel's observation that an overhead view "offer[s] a radical departure from conventional human-scaled landscape," as well as with the way she associates aerial photography (often but not always produced with wide-angle lenses) with concepts including "lack of horizon, abstraction, geometry, flatness, dehumanization, and deception of scale" (2011, 94). Still, there's something about these particular aerial photographs that makes the common association of the genre with a term like dehumanization feel a little off. Even though I can imagine these images as an antonym of portraiture, I can't help finding a decided (if distorted) humanness in them. Moreover, there is a leakiness in the pleasures provoked by Hanson's aerial views that strikes me as closely related to the values and possibilities of geopoetics, at least insofar as this term is concerned with couplings that translate between the scale of the human and the scales—large and small—of the more-than-human world.[4]

I attribute this leakiness to the way Hanson's photographs are presented not just as images of environments but also as images that participate in intertextual, multimedia environments. This presentation demands attunement to all that cannot be captured even in a wide, overhead shot. It asks us to think of slag heaps and access roads as creative marks made by human action, expanding (or at least reemphasizing) the many scales at which human-instigated landscape "writing"— here, inclusive of asemic transcriptions perpetuated by industrial and military infrastructures—operates in symbolic ways. This multimedia presentation also asks us to acknowledge how toxins hide from the overhead eye that has defined surveillance history.[5] The underground movement of toxins allows them to contaminate soil and groundwater; such contamination shows up in official texts even as it betrays the boundary lines found on official maps. When Hanson writes about choosing to emphasize aerial perspectives in his work, he writes about these and related ideas; in addition to addressing the history of military surveillance directly, he writes about how aerial photography "allows for the framing of relationships between objects that may seem unrelated on the ground, and [how] it permits access to sites with security restrictions" (1997, 5). He also writes about how this translates into the more ontological premise that when one takes to the air with one's equipment, "what otherwise cannot be pictured becomes available to the camera" (5). This impulse is congruent with the openness of geopoetic thought and with its emphasis on a situated politics of form.

In the presence of the example set by a project like *Waste Land*, the line some critics would have us draw between grounded, ethical responses and abstracted, aesthetic responses to various objects—including images and landscapes—blurs. Like the idea of the geopoetic, abstraction functions in plural, nuanced ways. It can be a synonym for something that exists only as an idea or something that has slipped free of all context. It can also be the antonym of representation. The human brain is good at filtering out details it perceives as noise, that is, at abstracting toward meaning. The human brain is also good at filling in missing details, that is, at converting abstract images into concrete images of some aspect of the world (think here of Gestalt principles in both art and psychology). In short, abstraction is not always and necessarily about holding human experience at a distance. It does, however, ask us to expand our idea of where the human lives in relation to a text or an image. I suggested at the outset of this chapter that it has often felt important to me as a poet to articulate ways in which form can hold deeply personal experience even as semantic content eschews the autobiographical. A color field painting can be read as a text that's "about" a painter's expert, bodily relationship to surface and texture, or it can be read as a text about the relationship between the scale of the canvas and the scale of the painter's and/or the viewer's body. A photographer's body is, of course, implicated in image production in similar ways, although it is somewhat less common for critics to talk about this fact. Part of the pleasure of using a wide-angle lens is precisely that what it produces *is not* exactly what the body doing the capturing saw. Part of the allure of grand landscape images, especially when their compositions contain visual clichés, is that a photographer took their body and equipment to a location that viewers are not necessarily likely to take themselves to. This question of the scalar relationship between a photographer's body and a set of images is linked to the methodological point that rhetorician Valerie Hanson makes in her study of communicative potentials linked to scanning tunneling microscopy, but which applies to all kinds of imaging processes (analog, digital, and biological): "Attending to imaging processes when analyzing the rhetorical work of an image can reveal aspects of rhetorical elements that frame both image-creators' and image-viewers' experiences of the information the images communicate" (2015, 166).

In addition to referencing legal, historical, and aesthetic features of the overhead view in his discussion of the choice to showcase aerial imagery in the *Waste Land* series, the photographer Hanson also admits a desire "to minimize [his] own exposure to these highly toxic environments" (1997, 6). He is concerned in a real, human way about the unknowable long-term effects that visiting *so many* toxic sites might have on his own body across time. Such a focus on collateral, systemically induced bodily damage suggests that the human scale is, in fact, present in these "scientific" aerial shots; it is simply present as context rather than in the images themselves. Indeed, many of the unnamed individuals whose bodies *have* been assailed by these particular sites engaged them in obtuse or abstract ways across swaths of time; they were neighbors as well as laborers. Dramatic images of their bodies would have poorly addressed the mundane nature of illness and debility, in addition to partaking of a trope that transforms mass negligence into stories of individual tragedy and/ or overcoming. Which is to say, a picture of a body—depending, of course, on how the pictures are taken, how they invite viewers to envision, and what viewers bring to them—doesn't necessarily present bodily implications more compellingly than an aerial image. Beyond highlighting the human-ness involved in inhuman aerial images, Hanson's fears about the dispersed danger America's most toxic sites continue to present to bodies that transect them also tie his machine-mediated visions of these landscapes to some core bodily insights from the field of disability studies. This is a field with a vested interest not only in ways an environment can limit or enable an individual but also in the cultural functions of assistive technology (a category in which we might include any camera, but especially cameras featuring lenses with focal lengths notably distinct from the focal length of a human eye and those that act in assemblage with flying machines to change a human's perspectival capacity).

The version of geopoetics that makes the most sense to me is linked, then, structurally (not incidentally) to insights that have been well articulated by disability studies scholars; in particular, this version of geopoetics takes for granted that human "capacity is not discretely of the body. It is shaped by and bound to interface with prevailing notions of chance, risk, accident, luck, and probability, as well as with bodily limits/incapacity, disability, and debility" (Puar 2017, 19). This version of geopoetics also casts landscapes, including urban and industrial land-scapes, in related terms—as capacious, but shaped by chance and the idea of chance, complex ecological and social systems, and the capacities and limits of the humans that interact with them.

* * *

Early 2013–Pittsburgh, Pennsylvania. I have been researching aerial imagery and working on the same long cycle of poems about emplacement and embodiment for more than a year. Fragments of landscape language clang in my mind, contradicting one another, but also binding me to this city. I now live two blocks from an abandoned Mazda dealership filled with ghost trees—saplings that grew up through the cracked floor and then found the place lacking light or water or room to root. Near-transparent leaves cling tenuously to their slim branches. There's a stillness to them that I have come to appreciate. I walk past these trees on my way to the cancer center where I am receiving chemotherapy to treat Hodgkin's lymphoma. Turn by turn, the drugs seem to be dismantling the mechanisms that control my short-term memory. Some days this only hurts my ability to do research reading. Some days, I have worse aphasia than I can bring myself to admit. While I have often chosen not to speak, either out of propriety or because I was not ready for the social fallouts language invites, the feeling of not having a fluent private language for the world catches me off guard. For obvious reasons, I remain interested in the abstract and the inarticulate, in poetics of fracture, fragility, and vulnerability. But something about the aerial imagery study feels decidedly *useful* to living through this, too. That should surprise me less than it does: I know better than most that humans frequently look to texts, images, and tasks to define (and, when necessary, redefine) ourselves and our places in the world. Likewise, I am better acquainted than most with the idea that "when we think, we look to technology, pulling from its forms and artifacts words to express the world" (Rickert 2016, 227). Snapshot-style images continue to appear in my mind, and likely because I have recently logged so many hours examining satellite images and photographs taken from air-planes and hot air balloons, these spontaneous images often comprise overhead, wide-angle views.

My study of Hanson's Superfund images has given me the language to say an overhead view can function variously even as it distances and abstracts; it can expedite unexpected intimacies, provoke research, or change our relationship to environmental contexts. That I have a language for engaging the complex historical trends that link technologies developed by government agencies to support military surveillance to recent decentralized practices such as the promotion of citizen science or counter surveillance of government activities feels relevant to the fact that modern chemotherapy, without which I would have no chance of survival, has its own wartime history (related to the observation of damaging effects mustard gas had on soldiers during WWII).

Infusion days are long. There's a roof deck attached to the blood cancer center, and when the spring weather is decent enough, after being lectured on the need for sunscreen despite the steel city's near-perpetual clouds (one of the drugs I am getting causes photosensitivity), sometimes I drag the IV pole that I'm attached to out to edge of the roof. I'm only four floors up, but there's some grand pleasure in just being outside. I begin calling the street that runs between this building and the main hospital *the canyon at the center of the medical complex*. But mostly I just stand as close to the sky as I can, listening to all the little sounds that make up what someone else might call silence. I don't try to name the sounds because I don't have the capacity. Eventually—usually very quickly, although time dilates, so I can't say exactly what that means—I stop trusting my ability to stand up, so I go back inside.

The sensory images I collect on the roof deck are mundane. They are filled with distance. I do not feel they exist on the scale of my body, and—given my body's recent betrayals—that's part of their pleasure. I began this chapter by asserting the value in "merely" describing some features of one way of sensing and engaging the world that is well described by the adjective geopoetic. One thing I've been describing in this chapter, however obliquely, is how a literal *accumulation* of mundane images in my body has changed both how I tune to the world around me and how I use language to engage that world. I've also paid attention to ways that carefully deployed abstractions can invite viewers or readers into acts of meaning making by facilitating new points of contact between bodied experience and the world at large. Where geopoetics invite something that might be worth calling wide-angle engagement, it has to do with these acts of accumulation and facilitation and with the difference between asking directly what something means and asking the kinds of questions that make new, contextual connections available to view. Flipping back through my notebooks from earlier in the year, I find the line:

Bodies suture :: geographic rupture.

I do not think that "bodies suture geographic rupture" is a promise geopoetic methods can make fully or entirely. I do think it is a proposition that invites continued exploration of potentials; if bodies belong to the world, then we have a duty to care for that world, as well as an ability to allow the world to comfort us and connect us with each other.

Notes

1 Where conversations about realism in photography tend to stray quickly into ontological territory (e.g., see Barthe's iconic text [1981] or the more recent engagements of Kember [2008] and Walton [2008]), the term immersion can suggest a water bath or language-learning class just as easily as it suggests virtual reality (VR) tech; it invokes an *engrossing* environment, emphasis on the adjective.

2 The distinction between "groundedness" and "authenticity" is one I first learned to make through discussions of Dydia DeLyser's "Authenticity on the Ground: Engaging

the Past in a California Ghost Town" (1998) in a Human Geography seminar in 2011 with Tom Sullivan at the University of Montana.

3 Histories of critical documentary in the United States are beyond scope here; they speak to the American nationalisms, the concurrence of westward expansion by Euro-American settlers, and the technological developments in photography and broader art histories. They include: infamous incidents, e.g., a presentation of US Geological survey photographs by William Henry Jackson that some rhetoricians argue was a major factor in the passage of the Yellowstone Park Act of 1872—this despite a lack of clarity about the actual historical role the images played (see Bossen 1982); organized multi-photographer campaigns such as the one that produced the Farm Service Administration (FSA) photographs of the 1930s; and movements such as the "New Topographics" of the 1970s, which push back on the romanticism of grand masters such as Ansel Adams by focusing on altered or processed landscapes (see Foster-Rice 2011).

4 Another way to imagine this link is to argue that the more-than-human styles of abstraction Hanson's aerial views instruct us in are closely related to the details of Eric Magrane's assertion, "[g]eopoetics can help us to make meaning of the world, at the same time resisting explanatory models that themselves preconfigure what we see" (2015, 97).

5 The historical relationship between aerial photographic techniques and wartime reconnaissance activities is well established, beginning with photographs taken from balloons shortly after photography's invention in the nineteenth century. The advent of dry plate photography in 1871 lent practicality and flexibility to the idea of aerial photography, with rockets, kites, and pigeons taking their places alongside balloons as pre-airplane-era camera carriers. In 1903, the Bavarian Pigeon Corps notably began using carrier pigeons for aerial reconnaissance. Their birds wore tiny breast-mounted cameras, which could be set to take automatic exposures at 30-second intervals.

References

Barthes, Roland. 1981. *Camera Lucida: Reflections on Photography*. Translated by Richard Howard. New York: Hill and Wang.

Bossen, Howard. 1982. "A Tall Tale Retold: The Influence of the Photographs of William Henry Jackson upon the Passage of the Yellowstone Park Act of 1872." *Studies in Visual Communication* 8 (1): 98–109. https://repository.upenn.edu/svc/vol8/iss1/10.

Cicala, Roger. 2011. "The Development of Wide-Angle Lenses." *Lens Rentals: History of Photography*, March 8. www.lensrentals.com/blog/2011/03/the-development-of-wide-angle-lenses/.

Daston, Lorraine, and Elizabeth Lunbeck, eds. 2011. *Histories of Scientific Observation*. Chicago: University of Chicago.

Dydia DeLyser, Dydia. 1998. "Authenticity on the Ground: Engaging the Past in a California Ghost Town." *Annals of the Association of American Geographers* 89 (4): 602–632. doi:10.1111/0004-5608.00164.

Farman, Jason. 2011. *Mobile Interface Theory: Embodied Space and Locative Media*. Abingdon: Routledge.

Flynn, Ciara. 2012. "Geographic Cure." *Thought Catalog*, April 26, 2012. https://thoughtcatalog.com/ciara-flynn/2012/04/geographic-cure/.

Foster-Rice, Greg. 2011. "Systems Everywhere': *New Topographics*, and Art of the 1970s." In *Reframing the New Topographics*, edited by Greg Foster-Rice and John Rohrback, 45–70. Chicago: Center for American Places at Columbia College.

Gabrys, Jennifer. 2016. *Program Earth: Environmental Sensing Technology and the Making of a Computational Planet*. Minneapolis: University of Minnesota.

Greider, Thomas, and Lorraine Garkovitch. 1994. "Landscapes: The Social Construction of Nature and the Environment." *Rural Sociology* 59 (1): 1–24. doi:10.1111/j.1549-0831.1994.tb00519.x.

Hanson, David. 1997. *Waste Land: Meditations on a Ravaged Landscape*. New York: Aperture.

Hanson, Valerie. 2015. *Haptic Visions: Rhetorics of the Digital Image, Information, and Nanotechnology*. Anderson: Parlor.

Harmon, Katherine. 2003. *You Are Here: Personal Geographies and Other Maps of the Imagination*. New York: Princeton Architectural.

Kara, Helen. 2015. *Creative Research Methods in the Social Sciences: A Practical Guide*. Bristol: Policy Press.

Kember, Sarah. 2008. "'The Shadow of the Object': Photography and Realism." *Textual Practice* 10 (1): 145–163. doi:10.1080/09502369608582242.

Kimmerer, Robin Wall. 2013. *Braiding Sweetgrass: Indigenous Wisdom, Scientific Knowledge, and the Teachings of Plants*. Minneapolis: Milkweed Editions.

Maddux, James F., and David P. Desmond. 1982. "Residence Relocation Inhibits Opioid Dependence." *Arch Gen Psychiatry* 39 (11): 1313–1317. doi:10.1001/archpsyc.1982.04290110065011.

Magrane, Eric. 2015. "Situating Geopoetics." *GeoHumanities* 1 (1): 86–102. doi:10.1080/2373566X.2015.1071674.

Maxwell, Glyn. 2012. *On Poetry*. Cambridge: Harvard University Press.

Palang, Hannes, Katriina Soini, Anu Printsmann, and Inger Birkeland. 2017. "Landscape and Cultural Sustainability." *Norsk Geografisk Tidsskrift—Norwegian Journal of Geography* 71 (3): 127–131. doi:10.1080/00291951.2017.1343381.

Parikka, Jussi. 2015. *A Geology of Media*. Minneapolis: University of Minnesota.

Puar, Jasbir K. 2017. *The Right to Maim: Debility, Capacity, Disability*. Durham: Duke University Press.

Pugh, Jonathan. 2009. "What Are the Consequences of the 'Spatial Turn' for How We Understand Politics Today? A Proposed Research Agenda." *Progress in Human Geography* 33 (5): 579–586. doi:10.1177/0309132508099795.

Rickert, Thomas. 2016. "Afterward: A Crack in the Cosmic Egg, Turning into Things." In *Rhetoric: Through Everyday Things*, edited by Scot Barnett and Casey Boyle, 226–231. Tuscaloosa: University of Alabama.

Sichel, Kim. 2011. "Deadpan Geometries: Mapping, Aerial Photography, and the American Landscape." In *Reframing the New Topographics*, edited by Greg Foster-Rice and John Rohrback, 87–106. Chicago: Center for American Places at Columbia College.

Sommer, Doris. 2014. *The Work of Art in the World: Civic Agency and Public Humanities*. Durham: Duke University Press.

Turchi, Peter. 2004. *Maps of the Imagination: The Writer as Cartographer*. San Antonio: Trinity University.

Walton, Kendall L. 2008. "Transparent Pictures: On the Nature of Photographic Realism." In *Photography and Philosophy: Essays on the Pencil of Nature*, edited by Scott Walden, 14–49. Malden: Blackwell.

Warf, Barney, and Santa Arias, eds. 2008. *The Spatial Turn: Interdisciplinary Perspectives*. Abingdon: Routledge.

Withers, Charles. 2009. "Place and the 'Spatial Turn' in Geography and in History." *Journal of the History of Ideas* 70 (4): 637–658. doi:10.1353/jhi.0.0054.

Wood, Denis. 2010. *Everything Sings: Maps for a Narrative Atlas*. Los Angeles: Siglio.

2

A COSMOLOGY OF NIBI

Picto-poetics and palimpsest in Anishinaabeg watery geographies

Kimberly Blaeser

We emerge from and live immersed in a particular cultural place and moment, as well as from and in a specific geographical and cosmological system of knowledge. Our inheritance of world and worldview informs our every experience; ultimately, our patterns of being transform into a lifeway. Both vision and experience create

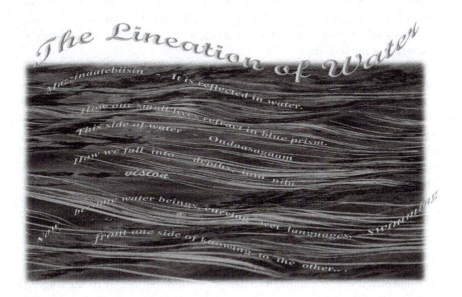

FIGURE 2.1   The Lineation of Water.

our aesthetic, our understanding and expression of beauty. Perhaps artistic desire itself arises from the factors of place.

How our eyes bend to water image

On White Earth Reservation in the northern Minnesota of my childhood, I grew up in rich water country. My family spent nearly every weekend of my early life fishing, swimming, skating, setting net, spearing, ice fishing, and harvesting wild rice. We knew Water as provider, as well as the home of the great underwater panther, *Mishibizhi*. Water didn't come from a faucet. We collected rain, pumped our drinking water from the belly of earth, or hauled it from springs in cream cans in the back of my uncle's pickup. Our daily household water source was a pail and a dipper.

Because water came through labor, we didn't waste it. Because our life was linked to water, we knew it as a relative. Because stories told how the cycles of women and moon were tide-driven, we understood Water's power and its mythic truth.

The imprint of years, of the cycle of seasons, in this watery geography, also formed my internal rhythms and impacted my perception. The impression endures. Even now it affects how I see, what I notice, and how I assess value. Years swing between freeze-up and break-up. The insides of my eyelids are etched with the images of fish swimming beneath the spearing hole of a winter fish house. And templates for beauty appear in watermarks on rocks, in the sound of ice cracking, in the shapes of mottled lily pads lifting in the fall wind. The minutiae of my days and nights awaken a particular kind of longing in my spirit, a territorial hunger or tenderness in my fingertips.

Thus, as a Native poet with a kinship consciousness formed among the White Earth Nation and northern Minnesota landscapes, various threads of my inherited cultural understandings always surface in my aesthetic and intellectual vision, in my desires, and in the kind of work I do. My experience of the natural world arises partly from and is imbued with a tribal context formed of ancestral connections, ideas of tribal continuance, and complex spiritual realities. These Indigenous metaphysical and epistemological groundwaters also seep into my artistic practice. The cultural nexus that feeds my creativity incorporates not only physical places, but also origin stories, ceremony, and ritual. Those native to the vast water country learn to honor *nibi* with songs, prayers, and offerings, and to live gratefully, aware of our dependence. Together we carry the burden of continuance.[2]

Anishinaabe cultural ecologist Melissa Nelson speaks of an "ecoliteracy," which involves "stories of kinship with our other-than-human family" as well as "unmediated time . . . alone in landscapes to . . . listen deeply to the language of the land" (2013, 216). In my understanding, that ecoliteracy involves not only an acquaintance with natural creatures, places, and their stories, but also a certain kind of embedded land knowledge that sometimes yields visual and verbal language and includes a deep awareness of relatedness that evolves into a moral compass of reciprocity. Relationship results in coded knowledge and carries responsibility.

As Okanagan poet Jeannette Armstrong claims, "All indigenous peoples' languages are generated by a precise geography and arise from it" (2016, 148). Indeed, if we think of language as "a systematic means of communicating ideas or feelings by the use of conventionalized signs, sounds, gestures, or marks having understood meanings" (*Merriam-Webster*, "language"), the ecolanguage we learn involves not only words but also natural sounds such as bird communications or the voice of thunder, as well as visual signs such as tracks or wave patterns. It includes the observed behavior of weather as well as that of mammals, fish, and amphibious beings. The language of environment or habitat inevitably reflects Indigenous wisdom gathered across generations and often includes the embedding of that familial understanding in stories—whether factual or metaphorical. Thus, the cycle of traditional stories, sometimes designated as tribal "myths," actually arises from or evolves as expressions of these crucial geo- or cosmo-logical understandings and therefore belies the implied connotation of the word "myth" as "imaginary or unverifiable" (*Merriam-Webster*, "myth"). Like tribal stories, many of the creative acts of poets and other Indigenous artists likewise arise from within and reinforce the geo-wisdom of Native nations.

Philosopher David Abram suggests Native songs and stories serve a "topographical function," but also "provide the codes of behavior for the community and its relationship to place" (1996, 175). Because Anishinaabeg people have lived for generations along and among the Great Lakes and river systems such as that of the Mississippi, our traditional worldview and vernacular—or what I call a "cosmology of *nibi*"—arise from these watery geographies. As Anishinaabe writer and linguist Margaret Noodin notes, "The center of *Anishinaabewkiing*, or Anishinaabe country, is the life-giving *gaming*, the 'vast water'" (2014, 1).

When Nelson characterizes the "hydromythology" or oral water stories of the Anishinaabe people as "imaginative ecosystems," she acknowledges the wide swath of life they encompass, including: origin stories such as that of the Earthdiver; migration stories of travel by water from the St. Lawrence River, following the sacred *Miigis* shell, to the Great Lakes and to the *maanomin* (food that grows on water); and stories involving the balance maintained by *Animikii*, the thunderbird, and *Mishibizhii*, the great underwater panther. This ecohistory traces decades of lives sustained by various waterways that supply everything from *gigoonyag* (fish), to *anaakanan* (woven reed mats), to *mashkiki* (medicine). Nelson notes how spiritual practices often center on the healing water drum (*dewe'igan or mitigwakik*) of the Midewiwin Society, and she points out that the Anishinaabeg clan system includes spiritual teachers of the Fish Clan and sometimes the "*niibinaabi* clan or 'water spirit' clan" (2013, 218).

The "cosmology of *nibi*" as a way of conceiving of the world must also involve the interdependent relationships between realms—between underworld/ underwater, earth, and sky. Obviously, water does not only sit on or under the earth; it moves through the whole system of beings, including plants, atmosphere, and all earth and animal bodies. Understanding this dynamic hydro-reality means entering into the motion and perpetual transformations of substances and spiritual entities beginning, for example, with the watery womb and including the moon-driven reality of tides and female cycles. Water literally sustains life, but it also becomes a sustaining symbol for the sometimes unknowable workings of spirits and nature. Many Anishinaabeg teachers and elders explain the high regard our people hold for water, claiming *nibi* is, as Nelson says, "imbued with great meaning and supernatural power"; it "*is* a manitou, and contains manitous" (2013, 217, emphasis in original).[3]

Thus, the ecolegacy of Anishinaabe peoples involves a moment-to-moment "reading" of land languages that insures sustainable practices and flourishing in a place, as well as the mythic and spiritual understanding of kinship. Taken together, these create a certain world or cosmological view—an Indigenous ontology. I believe this ontology also informs creative endeavors and forms what I understand as a cultural "geo-aesthetic."

Art as reciprocity

Among Native peoples, the importance of land narratives (oral and written, ancient and contemporary) have long been recognized. Anishinaabeg professor Brian McInnes claims these stories provide "a map to both the structure of the universe and our lives" (2016, 17). "We seek stories like a plant seeks sunlight," he writes. "They are vital to our growth and well-being in the world" (17). Like many other aspects of Native life, traditional literary arts are informed by a reciprocity or a reseeding of story. When our vital cosmos comes alive in tribal narratives and songs in ways that provide us sustenance, it awakens our desire, need, or responsibility to story the land and our kinship with it. In a relational universe we do not remain passive—we make return.

In *The Way to Rainy Mountain*, N. Scott Momaday clearly illustrates such a relationship when he describes the Kiowa creating a story at the base of Devil's Tower:

> There are things that engender an awful quiet in the heart of man. Devil's Tower is one of them. Two centuries ago, *because they could not do otherwise*, the Kiowas made a legend at the base of the rock.
>
> *(1969, 8, emphasis added)*

The perceived power or immanence of place can inspire or even compel artistic response in those invested in its workings. Indigenous cultures embody, interpret, translate, or "story" elemental voices, as the Kiowa do in Momaday's telling.

Inherent in this practice—that is, in the translation of perceived communications or embedded knowledge of place—is the underlying belief that "other" beings in the universe hold wisdom and somehow give voice to that foundational understanding, or alternately, sing the mystery. Also key to the process of such an exchange is a certain familiarity with the beings and patterns of local environments and a general practice of attentiveness among its inhabitants. Anishinaabe artist Bonnie Devine wonderfully expresses this kind of engagement in her work. Half of her diptych, "Letter to William" (from 2008), for example, includes the following language on an abstract close-up of rockface: "I have come to listen, believing the rock is filled with stories. I have come to read, believing the rock is a text" (reproduced in Penny and McMaster 2013, 94–95).

In an artist statement about the project, Devine remarks, "There is a story in the rock, not related to language or words but mute, elegiac, and undefended" (2016). These two declarations taken together highlight the mindset and performance of a geo-aesthetic as I understand it—a geo-aesthetic that includes an attempt to rein-scribe Indigenous land histories and thus simultaneously to honor the sovereignty of Anishinaabeg intellectual traditions.[4]

Expressing a similar philosophical understanding about embedded stories, Brian McInnes discusses Canadian Anishinaabeg teachings and underscores Native art and artists' fundamental calling in relationship to place. The element(s) of "First Nations' stewardship of the land," he claims, include(s) "maintenance of these places *and their stories*" (65, emphasis added). For the sake of continuance of our mutually constitutive community, our human care of place includes the bond to herald it. To reinforce knowledge of, relatedness to, and responsibility for our geo-community, we "teach" place through literary and artistic expression. In this context, art can serve as reciprocation, the return we make in our pact of kinship—a crucial tool for imparting sustainable wisdom.

Indeed, Indigenous oral and material traditions have long served an interpretive or teaching role among Native peoples. But with some traditions—weakened or no longer practiced universally—the crucial lessons they embody could go unheard by a portion of tribal members. Here art might serve as a conduit of valuable cultural knowledge or even be used to revive or reanimate traditions. Curator and Native art historian Bruce Bernstein claims, "Native art plays a central role in health and endurance. . . . Art is why tribes persist, as it strengthens Native American aspects of life (2012, 30).

Art as gesture

If Native arts encode Indigenous knowledge, they do it largely through their enact-ment of the cosmic fluidity. Tribal compacts respecting sustainability arise within a culture aware of the inherent overlapping of beings and energy. As Anishinaabeg, we know the *wiishkobaaboo* that spills from the maple tree in spring depends upon the root system of the tree, the stored moisture in the earth, the fluctuations in the spring weather—but really it depends upon a much longer cycle of happenings, of relationships and exchanges made throughout the year. It depends, for example, upon beings such as *giizis* (our sun Grandfather). The braided and sequential series of events and conditions involved in the production of the water-based gift of sap can best be expressed through "language" patterns that honor and enact an aware-ness of this elemental interdependence.

Native geopoetics often succeeds in voicing this intermingling of science and kinship because it extends itself across ways of knowing, from the empirical to the imaginative, and it employs a wide swath of creative methods. Jaune Quick-to-See Smith describes how in "old-times" Native artists were "the keepers of memory, the recorders of events, the markmakers of prayers, and the shaman who brought the unseen world into view" (2009). Geopoetics sits easily in this aesthetic tradition, and, in its frequent quest to express the ineffable or the ultimately unsayable, the artistic "currency" of place-based tribal arts often involves symbol and gesture. Lakota artist Colleen Cutschall, for example, writes about "drawing where the seen and unseen come together to form the narrative" (1996, 65). In my conception of this praxis, the visual and lingual cues employed might involve concrete physical details of place: the loam and lichen, the steam rising from snow. But, they also point toward an animate, multilayered, and intricately (some might say rashly) interconnected universe. As Laguna writer Leslie Silko describes it, the world and ourselves in the world are "part of an ancient, continuous story composed of innumerable bundles of other stories" (1996, 31).

In my own creative work, I endeavor through image, language, presentation, and gesture to insinuate these complex bundles within bundles of relationships of my known culture and cosmology. Recent work, for example, incorporates layers of text and image in what I call "picto-poems." Inspired by both Anishinaabeg pictographs and Native American ledger art, these sometimes involve an ekphrastic pairing of poems and image, sometimes layered palimpsestic or collage pieces. Often my poems and picto-poems use the reflections and refractions of water or various eco-signs from animal markings to rock formations. I write the visual realities of place in order to gesture toward palpable but invisible imprints of history or story—to imply the unseen or sometimes unknowable. A "true" or successful rendering of natural places in my work comes alive as Native spaces. Hence, as in the following piece, the water animals of my experience become the Earthdivers of tribal story.

The well-known Earthdiver myth reflects a tenet central to Native understanding—our reciprocal relationship to this earth we call Turtle Island. Common to peoples all over the world, the key elements of the story include a flood and the need to reach the bottom of the water to gather grains of sand from which to build a new earth. Though versions vary with the teller, associated Anishinaabeg teachings emphasize the significance of all creatures—small or large beings—and our responsibility to work together in the continuing creation and recreation of the world (distinguishing the account from biblical creation stories in which the world is created for human beings, and is ordered hierarchically). Anishinaabeg tellings in particular feature the self-sacrificing action of one of the smallest of water beings—Wazhashk, or muskrat.

My picto-poem wants, in its allusiveness, to emphasize our human insignificance and, at the same time, our relatedness, our belonging to something more vast than ourselves—to the immense bodies of water, to story, to all time. In the following image, I use the surreal rendition of "water beings," repeated like layers of existence—like retellings or echoes of stories in order to untether the reader/viewer from "reality" or so-called "realism." The piece invites a reimagining of our own multiplicity, our "dream" selves, perhaps even a realignment of our "mythic quest." The doubled language of the bilingual poem assists in creating this "rashly" plural reality, while ripples and the wake generated by the swimming being(s) insinuate ongoing motion. In another crossover, the story text, by itself becoming image, further erases the idea of supposed separations.

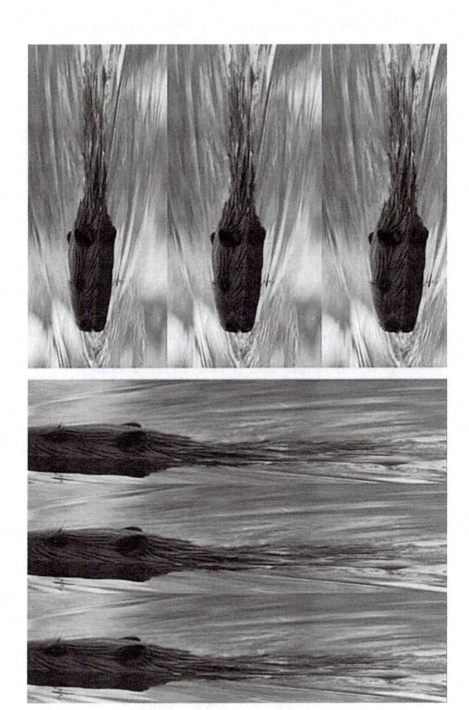

FIGURE 2.2 Dreams of Water Bodies.

Dreams of Water Bodies

Wazhashk,
small whiskered swimmer,
you, a fluid arrow crossing waterways
with the simple determination
of one who has dived
purple deep into mythic quest.

Belittled or despised
as water rat on land;
hero of our Anishinaabeg people
in animal tales, creation stories
whose tellers open slowly,
magically like within a dream,
your tiny clenched fist
so all water tribes.
might believe.

See the small grains of sand—
Ah, only those poor few—
but they become our turtle island
this good and well-dreamed land
where we stand in this moment
on the edge of so many bodies of water
and watch *Wazhashk,* our brother,
slip through pools and streams and lakes
this marshland earth hallowed by
the memory
the telling
the hope
the dive
of sleek-whiskered-swimmers
who mark a dark path.

And sometimes in our water dreams
we pitiful land-dwellers
in longing
recall, and singing
make spirits ready
to follow:
*bakobii.**

*Go down into the water.

Nibii-wiiyawan Bawaadanan*

Wazhashk
agaashiinyi memiishanowe bagazod
biwak-dakamaadagaayin
mashkawndamyin
googiigwaashkwaniyamban
dimii-miinaandeg gagwedweyamban.

G'goopazomigoog
ninii-chiwaawaabiganoojinh akiing
ogichidaa Anishinaabe
awesiinaajimowinong, aadizookaang
dash dibaajimojig onisaakonanaanaawaa
nengatch enji-mamaanjiding
gdo'bikwaakoninjens
midash kina Nibiishinaabe
debwe'endamowaad.

Waabandanan negawan
aah sa ongow eta
maaaji-mishiikenh-minis
minwaabandaan aakiing maampii
niigaanigaabawiying
agamigong
Wazhashk waabamang, niikaanaanig
zhiibaasige zaaga'iganan gaye ziibiinsan
mashkiig zhawedaamin
mikwendamin
waawiindamin
ezhi-bagosenimowaad
ezhi-googiiwaad
agaashiinyag memiishanowewaad
bagazojig
dibiki-miikanong.

Nangodinong enji-nibii-bawaajiganan
gidimagozijig aakiing endaaying
bakadenodamin
dash nagamoying
jiibenaakeying
noosone'igeying
bakobiiying.
*Translation by Margaret Noodin

Art as the ritually necessary

"Dreams of Water Bodies" strives, through its various ambiguities and gestures, to reanimate and make viable an Anishinaabe story, to link tribal "myth" to contemporary experience, and ultimately to invite a discovery or contemplation of the fundamental workings of the cosmos. These movements in my work seem to align it with a longer visual artistic tradition among woodland artists that attempts to image cosmic relationships. Art history professor Ruth Phillips, for example, writes about the workings of Anishinaabe art in regard to its focus on the spiritual and cultural understandings of relatedness, and notes:

> Awareness of . . . cosmic structures enables an individual to center him or herself in relation to the forces that animate the universe, to live in the world harmoniously, and to achieve the experiences of connection with the animating spiritual powers through which blessings of power are obtained. Art helps create this awareness and produce the ritually necessary experience of centeredness.
>
> *(2013, 62)*

Other Indigenous authors and artists have likewise flagged the fruitful, sometimes transformational, role of story and art in tribal spiritual matters. Métis artist Dylan Miner, for example, compares the work of artists to that of medicine people, saying, "Like the *mshkikikwe* or *mshkikiwinini*, the artist and storyteller observe the world around them in ephemeral ways that create meaning from an otherwise unintelligible existence" (2013, 318).

Phillips's description of classic Anishinaabeg visual representation explains in some detail the way cosmological underpinnings may become manifest in art, noting, for example, layout and certain key relational factors between elements and realms:

> Cosmic space, in Anishinaabe and other shamanistic world views, is conceptually aligned along a central axis, sometimes imagined as a tree, that links an upper sky-world zone with earth and with an equivalent zone of power under the surface of earth and water. These zones may be visually mapped as a kind of vertical layering.
>
> *(59–60)*

She writes, too, of the way "the cosmos is conceptualized as a circular space centered on the world axis" with the "spatial field" often "divided in half to reflect the interdependent nature of . . . sky/earth, male/female, day/night, light/dark, good/ evil or creation/destruction."[5]

These and other observations of Anishinaabeg style and content all hearken back to the substrata of worldview. We create out of an awareness, imprint on our story or art our way of seeing and being in the world. Although I recognize how I use visual cues in my own work to suggest cosmic realms and the dynamic crossings or relationships I story between them, I come to these visual embodiments intuitively. I actually create toward a sensibility—by hearkening after a certain slant of knowing. Embedded understandings born of intergenerational investment in place feed the process of discovering meaning in and through creative acts.

In the "water bodies" picto-poem shown earlier, both the visual and the lingual bodies of the text display an *axis mundi*, but they also imply the link, or imminent crossing of the depicted zones. Likewise in the image "A Crane Language" (to follow, fig. 2.3, top), realms and the experience of their overlapping animate the picto-poem. The process by which I arrived at this figuration perhaps best illustrates our elemental immersion in cosmological understandings and indeed the insistent insinuation of these foundational perspectives into each small situation of our lives.

This particular creative sutra began with a Great Blue Heron flying through the blue of spring skies. During a paddle with my son in the Horicon Marsh in Wisconsin on Mother's Day in 2012, we spent a shimmer of time watching the heron stalk fish, snatch them, and swallow them whole. It flew with great pomp in liftoffs and landings, sometimes with a yellow-bellied *giigoonh* still carried in its beak. Gavin turned the canoe while I panned my camera, making photos—we, the bird, the world, all circling something unnameable.

Because the feel of the encounter was rapturous, the first printed image seemed an anemic half rendering. I didn't know why. The position of the bird was stunning—huge wingspan and wingtips curving up, neck in a classic curve, legs extended, and feet pointing straight out to forever. What the camera "captured" was postcard beauty, but what we became in that moment was transcendent. I realized we don't ever see only what is before us: We see what we bring to the experience. What is literally before us, we always experience in the context of our own understandings and beliefs. That heron had flown through the territories of Anishinaabeg ontology and mythology as *ajijaak*, the echo-maker and totemic figure for the crane clan from my cosmology of *nibi*.

How could I make the image suggest the context of my layered reality? I employed palimpsest and gesture, making symbolic use of Anishinaabe floral beadwork, together with snatches of language including the bird's name in Anishinaabe-mowin—*Zhashagi*. The woodland beadwork itself already embodied Anishinaabe symbolic flourishes, the floral circles with their four quarters suggesting among other things a medicine wheel with the four cardinal directions. But referential layers include other teachings, such as that about *binesiwag* (birds) who become symbolic messengers or intermediaries between realms. Their transcendence of boundaries reveals the porousness of divisions such as that between water and sky or the physical and spiritual. Indeed, in the picto-poem, the sky layer and the floral earth are one and the same—their realms meshed. Visually in the palimpsest, the extended body of the totemic *ajijaak* and the alphabet of its name, *Zhashagi*, lay across, or join, the realms above and below.

The lines of the poem, too, story my search to understand our encounter. The language suggests the perspective "ancient" and beyond "reason" from which

I experienced the moment, and presents the *Zhashagi* episode as inflected with a tribal context. Taken together, the image and the poem attest to the "knowledge" that Native spaces still exist in our everyday landscapes, to my belief in how readily we reinhabit them.

In thinking about the complex interaction that gave rise to the picto-poem "A Crane Language," the phenomenon Gerald Vizenor terms "transmotion," as well as his characterization of its "tropes" in literature, provides insight. He describes the psychic leap from physical to spiritual perceptions this way:

> Natural motion is a heartbeat, ravens on the wing, the rise of thunderclouds and the mysterious weight of whales. Transmotion is the visionary or creative perceptions of the seasons and the visual scenes of motion in art and litera-ture. The literary portrayal and tropes of transmotion are actual and visual images across, beyond, on the other side, or in another place, and with an ironic and visionary sense of presence. The portrayal of motion is not a simu-lation of absence, but rather a creative literary image of motion and presence.
>
> *(2015, 7)*

Similarly, Phillips has described the way Anishinaabeg art representations include "interrelationships of the forces that animate the cosmos" (59). The transmotion—the tracing of interconnected energies—arises out of an Anishinaabeg worldview that both recognizes the inherent literal and figurative crossings of beings, realms, and spiritual presences and manifests them in its artways.

Both the "water bodies" and "crane language" picto-poems also derive artistic "meaning" from an implied sense of their being part of an ongoing process. Despite their textual and visual stopped-time reality, they are not static, but instead seem *in medias res*. Through its allusiveness, the language of the poems implies an expansive mythic and historical context. Each image shows an action in progress, suggesting the potential continuation of movement. Because this "transmotion" gestures toward a presence that remains unnamed, it invites a certain transcendence of the merely physical. For example, in "A Crane Language," when the poem speaks of the "ancient light" that "enters," this presence might be understood as arising from the "fissure," or as Cutschall says, from "where the seen and unseen come together" (65).

Recognizing how Native creative works "allow us to envision the possibility of things not ordinarily seen or experienced," Diné storyteller Yellowman calls this place of un/knowing "the intangible part of our thinking mind" (quoted in Beck, Francisco, and Walters 1990, 61). Thus, the arrival point in these works is not a singular place or moment, but rather a way of seeing or being. However we might characterize this particular pathway or vein of understanding, ultimately the picto-poem—with Zhashagi flying through a sky of woodland beadwork—disrupts expectations. "Reading" the image or the poem requires the reader/viewer to inhabit "an other" perspective, one alive with Anishinaabeg geo-knowledge and aligned spiritual "light." It creates a pathway toward this alternate way of perceiving, toward a moment of transcendence.

↔◊↔◊↔

A Crane Language

From where behind you ancient light
enters. Curves of this woodland sky,
a fissure in the propriety of reason.
Stitched here is first translucence. Sun
turning the hose of your throat to a vessel
of fire. Body angled like a sundial
against the cup of all reluctance. Our rock cold
clan longing. This silence. Released now
from the weighted anchor of time.
Your voice—a torch of language. *Zhashagi.*

The flame of your tongue gives light.

Resistance and the reassertion of an ancient cosmology

Just as traditional arts like the Ojibwe pictographs often had accompanying songs or stories, visual poetry and my picto-poems likewise employ both verbal and visual elements to convey key understandings about Anishinaabeg cosmologies, belief systems, or foundational community practices related to sustainability and spirituality. In this method, epistemological concepts thus receive dual reinforcement or their complexity rendered more fully. I apprentice myself to tribal art and story because I believe they continue to play a significant role in creating a context for tribal survivance.

As scholar Dean Rader notes in writing about Indigenous resistance in literary and other creative works, the artistic "language" is "not simply an aesthetic, but also an ethic" (2011, 53) and becomes "both a measure and a means of sovereignty" (53). The sometimes idiomatic and imagistic language of art often employs both "compositional" and "contextual" resistance (53). What we create as Native artists often strives to be both "affective"—or beautiful as art, and "effective"—doing something in the world.[6] Paul Chaat Smith, Comanche author and art curator, says, "Artists are deeply respected in the Native World. We ask of them just two things: 1) make fabulous art, and 2) lead the revolution" (2012, 215). Chaat Smith's statement may be playful, but his comments align with the previously quoted declarations by Cutschall and Quick-to-See Smith. Many other Native artists and intellectuals, from Edgar Heap of Birds, to Jolene Rickard, to Victor Masayesva, have also pointed to the key role art plays in the survival and flourishing of Native cultures. Talking about the relationship between "tradition, art, and Indigenous sovereignty," Rickard, for example, claims artistic traditions are "a reinvestment in a shared ancient imaginary of self and a distancing strategy from the West" (2011, 472).

One need not look far in the current news to realize contemporary "Western" teachings and conditions sometimes run radically counter to traditional

FIGURE 2.3 Ancient Light and Of the many ways to say: *Please Stand.*

Indigenous belief systems, especially when we consider ecological ideals. Tenets of Indigenous wisdom on sustainability often directly conflict with generally accepted or mainstream value systems and cultural lifeways. Communal values, for example, may be set against capitalist ideals. Likewise, Native notions of reciprocity, kinship, and the animate water or earth do not align with resource "ownership" or the labels of "inanimate" and "thing" often used when categorizing elements of the natural world. The "rights" and "protection" afforded humans in our current capitalistic empire are seldom extended to non-human beings or elements.[7]

Actions following from these different understandings frequently result in cultural clashes respecting environment issues. Native activists undertake many actions in response to potential threats to the places and relationships we value—from protecting the water resources at Standing Rock to working to prevent development of sacred sites such as effigy mounds. In her essay "Places to Stand: Art after the American Indian Movement," scholar Jessica Horton claims, "The vitality of contemporary struggles rests on uncovering historical ground" (2017, 30). Part of the remembered historical ground in regard to environmental issues includes past untenable actions resulting in pollution and destruction, but it also includes basic epistemological concepts underscoring core values and practices. The African American author and activist bell hooks has noted how connecting art to "lived practices of struggle" helps to create a "genealogy of subjugated knowledges" (1991, 59). According to the traditional teachings of Anishinaabeg people, for example, "water or *niibi* is a primary sacred element in life and therefore must be cherished as an essential relative, elder, and teacher" (Nelson 2013, 216). Today Anishinaabeg have Water Walkers working to change destructive patterns in the human relationship with water, and many of us become Water Protectors in whatever ways we can—sometimes through art and poetry.[8]

Eloquence of Aki

How do we translate the flashing fins
of poisoned fish? What other alphabet do you know
to spell *contaminated waters*?
Like banned books words still burn
on my tongue—*reciprocity, sacred,*
preservation, earth, tradition, knowledge, protect.
Even the vellum of *justice* disdained,
crumbled in quick greedy fists.
Meanwhile we gather here,
descendants of *ajijaak* and *maang*
lift our ancient clan voices in longing,
for a chant of restoration
in a Faustian world.

The "ecocritical" in my oeuvre often engages with specific contemporary struggles for clean water and focuses on the physical and spiritual impact of environmental actions. Recently, in response to threats to the Boundary Waters Canoe Area Wilderness (BWCAW) from potential copper mining, I became a plaintiff in a lawsuit filed by a group called Northeastern Minnesotans for Wilderness against the United States. The declaration I wrote actually included passages extracted from creative works because these "creative" works arise out of an ethic of reciprocity, kinship, and sustainability. (Shouldn't our laws and our art both embody our foundational beliefs?)

The picto-poem "Of the many ways to say: *Please Stand*" (see previous image and later text) might also be categorized as "documentary" poetry, making use as it does of historical and political details and newspaper accounts related to the debate of a proposed open-pit iron mine in the Penokee Hills of Wisconsin. But the poem simultaneously alludes to a larger cosmological view and to Native philosophical wisdom—here specifically, to the practice of making decisions while considering seven generations in the past and seven generations into the future. The Penokees are the headwaters of the Bad River, and various lakes and streams in that ecological region flow into Lake Superior (one of the world's largest bodies of fresh water). Protecting this water resource means safeguarding the future, period.

In in addition to "speaking truth to power," the tangible action the poem calls for is resistance: standing up for those impacted by environmental racism, standing against the companies and legislators who threaten our land and water, standing with those who oppose environmental degradation. Of course, the poem also contains multiple allusions as well as pointed indictments. It sarcastically gestures, for example, to the convoluted paths of corporate and capitalist rationalizations or the supposed complexity of contemporary situations through the use of a mathematical equation with multivariable functions. Ultimately, however, it upholds the "simple math" of survival. Gestures in the poem link the current environmental case to others in the long history of untenable actions such as unsafe uranium mining and the environmental contamination caused by atomic bomb detonations.

The kind of historic contextualizing of Indigenous resistance I undertake in the picto-poem matters in nations such as America where colonial forgetfulness has become a middle-class pastime. Tommy Orange's novel *There, There* effectively underscores the way Native people "continue to be slandered despite easy-to-look-up-on-the-internet facts about the realities of our histories and current state as a people" (2018, 7). Indeed, as Horton suggests, the appropriate historical context for our current state is "a vast colonial story implicating indigenous bodies and territory for more than five centuries" (2017, 30).

↔◊↔◊↔

Of the many ways to say: *Please Stand*

I. ∂,Partial Differential Equation

What we erase from polite conversation. Bodies on fire. The historic cleansing of the landscape, the sweep of humanity west, west, west. Environmental r ism.

All things being equal, things are never equal. Think of scope. Like the reach of the imperial. Or consider variables. Value. Or commodity. Convenience policy, a tally mark across generations. Uranium mining. Atomic bomb detonation at White Sands. A complicated table of fallout factors. Plume of greatness.

Shift. Angles and perspectives. Ways of seeing. *Seven generations into the past; seven generations into the future.*

Or how to solve for survival.

II. *Zongide'en*, Be Brave.

Another partial differential equation. Let's say a corporation proposes a mine. Variables include Tyler Forks. Bad. Potato. Rivers. A 22-mile, 22,000-acre strip of land. Jobs. *Maanomin.* Open pit. Exceptional or outstanding resource waters. Legislation. Iron oxide. Fish. Blasting and pulverizing. New legislation.

The functions depend upon the continuous variables. Fluid flow, for example. And changing laws. Somewhere along the granite line, someone enters. Let's say they have put down one life and taken up another: the solution of the PDE.

Warriors (walleye, Indian, new-age) face arbitrary functions. Changing laws. Guards. Guns. If life is stretched over two points. It vibrates. We cannot measure that vibration in this generation.

We can sing it, or make it into light.

My new work in picto-poems and palimpsest alludes to many aspects of an Anishinaabe ontology, particularly one that elevates the sacred elemental being of water. It underscores: concepts of motion, repetition, or cycle; an understanding of overlapping or shared realities; totemic animals; autonomous reciprocal relationships; and an emphasis on community or kinship. In terms of form, I imply multidimensionality by literally employing several senses, by joining—sometimes juxtaposing or layering—text and image in a palimpsest of mutually referential elements. I use symbolic visual cues including reflection and refraction of image, shadows, and implied movement. I also attempt visual disruption through advertising absence and employing non-representation images. Finally, I make use of concrete poetry, unexpected juxtapositions or overlaps of language and image, and interactive textual allusions. These artways arise out of and return me to a sustaining cosmology of *nibi*.

Wellspring: Words from Water[9]

A White Earth childhood water rich and money poor.
Vaporous being transformed in cycles—
the alluvial stories pulled from Minnesota lakes
harvested like white fish, like *manoomin*,
like old prophecies of seed growing on water.
Legends of Anishinaabeg spirit beings:
cloud bearer Thunderbird who brings us rain,
winter windigo like Ice Woman, or *Mishibizhii*
who roars with spit and hiss of rapids—
great underwater panther, you copper us
to these tributaries of balance. Rills. A cosmology
of *nibi*. We believe our bodies thirst. Our earth.
One element. *Aniibiishaaboo*. Tea brown
wealth. Like maple sap. Amber. The liquid eye of moon.
Now she turns tide, and each wedded being gyrates
to the sound, its river body curving.
We, women of ageless waters, endure;
like each flower drinks from night,
holds dew. Our bodies a libretto,
saturated, an aquifer—we speak words
from ancient water.

Notes

1 Survivance, a neologism from the critical work of Gerald Vizenor, brings together ideas of survival and resistance, or some say survival and endurance. The word is now commonly used in Indigenous literary circles.

2 I occasionally employ specialized use of language ("story" as a verb, for example) or philosophical underpinnings (the "making" of photographs as an engagement not bound up in possession) in this chapter. I endeavor not to pause too frequently to explain the language choices, hoping the literary process of world-making will suffice in providing context.

3 Manitou, in this instance, means more than "spirit" or "god," which are words by which it is often defined. Rather, it involves the broader understanding Basil Johnston suggests: "mystery, essence, substance, matter, supernatural spirit, anima, quiddity, attribute, property, God, deity, godlike, mystical, incorporeal, transcendental, invisible reality" (1995, 242).

4 Although it is beyond the scope of this essay to discuss at length the complex issues of tribal sovereignty, if we think about the distinctions offered by Joseph Kalt and Joseph Singer between "de recto sovereignty, de jure sovereignty, and de facto sovereignty (i.e., sovereignty by moral principle or right, sovereignty by legal decree or legislative act, and sovereignty in practice)" (2004), acts of self-determination in art can be undertaken as and understood to be de facto sovereignty (often reinforcing a belief in de recto sovereignty and in resistance to the denial of de jure sovereignty by settler colonialism). For example, the reassertion by artists of tribal nations' place names, the conscious use of Indigenous language, and the privileging of Native foundational stories and teachings

challenge erasure and reassert tribal knowledge, including insight into sustainable wisdom (which is of particular importance in the practice of geopoetics).

5 Phillips actually uses the phrase "fundamental binaries" to characterize these elements (sky/earth, male/female, etc.), but despite her acknowledging their "interdependent" reality, the statement doesn't wholly fit with my understanding. Thus, I eliminated that phrase in my restatement of the idea.

6 I am indebted in this understanding to Seamus Heaney's *The Redress of Poetry* in which he locates in poetry both its "joy in being a process for language" and its "power as a mode of redress . . . an agent for proclaiming and correcting injustices" (6).

7 Other countries have recognized the rights and "personhood" of natural features. For example, according to the 19 June 2017 issue of The Conversation, "New Zealand's Whanganui River is a person under domestic law, and India's Ganges River was recently granted human rights. In Ecuador, the Constitution enshrines nature's 'right to integral respect'." See https://theconversation.com/when-a-river-is-a-person-from-ecuador-to-new-zealand-nature-gets-its-day-in-court-79278.

8 A group of Anishinaabe "grandmothers" began walking around bodies of water including the Great Lakes in order to honor *nibi* and raise awareness about the need to protect our waterways. Among the more famous of the participants is Josephine Mandamin. The Water Walkers have traveled over 10,000 miles by foot in Canada, the United States, and Central America. As of this writing, the movement is ongoing.

9 The image for this picto-poem is available at http://lithub.com/new-poetry-by-indigenous-women/.

References

Abram, David. 1996. *The Spell of the Sensuous: Perception and Language in a More-Than- Human World*. New York: Random House.

Armstrong, Jeannette. 2016. "Land Speaking." In *Introduction to Indigenous Literary Criticism in Canada*, edited by Heather Macfarlane and Armand Garnet Ruffo, 145–159. Peterborough: Broadview Press.

Beck, Peggy V., Nia Francisco, and Anna Lee Walters. (1977) 1990. *The Sacred: Ways of Knowledge, Sources of Life*. Reprint. Flagstaff, AZ: Northland.

Bernstein, Bruce. 2012. "Expected Evolution." In *Shapeshifting: Transformations in Native American Art*, edited by Karen Kramer Russell, 30–43. New Haven: Yale University Press.

Chaat Smith, Paul. 2012. "Famous Long Ago." In *Shapeshifting: Transformations in Native American Art*, edited by Karen Kramer Russell, 213–221. New Haven: Yale University Press.

Cutschall, Colleen. 1996. "The Seen and the Unseen to Form the Narrative. . . " In *Plains Indian Drawings, 1865–1935: Pages from a Visual History*, edited by Janet Catherine Berlo, 64–65. New York: Harvey N. Abrams.

Devine, Bonnie. 2016. "Artist Statement" for "The Transformation of Landscape in Canada: The Inside and Outside of Being." www.cacnart.com/bonnie-devine.

Heaney, Seamus. 1990. *The Redress of Poetry*. Oxford: Clarendon.

hooks, bell. 1991 "Narrative of Struggle." In *Critical Fictions: The Politics of Imaginative Writing*, edited by Philomena Mariani, 53–61. Seattle: Bay Press.

Horton, Jessica. 2017. "Places to Stand: Art After the American Indian Movement." *Wasafiri: Indigenous Writing and Literary Activism* 90 (Summer 2017): 23–31.

Johnston, Basil. 1995. *The Manitous: The Supernatural World of the Ojibway*. New York: Harper Collins.

Kalt, Joseph, and Joseph Singer. 2004. *Myths and Realities of Tribal Sovereignty: The Law and Economics of Indian Self-Rule* (Harvard Working Paper Series, RWP04-016). https://scholar.harvard.edu/files/jsinger/files/myths_realities.pdf.

McInnes, Brian. 2016. *Sounding Thunder: The Stories of Francis Pegahmagabow*. East Lansing: Michigan State.

Merriam-Webster Online, s.v. "Language." Accessed January 24, 2019. www.merriam-webster.com/dictionary/language.

Merriam-Webster Online, s.v. "Myth." Accessed January 24, 2019. www.merriam-webster.com/dictionary/myth.

Miner, Dylan. 2013. "Stories as *Mshkiki*: Reflections on the Healing and Migratory Pathways of *Minwaajimo*." In *Centering Anishinaabeg Studies: Understanding the World Through Stories*, edited by Jill Doerfler Niigaanwewidam James Sinclair and Heidi Kiiwetinepinesiik Stark, 317–339. East Lansing: Michigan State.

Momaday, N. Scott. 1969. *The Way to Rainy Mountain*. Albuquerque: University of New Mexico.

Nelson, Melissa. 2013. "The Hydromythology of the Anishinaabeg: Will Mishipizhu Survive Climate Change, or Is He Creating It?" In *Centering Anishinaabeg Studies: Understanding the World Through Stories*, edited by Jill Doerfler Niigaanwewidam James Sinclair and Heidi Kiiwetinepinesiik Stark, 213–233. East Lansing: Michigan State.

Noodin, Margaret. 2014. *Bawaajimo: A Dialect of Dreams in Anishinaabe Language and Literature*. East Lansing: Michigan State.

Orange, Tommy. 2018. *There, There*. New York: Knopf.

Penny, David W., and Gerald McMaster, eds. 2013. *Before and After the Horizon: Anishinaabe Artists of the Great Lakes*. Washington, DC: Smithsonian National Museum of the American Indian.

Phillips, R. 2013. "Things Anishinaabe: Art, Agency, and Exchange Across Time." In *Before and After the Horizon: Anishinaabe Artists of the Great Lakes*, edited by David Penney and Gerald McMaster, 51–69. Washington, DC: Smithsonian.

Quick-to-See Smith, Jaune. "Star Wallowing Bull: Born with a Gift." (blog). Accessed April 28, 2009. www.starwallowingbull.blogspot.com.

Rader, Dean. 2011. *Engaged Resistance: American Indian Art, Literature, and Film from Alcatraz to the NMAI*. Austin: University of Texas.

Rickard, Jolene. 2011. "Visualizing Sovereignty It the Time of Biometric Sensors." *The South Atlantic Quarterly* (Spring): 465–486.

Silko, Leslie. 1996. *Yellow Woman and a Beauty of the Spirit: Essays on Native American Life Today*. New York: Simon & Schuster.

Vizenor, Gerald. 2015. "The Unmissable: Transmotion in Native Stories and Literature." *Transmotion* 1 (1). https://journals.kent.ac.uk/index.php/transmotion/article/view/143/604.

3

TERMA

A dialogue

Sameer Farooq and Jared Stanley

In 2016, Sameer Farooq, a Toronto-based visual artist, and Jared Stanley, a Reno-based poet, were commissioned by The John and Geraldine Lilley Museum of Art to be the first exhibiting artists in their new museum space. Farooq and Stanley met the previous year while teaching in the MFA in Interdisciplinary Art at Sierra Nevada College, where they became fast friends, tied together by their mutual obsession with museum display strategies, an interest in the spiritual dimension of art, a love of Orhan Pamuk's *A Modest Manifesto for Museums*, and a shared sense of the speculative potential of reordering, scrambling, overlaying, and wandering through the ways history is expressed in physical and imaginative space.

Sameer Farooq is an artist whose interdisciplinary practice aims to create community-based models of participation and knowledge production in order to reimagine a material record of the present. He investigates tactics of representation and enlists the tools of installation, photography, documentary filmmaking, and writing, as well as the method of anthropology to explore various forms of collecting, interpreting, and display. The result is often collaborative work that counterbalances how dominant institutions speak about our lives: Farooq's work presents counter-archive, new additions to a museum collection, or a buried history made visible.

Jared Stanley is a poet and writer who often works with visual artists. His primary interest is in the intersection of lyric poetry, the history of landscape and land use, and the vernacular, ever-shifting ground of language as it changes as a result of migration, environment, and technology. His work has developed from an initial, book-centered writing practice into a formally expansive series of projects that take the materials of reading as a starting point. These projects ask fundamental questions about how typographical forms and environments of reading shape the reader's experience of a text.

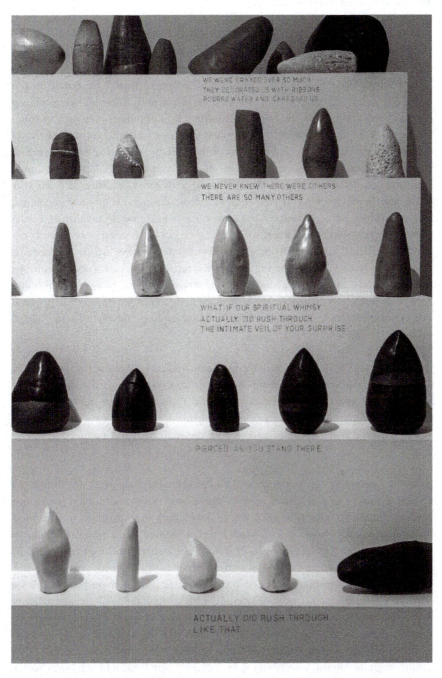

FIGURE 3.1 Sameer Farooq & Jared Stanley. *Terma, Images from the Ear or Groin or Somewhere* (If it were possible to collect all navels of the world on the steps to ASCENSION). Exhibited at the John & Geraldine Lilley Museum of Art, Reno, NV.

Source: Photograph by Ann Ploeger.

The result of our collaboration is a speculative artist's museum entitled *Terma, Images from the Ear or Groin or Somewhere*, which opened in January 2019. Located on the boundary between ethnography and science fiction, *Terma* takes its title from a Tibetan Buddhist tradition in which "hidden treasures" are buried in the earth, sky, water, and the mind. When retrieved by adepts, either by being dug up or meditated upon, these treasures are interpreted as important messages in troubled times. Farooq and Stanley use this practice as a metaphor to consider the ways museums (especially those dedicated to anthropology and archaeology) generate narratives about and taxonomies of objects and language. UNR was the initial incubator space for this project, which Farooq and Stanley envision as a ten-year collaboration presented across many gallery spaces. Through a long-term generative dialogue, they have constructed a museum from the ground up, making and conceptualizing its objects, narratives, spatial parameters, classification systems, and textual encounters. Ultimately, the project is an excavation of two artists' imaginations and an opportunity to reframe how museums organize the past spatially. In it, we ask: *How can we imagine and unearth the world differently?* What follows is a dialogue between Farooq and Stanley on the spatial construction of museums, their personal histories with museums, and the ways objects and language interact to create a lyrical method of moving through space.

JARED STANLEY (JS): When we first met, you told me you wanted to work with clay. In particular, you were talking about how you wanted to bring together clay from Pakistan and Cape Breton, Nova Scotia. Can you call to mind why these places were important to you in your investigation of this most earthy of media? Why clay? And is it important to you that the material has a connection to the earth?

SAMEER FAROOQ (SF): At the time of the project you mention, I was thinking a lot about notions of pluralism, a concept introduced roughly a century ago. In our current climate of cultural politics, where broadening boundaries of inclusiveness live alongside increasingly polarized societies, it seemed like a fitting moment to reflect on the concept of pluralism. Pluralism holds that distinct groups can exist side by side while considering qualities from other groups as traits worth adopting. In this way, diverse groups do not erase difference or disagreement between them, but rather they manage conflict peacefully, with all maintaining an understanding that a democratic society can be enriched by difference. I wanted to test this notion through material response: *How can ideas of cultural integration exist as objects?* Clay seemed to be the perfect medium because I could orchestrate the meeting of two distinct regions—in this case, from my birthplace (Cape Breton Island, Nova Scotia, Canada) and from my ancestral background (Lahore, Pakistan). What is interesting about clay is that it is completely stubborn and will act in whatever way it wishes. Just last week I had 15 items explode inside a kiln because the conditions were slightly off. So, unlike political speech, there is no way of smoothing over the behavior of clay. Rather than being fully resolved, my "plural objects" aimed to complicate seamless notions of pluralism by encouraging distinct materials to suggest surprising forms. I was curious how different clay bodies would actually react to each other when forced to coexist. I wanted their unpredictable reactions to challenge the philosophy of pluralism.

Unfortunately, this project has yet to be realized because, with current clay production monopolies and the difficulty of moving earth across borders, there have been many delays!

This project exists within the context of my ten-year project of building speculative museums that illuminate the blind spots in traditional ways of collecting. My earlier works focused on looking with the eyes open—on working with found objects, groupings of artifacts, display implements in museums, etc. I used this strategy to perform a sort of institutional critique. Yet I felt *language* was missing. In collaborating with you, Jared, I am surprised at where language goes within this space. In many ethnographic and anthropological museums, language insists on governing how objects and images are viewed, often as labels or wall panels. In *Terma*, we are proposing to build a museum through an internal excavation. On the subject of this project specifically, I am compelled to ask you a number of questions: What are the potential shifts you envision in the way language can be used within a museum space? Where could language live? How does it sound different? What is the linguistic equivalent to artifacts such as pottery fragments, a collection of marble heads, or a pile of fossils in the corner of a room?

JS: A museum is like lots of public spaces in that the free movement of bodies and minds is subtly directed by language and the visual rhetoric of signage. I'm a poet, so I'm concerned with freedom and meaning, especially when it is used for control in public space. I'm excited by the way language moves along the edges of genre. I think of signage as a spatial genre of writing, so I use the spatial metaphors of "movement" and "edges" intentionally—that is, I employ them because I want a vocabulary that is familiar with boundaries but nevertheless ignores them or uses their forms against them. I feel a great potential in the way language moves through the body of a viewer.

But this feeling abuts the conventional use of signage, which often condenses language to produce simplified meaning. And that makes sense in some scenarios. If one needs to turn right on a certain street, the name of the street is all one needs on a sign. The street's history, its orientation to a watershed, etc.: These are not important to a person in the enclosed space of a car hurtling at flesh-destroying speed. The simplifications that signage has brought to all kinds of landscapes have been extreme: We've come to expect such clarity and simplicity of language everywhere. Politicians now make speeches whose content is summarized on large banners or screens behind them, dwarfing the speaker.

And this kind of thing happens in museums, too. Wall text regarding a Rousseau painting reads "By placing the trees along a diagonal axis, [Rousseau] has conveyed a sense of the wind, in spite of the painting's static and naive style" (National Gallery of Art 2018). Really? If the wind is conveyed in the painting, why do we need wall text to tell us so? It smacks of condescension and disrupts a viewer's attention. I feel a strong grief, in our broken world, about the way language is used to control and narrow the attention.

And it's against this backdrop of distinctly (and deadly) narrowing distillation that the prospective language in *Terma* operates. We are doing a poet's work in space by expanding meaning through compressing language—by opening language's interaction with the nonlinguistic objects in our museum. As I compiled an archive of potential language for *Terma*, I wrote exploded aphorisms, fragments of text, cataloging systems, red herrings, and meditative phrases. When we use words such as "excavation," "extraction," or "archive" in this work, we are not using them metaphorically: We are using them literally.

When you first brought the idea of terma to my attention, Sameer, the notion made immediate sense to me: The monk's practice took into account my interest in the earth; it had a kind of serendipity, the possibility that one might stumble on some terma; it allowed us to put our trust in the universe; finally, it had a clear mystical dimension. Add to that the fact that terma as a practice and tradition was both an embodied notion (i.e., finding buried wisdom in the earth at a crucial time) and a mental notion (i.e., that one could "excavate" one's mind for terma). To me, all of these ideas are about ways of "finding" oneself by digging or inserting oneself into the earth. I have always thought of this project as having some very elemental relationship to Ana Mendieta's *siluetas*, in which the artist made "portraits" of her body by creating depressions in the earth. Similarly, our project is a record of our own "excavations" in both real earth and that more elusive plane that Henri Corbin calls visionary geography (1977, 24). Turning back to you, Sameer, could you talk about how our museum brings together the disparate objects of our collective excavations of terma? Is our museum a new site?

SF: I have been so compelled by the concept of terma. When I first came across it during a Buddhist retreat, it presented such a strong perspectival shift in which body, earth, and time were intertwined in a way that I was highly unaccustomed to.

As we mentioned in the introduction to this dialogue, the word "terma," meaning "hidden treasure," comes from Tibetan Buddhism and refers to teachings that have been deliberately concealed by monks to be discovered by adepts in troubled times as a source of renewed wisdom. Terma can take the form of a physical object that can be buried in the ground, lodged in a rock or crystal, secreted in a tree, hidden in the water, or placed into the sky. It can also be "mind terma": An object concealed in the mind of a *tertön* to be later revealed through a series of codes. I began to be fascinated by the visual quality of these objects/concepts: What does something have to look like if it is to be buried in the sky? For me, as a visual artist, what terma suggests really puts my imagination into overdrive.

In particular, the notion of mind terma is fascinating. I am reminded of a poignant statement curator Massimiliano Gioni wrote on the occasion of the 51st Venice Biennale: "Besieged as we are by the profusion of images that characterizes our present, we might be tempted to close our eyes to realize that pictures have always been swarming behind our eyelids" (2013, 24). Never mind if this is a form of terma (it probably isn't); it serves as an encouragement to excavate the ephemeral. Why is it that we believe in artifacts made from solid materials, but we distrust artifacts of the imagination? I began to get very curious in my own internal image stream, as well as in the streams of others; I wondered if there were patterns or overlaps within our vivid internal lives.

As a guiding principle for an imaginary museum, terma is an excellent metaphor as it allows objects to move beyond traditional object-based ethnography and encourages innovative collecting practices to be applied to the artifacts on display. Through building a museum fueled by our collective imagination, I am hoping to gain a better understanding of the forms and patterns that come forward through the process of meditation and collaboration. For me, our collaboration is also a move toward decolonizing the museum through the creation of a series of objects by which we acknowledge a deep subjectivity. The selection was not based on an established set of rules, expertise, and "best practices." Instead, this museum becomes home to an imaginary site; the project calls into question the institutional production of truth, while at the same time proposing a future archeology fueled by the ideals and imaginations of its builders and spectators.

In our early conversations, Jared, you spoke about the idea of overlapping geographies. Rather than simply bringing our earlier memories from the San Francisco Bay Area and Cape Breton Island together, I think you referred to our process as a form of "metaphysical geography." In your own writing, I've enjoyed reading how you've taken similar cognitive leaps (and then crashed back down to Earth again), ones that language makes possible. For example, you write that standing in the desert in Nevada can cause you to stand in the desert in Afghanistan. How do these acts of teleportation work in a museum space?

JS: In much the same way as the first line of a poem is a passage into the work of the poem, I hope *Terma* will operate as a threshold into an imaginary world, a pretend world. My childhood museum, the Oakland Museum of California (OMCA), was a "place of the imagination" in the grandest sense of the word. A beautiful, brutalist ziggurat-like structure, it was my mind's imaginative template for each paradise I encountered in reading. The OMCA, was (and is) the most beautiful place in the world. This kind of overlay, the reading into and out of a place that was also a kind of book, is the foundation of my explorations of the relationship between objects, language, and space, i.e., my poems.

But the museum's invitation to "reading" was more complex than this. When I consider the anthropological displays at the OMCA, the dioramas of the displaced peoples of California, I encounter a triumphant imperial museum's narratives about gone worlds—Edenic, pre-contact paradises—with all the murderous sentimentality such displays deploy. All of the usual tropes were on display, presenting the people of California as primitive and tragic, as specimens under glass. My experience of looking and reading has been shaped by these two contrasting experiences of a place that inspires the imagination and also shapes one's ideas.

I weigh these contrasting visions of how a museum encounters a viewer's imagination. In the first, I (the viewer) brought the reading I did outside the museum *in*, that is, I brought what I read to bear on the museum space. I made the beauty of the architecture congruent with the imagination I brought to the place. In the second, the curatorial bias of display formed a kind of sentimental, tragic myth about the settlement of California. It controlled my imagination, narrowed its response.

I hope that the language and objects of *Terma* will work in concert with the imaginations of our viewers. I hope what they encounter therein will not transport them to some place we have made, but rather will trigger the viewer to bring something of themselves into an in-between, third space. This is what's so wide open: The way we bring the entirety of ourselves to a place like a museum.

Meeting you has changed my life, Sameer. Our geographical separation made our meeting, in space and in time, both highly unlikely and, in retrospect, completely inevitable. I often think of our meeting in terms of luck, and I wonder if our meeting itself wasn't itself a kind of terma or mystical excavation. I keep thinking of museum audiences, of how we "enact," for them, the archive of "terma." I wonder about the phenomenology of the museum space, how our audience members move through, interact with, view, and finally turn visitors into poems themselves. Our museum does that! How do you think of the audience as a recipient (or co-performer) of *Terma* via our terma?

SF: I agree! How did we manage to meet? In terms of the public, I always imagine any speculative museum that I have been involved in as more of a provocation than as a set of expertise unfolding over a space. A major question in my own practice, which is relevant to our current project, is: *In our attempts to create historical narratives within museum space, what gets lost in the capture?* The public has always been a fundamental part of filling in this gap. My hope is that, through presenting a series of objects and mechanisms of display, the audience is asked to imagine what other possible artifacts are missing from this collection. Their imaginations, memories, and subjective experiences all become essential to completing the museum, and I have tried to construct previous museums with this prompt in mind. Instead of seeing museums only as hierarchical structures whose gates are upheld by experts in the field, I see museums as also being mechanisms for community-based knowledge production, with the public becoming collector, interpreter, and curator.

Words can make tangible the wilds of the imagination and, in a sense, create an artifact that others can see and, in turn, onto which they can map their imagination. I've found that the translation of an inner image stream into external images very difficult, and because language is very good at preserving a time-based sequence, it is a great medium to work with when I'm engaging in contemplative practices.

In traditional museums, words and things interact on many levels: museum labels, didactics, maps, instructions, and words inscribed on objects themselves. In a speculative museum such as *Terma*, what is an interesting challenge is to see how words can indeed become things. Perhaps interior spirits and processes will open up the possibility for language to operate in a much different way than in traditional museum spaces. Of course, your question makes me think of Foucault's *The Order of Things*, in which, in his famous introduction, he speaks about Borges's description of a Chinese Encyclopedia that details an incongruous list of what animals can be divided into. There is one line that I love:

> In the wonderment of this taxonomy, the thing we apprehend in one great leap, the thing that, by means of the fable, is demonstrated as the exotic charm of another system of thought, is the limitation of our own, the stark impossibility of thinking that.
>
> *(1970, xv)*

Could the "thought forms" of our imaginations present a similar impossibility as told through words and things? Can an idiosyncratic list collapse traditional museology while presenting new objects of our imagined time?

As Theosophists such as Annie Besant and CW Ledbetter hoped, are there universal forms that we all share? Do these forms show up in objects and words? Or, is this just an impossible task that will make us obsessive and exhausted?

JS: I love the syncretic romance of theosophy—the desire to see oneself and one's ideas refracted in all traditions, to see the rhymes between traditions, places, and times. It's a (perhaps world-ending) tragedy that the evidence of the rhyme is more available now than ever, and yet the ability to see it is further away and more impossible than ever. So many are afraid of rhyme, because the rhyme is difficult, indifferent. And nevertheless ... rhyme is the poet's way to the universal. Why, for instance, do I turn, in this age, to the last poems of the exiled Russian Jewish writer, Mandelstam, to the poems of the steppe's black earth? Is it because I rhyme with the poems? Well, yes, but why? Mandelstam's life and suffering are completely alien to me. And yet, in the spirit of his own dreaming, toward Ovid, toward Dante, toward Constantinople, he presents the pain of the universal—which for him was cultural, cosmopolitan, Greco-Roman—confronting another universal, that is, the Soviet State, with its totalitarian scientism and its absurd lies. We don't live in that world, but his loss rhymes so strongly with our own bloody-minded impasse.

I'm tempted to dwell on the irony that the only people who seek a kind of universality are branded eccentrics, cultists, heretics—in short, heroic figures, or what Robert Duncan, in his frightening way, calls "predators of the marvelous" (2014, 541). And to answer your question: Because we are eccentrics, the answer must absolutely be *yes*, there are universal forms, and *yes*, our attempts to map them must absolutely be spiced with the self-knowledge of our absurdity and frustration. We are on a fool's errand because to be anything other than a fool or a naïf in this world is to be crushed by the cancerous fantasies of capital.

One thing that seems crucial to our collaboration is that we both harbor deep interests in the place the other comes from. As you know, I am something of a Canadaphile, and you are an honorary California Girl. Additionally, I live a life that is very rooted in Reno, and you do projects all over the world, traveling extensively. Personally, I love the way our differing "lines of flight" give us complementary ways of entering into collaboration. Could you talk about your researches in museum display in an international context? How is the "international style" of display disrupted by local conditions? I'm thinking in particular of Diego Rivera's Anahuacalli Museum in Mexico City as a marker of an idiosyncratic interpretation of Mexicanidad, or the *Museum of Innocence*, which is a deeply personal museum about Pamuk's Istanbul (Pamuk, 2014).

SF: Over the past few years, my work has been facilitated by a fellowship that I received from my provincial arts funding body to study museum display as an aesthetic medium. This fellowship offered me an incredible opportunity to take a bird's eye view of how museum collections were structured (rather than getting lost in the specific histories of objects). Exhibition design is a complex space of social codes, ideological agendas, and haphazard protocols, and you very quickly learn that museum design is much more about a self-portrait of the institution and a representation of the social codes and limitations of specific moments in history. The fingerprints of the institution will always be on the material.

What has been interesting about visiting and photographing "memory museums" (ethnographic, anthropological, historical, etc.) worldwide has been witnessing a decrease in the experimentation of installations in larger, more "professional," and better-funded institutions, in contrast with the highly experimental moves that smaller, more community-based institutions have enlisted. Specifically, in the larger, nationally funded institutions, a series of professional formulas or "best practices" have entered into each museum space, largely collapsing site-specific response in favor of a (highly European influenced) international style. It is worth noting that this has not always been the case. In the early part of the twentieth century, wildly experimental moves took place in museum display, such as in Frederick Kiesler's *Surrealist Gallery* (exhibited at the Art of This Century Gallery in New York in 1942), where lighting was engineered so that half of the paintings were lit half the time and where every two minutes attendees heard the recorded sound of the roar of a train. Another example is El Lissitzky's *Abstract Cabinet* (exhibited at the Hanover Provincial Museum in 1927), where moveable partitions and reflective materials were used as the backdrop to display and activate artworks.

I have visited more than 80 museums globally over the past few years to study their mechanisms of display—from the Egyptian Museum in Cairo to the

Pyramid Lake Paiute Tribe Museum in Nevada—and have been most heartened by smaller institutions that needed to take much larger creative license to address a gap in funding or to confront their sheer smallness and isolation. In these spaces, I would see ingenious display strategies (repurposing found furniture, carving out actual walls in which to place artifacts, etc.) that were created in response to the urgent need on the part of these small institutions to mark a place in our historical imagination. At the same time, I've found great humor in larger, more "professional" institutions and their difficulties keeping up with the archival imperative. For example, in most archaeological museums there is a pile of large rocks or fossils in the corner that nobody knows quite what to do with and that are too costly to be housed in any display case. In each of these cases, museum display is highly affected by external conditions: geography, political climate, proximity to the public, etc.

What is unique about both the Anahuacalli Museum in Mexico City, designed by the Mexican artist Diego Rivera, and the museum the Nobel laureate Orhan Pamuk opened in Istanbul after the success of his novel of the same name, *The Museum of Innocence*, is that they are both memory museums created by artists and, in essence, are total works of art. They also each share a deep homage to the places both artists called home, allowing the strategies of the museum to help the artists reckon with their own relationships to their national histories. Walking into the Anahuacalli Museum, with its large, porous, volcanic rocks forming a temple around Rivera's collection of pre-Columbian artifacts, is deeply moving. Through the materials, lighting, and display mechanisms, there is little care for taxonomy or distinction; instead, Rivera encourages the public to relate to the artifacts on a metaphysical level.

I see *Terma* as a gesture related to decolonizing the museum. More specifically, I see it as a highly subjective and intimate approach to challenging the ways in which objects have been collected. In this way, the traditional museological machine gets disrupted at an early stage. In many ways, you and I, Jared, wish to subvert a "logical" impulse or to question best practices among museum professionals. At the level of taxonomy and the narratives inside of museums, can you imagine your approach being an extension of this decolonizing tendency?

JS: Yes, but always with a sense that English is an imperial language, and as such its colonial history is embedded in its vocabulary. Therefore, we can recognize English as being an imperfect instrument, or, perhaps more profitably, an ironic, subtle, sublime instrument. In a nod to Caliban in Shakespeare's *The Tempest*, I can say the word "khaki" and laugh my ass off (bitterly) every time. My love of English is analogous to what Flaubert said of the mystic: "I am a mystic at heart and I believe in nothing" (1953, 167). In this sense, I am a poet *devoted to* English, and I do not believe in that language, which I nevertheless love. It's absurd.

If one of the tenets of a colonial way of making space in the museum is the hierarchy of expertise that language (as signage) would tend to imply, then yes, we are decolonizing. I'm glad that you mentioned narrative, Sameer, because narrative is hierarchical. Our museum is lyrical as opposed to narrative. We've talked many times about the way a lyric approach to a museum might change how a viewer moves through the space: Where a narrative implies an "orderly" progression through time and space, a lyric museum is about another sense. The philosopher Jan Zwicky says it well: "The eros of philosophy is clarity. The eros of lyric is coherence" (2015, 3). In this statement, I take coherence to mean something akin to your statement about "universal forms": That a lyric museum is less concerned with creating a clear narrative and assigning fixed roles to subjects, instead being devoted to finding a kind of unity or coherence in proliferation, in the interaction of specific objects in the museum space. Zwicky goes on to quote Heraclitus, at his most mellifluous: "The fairest order in the world is a heap of random sweepings" (2015, 42). In a funny way, the "heap" is the coherence. A heap is a form, and there is formal pleasure in museums that appear to be mere heaps. Some museums gain their charm from the collision and touch of form-on-form.

I have been rewatching Parajanov's *The Color of Pomegranates*, a lyrical film that tries to enact the biography of the Armenian/Azeri/Persian/Georgian poet, Sayat Nova, by avoiding narrative altogether. Instead, Parajanov focuses on constructing gorgeous, wild, and mystical tableaux in which the ruins of Armenia—indeed, the very rubble and dirt of the Caucasus—are a part of the wholeness of the poet's fragmentary visions. The only connective tissue in the film is intertitles, with paraphrases from the poet and a voiceover of various actors reading the poet's verses. The lack of narrative focuses the viewer, quite intensely, on the poet's specific visions.

SF: *Terma* proposes to build a museum through an internal excavation. Where do your images come from? What does your process look like to get to this material?

JS: The way images in poetry serve as metaphors, just as the way in which music in poetry acts as a metaphor, is for me an endless source of pleasure, because the misprisions in the metaphor create so much opportunity for slippin' and slidin'. I'm an empirical writer—my images come from without—and the poems are the residue, a byproduct of compiling and combining the images together with the kind of internal logic of accretion. My poems are, in this sense, tiny museums. The poet Laura Jensen says in her poem entitled "Animal," "I am a cousin / of the crow's collection" (1977, 14). Don't you feel like that?

References

The Color of Pomegranates. 2018. Directed by Serge Parajanov. New York: Criterion.

Corbin, Henry. 1977. *Spiritual Body and Celestial Earth*. Princeton, NJ: Princeton University Press.

De Certeau, Michel. 2011. *The Practice of Everyday Life*. Translated by Steven Rendall. 3rd ed. Berkeley: The University of California Press.

Drucker, Johanna. 1998. *Figuring the Word*. New York: Granary Books.

Duncan, Robert. 2014. "To Speak My Mind." In *The Collected Later Poems and Plays*, edited by Peter Quartermain, 541. Berkeley: The University of California Press

Flaubert, Gustave. 1953. "Letter to Louise Colet." In *The Selected Letters of Gustave Flaubert*, edited by Francis Steegmuller, 167. New York: Farrar, Straus and Young.

Foucault, Michel. 1970. *The Order of Things: An Archaeology of the Human Sciences*. Translated by Alan Sheridan. New York: Pantheon.

Gioni, Missimiliano, and Natalie Bell. 2013. *Il Palazzo Enciclopedio = The Encyclopedic Palace: Biennale Arte 2013*. Venice: Marsilio Editori.

Jensen, Laura. 1977. *Bad Boats*. New York: The Ecco Press.

Lippard, Lucy. 1995. *Overlay: Contemporary Art and the Art of Pre-History*. New York: The Free Press.

Mallarme, Stephane. 1996. *The Book, Spiritual Instrument*. Edited by Jerome Rothenberg and David Guss. New York: Granary Books.

National Gallery of Art, Wall Text (Rousseau, "Surprised!" 1891).

Pamuk, Orhan. 2014. "Manifesto." Masumiyet Müzesi. https://tr.masumiyetmuzesi.org/page/manifesto.

Pamuk, Orhan. 2012. *The Innocence of Objects*. New York: Abrams.

Staniszewski, Mary Anne. 1988. *The Power of Display: A History of Exhibition Installations at the Museum of Modern Art*. Cambridge: MIT Press.

Thondup Rinpoche, Tulku. 1998. *Hidden Teachings of Tibet: An Explanation of the Terma Tradition of Tibetan Buddhism*. Edited by Harold Talbott. Boston: Wisdom Publishing.

Zwicky, Jan. 2015. *Alkibiades Love: Essays in Philosophy*. Montreal: McGill-Queen's University Press.

4

ALL VISUALS HAVE SOUND

The verbalization of geography and the sound of landscape

Cecilie Bjørgås Jordheim

> The hills are alive with the sound of music
> With songs they have sung for a thousand years
> The hills fill my heart with the sound of music
> My heart wants to sing every song it hears.
> > Rogers and Hammerstein,
> > *The Sound of Music*, 1959

What would an encoding of nature and geography sound like?
Does music always mimic nature, or is it an abstract, autonomous form?
Is it possible to consider nature and sounds in text—that is, auditivity—as isomorphic?[1]

Because I primarily create visual scores, my work has a significant crossover with visual and conceptual poetry. I root these scores in geopoetics and create them using ecological and geographical lines and fragments. I often collaborate with musicians, and the work becomes part of a performance in which my visualizations are translated into musical improvisation. By doing so, my work questions if there is a direct connection between language and the world—between topography, typography, text, architecture, and sound.

Furthermore, my ideas arise out of a dedication to interdisciplinary work, and I create installations, visual scores, and concrete poetry that move between genres, fields of art, and artistic mediums. Despite the range of work I produce, it is united by a common theme: All visuals have sound.

My path to concrete and visual poetry started in the field of analog Super 8mm and 16mm film. I started out as a painter at an art and film school, where the line between the two fields was fluent. I was amazed that the tactile, hands-on film was either 18, 24, or 32 paintings a second. (Just imagine the production!) For me, the introduction to traditional narrative filmmaking led to a skepticism about the pairing of sound and image. There was a clear hierarchy, with sound at the top: Musical soundtrack or Foley (that is, sound effects created for film) overpowered the image time after time in traditional films. Sound manipulated the viewer to feel and think the way the filmmaker intended. To me, the expression was forced. There was no direct or isomorphic connection between the sound and the image. This led me to the question, What if the two media—sound and image—could be on equal footing? What if we considered what they had in common, such as repetitive patterns or arrangement of contrasts and complex structures, in the way they related to the world?

When I started working with visual soundtracks on 16mm film, I not only found the commonalities between sound and image, but I also discovered how to look at other notational systems and ideas of interdisciplinarity in sound production, not to mention in my art production in general.

In this chapter I will present four projects that are rooted in a common geopoetics. My aim in these projects was to combine seemingly separate and different systems of text, notation, and movement in order to find a common sensory connection.

Before elaborating on how my art projects seek interdisciplinarity, I want to provide a brief history of Western thinking about the connections between art and science in order to demonstrate how we are in an age of interdisciplinarity, distinct from previous ages in which disciplines were often more discrete.

Making the ungraspable graspable

The fascination with the untouched and the accompanying desire to conquer it have been present in many cultures. In Norway, for example, we have seen this fascination in adventurous polar explorers such as Fridtjof Nansen (1861–1930) and Roald Amundsen (1872–1928). Some conquer by climbing, crossing, or diving down; others systematize. We develop systems to make the ungraspable graspable— to make something abstract and elusive concrete and solid. Systems and codes for communication and classification are part of our natural cognitive development. These are taught at an early stage, through parenting and education, so they can be understood, executed, and passed on.

As part of civilization and intellectualization, codes are a contract with society, where one agrees on the parameters that a particular society uses. Systematizing and coding help us cope with a constant stream of information efficiently; they facilitate our ability to navigate more easily through information and to gain control, reproduce meaning, and master new fields of knowledge. As Foucault noted:

> The fundamental codes of a culture—those governing its language, its schemas of perception, its exchanges, its techniques, its values, the hierarchy of its practices—establish for every man, from the very first, the empirical orders with which he will be dealing and within which he will be at home.
>
> *(1970, xxii)*

Gestalt psychology is a philosophy of mind that originated in the Berlin School of experimental psychology. It describes the mind's ability to find a global whole and connect everything the eyes see or the ears hear. In this theory, humans endeavor to interpret incomplete visual information and seek to perceive things in the easiest way; they yearn for stability and wholeness (Heim 2008, x). Gestalt psychology has taught us that the brain will seek interconnectivity ("Gestalt Psychology" 2018). It tries to explain how the brain groups fragments of sensory information to create an order, a system—to maintain meaningful perceptions in a seemingly chaotic world.

The Greek philosopher, mystic, and mathematician Pythagoras (580–500 BCE) tried to translate the whole physical world into numbers. Indeed, he is widely accepted as the father of modern numerology. According to ancient sources, Pythagoras was the first to discover the connection between the sound from celestial bodies and those from terrestrial bodies by placing the seven known planets into a notational system. Pythagoras researched the connection between the pitch and the length of the vibrating string in a monochord, and he expanded this connection into astronomical relations (Proust 2011). Pythagoras concluded that the planets were not randomly placed in the solar system; the distance corresponded to a harmonic ratio of numbers. The translation from the planets' position to music was based on arrhythmic methods. Every planet had its place in the solar system, and the relations between them were reportedly perfect and harmonic. In this way Pythagoras established a precise connection between everyday reality (mathematics and astrology) and the divine (music).

Music, essentially, was a new mathematical system.

According to Pythagoras's cryptic, esoteric theories, music is its own argument for humanity's divinity and intelligence. In his view, humanity has the capacity to find spiritual and perfect harmony within itself; humans merely need to understand the planets' perfect mathematical harmony and embrace celestial directness through music.

Art and science united

In contrast to music (with its inextricable connections with mathematics), the prevailing opinion was that visual arts were considered a craft: From pre-Renaissance Europe until the rise of Humanism (ca. fourteenth to sixteenth centuries), the individual artist was in no way elevated to a place of esteem in the way they often are in the Western world today (Jakobsen 2011). Art and science were first united in Filippo Brunelleschi's (1377–1446) discovery of the linear, or scientific, perspective (Janson and Janson 2004, 60), and Leonardo da Vinci (1452–1519) based his declaration that painting was mathematics on this discovery (Kristeller 1996).

With the Renaissance European connection between art and science came an increase in prestige for the art disciplines. Later, these humanist thoughts from the Renaissance inspired the eighteenth-century Age of Enlightenment, where the revolution in natural science created a new worldview; individuals in this later age used mathematical tools from Antiquity to make achievements in natural science.

Unlike today, where we often use the term "interdisciplinary" to make connections between disparate fields, specialization of disciplines did not exist in the Enlightenment period: Philosophers, writers, artists, and scientists assembled to participate in ambitious discussions in which they showcased their verbal brilliance, turning their conversations into a form of entertainment (Jakobsen 2011).

It was during the period of Romanticism in the late-eighteenth century that intellectual life became fragmented and split into disciplines; scientists became specialized and departed from each other. Both philosophy and art became professionalized, and the interactions between disciplines were notably reduced. By the long nineteenth century and through the two world wars and the Cold War, the interdisciplinarity and connections on which Pythagoras had founded his thoughts had dissolved.

As humans have evolved, they have developed technology and other tools to interpret nature. Moreover, humans have developed ideas and institutions based on sensory experience of phenomena. In the more recent history of humankind—perhaps in the last few decades—the scientific environment has seen a shift toward looking at art as a system of knowledge. In order to bring science into the human experience, making big science intimate, the world's largest particle physics laboratory, CERN, has offered artist residencies since 2011. Here, artists can co-create, make sense of these experiences of phenomena, and benefit the scientific community. As one of the artist-in-residence, Haroon Mirza, observed:

> Scientists at CERN are beginning to talk about this idea of science heading towards a brick wall. They're starting to accept there are some things we just aren't able to understand; something that artists encounter every day.
>
> *(quoted in Bello 2017)*

CERN has accepted that the scale and speed of the impact of human activity on the planet has transitioned into something that we cannot easily comprehend through mathematics or instruments (Bello 2017). Artists question how we can make sense of the experience and give new interpretations of the world and reality.

In 1922, philosopher Ludwig Wittgentein remarked in *Tractatus Logico-Philosophicus* on the need to reveal and define the borders of science. It's the philosopher's task to render clear insight that cannot be verbalized (Johannessen 1989, 17). Thus, in his endeavor to make sense of the world, Wittgenstein chose to communicate using the philosophy of mathematics and metaphysics. In his philosophical writings, Wittgenstein provided new insights into the relation between the world, thought, and language—and thereby into the nature of philosophy (Biletzki and Matar 2018).

We encounter Wittgenstein's thoughts on isomorphic connections between grammatical relations of the world and facts about nature in some of his writings:

> The gramophone record, the musical thought, the score, the waves of sound, all stand to one another in that pictorial internal relation, which holds between language and the world. To all of them the logical structure is common.
>
> *(Wittgenstein 1999, 32, statement 4.014)*

Lost in translation

Through all times, human knowledge has been transferred and handed down both orally and through human participation in practical processes. Later, markings and writings facilitated the knowledge transfer. Notation made it possible to archive music before the phonograph was invented. The phonograph introduced a new era in archiving and preserving music; it reduced the reproduction's inaccuracies through the tangibility of the source, making it possible to listen to and repeat the sounds. Meanwhile, notation still needs a key, a system on which the creator and the interpreter can agree, so that the work can be reproduced according to the creator's intentions.

One of the earliest surviving notations of music, known as the *Epitaph of Seikilos* (first century CE), is an excellent example of a system with a lost key. The epitaph is found on a tombstone near Aidin, in present-day Turkey. In addition to musical notation, the stone's inscription reads: "I am a tomb stone, an icon. Seikilos placed me here as an everlasting sign of deathless remembrance" (Sohma 2016). Historians have had considerable difficulty deciphering the music noted on the stone, but Seikilos appeared to believe that the melody would be preserved and archived forever. Such a conviction shows his apparent lack of historical awareness respecting shifting systems of communication and a corresponding belief in a universal system—one unbound by time, spared from discontinuity and cultural change. Although there are examples in texts from the early Renaissance where the Greek notational system is added, according to the National Museum of Denmark, there is no evidence that the Greek notational system survived from Antiquity into the Middle Ages ("De første noder?" 2019). Attempts have been made to transcribe Seikilos's short melody into a Western notational system, but they have failed as the contextual codes are no longer present. With no key to the notation, and thus to the cultural contract, a definitive reconstruction/recreation of the melody is impossible.

The constructed systems, such as the notational system, are only graspable for those holding the key; systems are not necessarily universal. Although language is a human construct that is marked by inaccuracies, it is nevertheless a bearer of meaning in society.

The notation of Norwegian folk music in the era of Norwegian Romantic nationalism (the late-nineteenth century) is a prime example of the inaccuracies that can occur when translation from the oral to the written is colored by social and cultural codices. One of the men traveling in Norway in the 1840s to collect folk music was the Norwegian composer and organist, Ludvig Mathias Lindeman

(1812–1887). Among other places, Lindeman collected tunes from the mountainous region of Telemark in Southern Norway. Because the region was remote and thus seemingly untainted by Danish and Swedish culture, the cultural elite deemed these folk tunes to be the most authentic.

Lindeman's systematization was tinted by his background, and he adjusted where it seemed inappropriate. To a classical scholar at that time, the rhythm did not seem to fit into a normal time signature, and the music's microtonality would seem "off" or out of tune, not corresponding to the aesthetic musical ideal of the time (Ressem 2016). Lindeman arranged the folk tunes for piano, forcing the melodies into the Western tempered scale, contrary to the musical traditions in the region. This led to inaccuracies, which means the notation he produced does not give an exact picture of the original tune.

Even today the notation for traditional fiddle is a mere guideline; to crack the codes for playing it correctly, one needs to know the local history of the tune as well as the story, temper, and personal blemishes of the author. One can say that in classical notation, the connection between written music and what is performed is more direct. The conductor's interpretation is, by comparison, another translation that contributes to the final result.

The Age of Enlightenment's model for divine role models came from Antiquity in art, and as a result, the expressive and the domestic became increasingly important in Norwegian Romantic nationalism. The vernacular, the mother tongue, and folk culture contributed to building a bridge between the nation and state in a time of secession, as Norway started to free itself from Denmark (1814) and later Sweden (1905). In visual arts, this period is recognized by an idealization of rural motifs, such as the paintings of mountain scenery and the midnight sun from Lofoten and Steigen.

As Peter Blom, Governor of Buskerud County and Member of the Norwegian Parliament wrote in 1827, "Lofoten is devoid of any form of natural beauty" (Blom 1832). His view of Norway's Lofoten islands and their surroundings was not rare at the time. However, in the era of Norwegian Romantic nationalism this changed drastically, with painters such as Christian Krogh, Peder Balke, Theodor Kittelsen, Gunnar Berg, and Otto Sinding drawing inspiration from the area. They documented the landscape on canvas and spread their art—and their perception of the archipelago—all over the country.

In the desire to portray the uniqueness of the Norwegian folk culture and the Norwegian landscape, and thus serving the Romantic project, this led to a decrease in realistic presentation of nature: Artists often chose to emphasize the motif's dramatic effect and humanity's relation with the forces of nature, as shown in Peder Balke's (1804–1887) interpretation of Stetind in the painting *The Mountain Stetind in Fog* (1864). His unique presentation of the landscape was a result not only of his own self-reflectiveness and the circulating Romantic tales of genius but also of the tendency of humans to infuse nature with their own temperaments (e.g., dramatic landscape, quiet meadow). Even though historical consciousness was developed at that time, the period lacked many keys and codes for arriving at an exact translation.

The art projects

The projects I turn to now are rooted in a common geopoetics. As I stated earlier, my aim in these projects was to combine seemingly separate and different systems of text, notation, and movement in order to find a common sensory connection. The four projects are as follows:

1   *Horizon. No Horizon (The poetry of) Repetitive Vertical Movement (at sea)*, 2010. In this project, I translated a boat's movement at sea into readable, visual scores through drawing, with the possibility of having these scores performed by voice.
2   *We Built This City (On Rock and Roll)*, 2012. I translated lyrics by the rock band Starship into a graphical urban geography/architecture.
3   *Black Walnut Grove*, 2016–present. I placed staff paper (blank sheet music) in a rural abandoned area near Kemptville, Canada, and let it become altered by wind and rain.
4   *From the West Fjord*, 2011. I translated the shape of a mountain in Steigen, Norway, into a score, which was later performed by a string quartet.

In the project *Horizon. No Horizon*, the key connecting nature and music is the decoding of the letters A and e, while the key in *We Built This City* is the black-lining

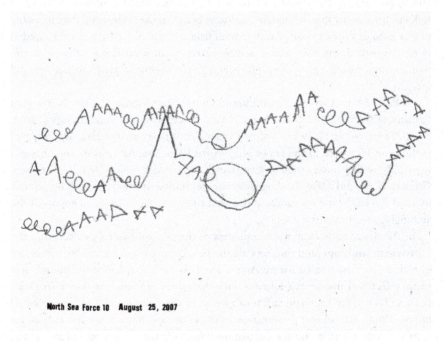

North Sea Force 10   August 25, 2007

FIGURE 4.1   *Horizon. No Horizon. (The poetry of) Repetitive Vertical Movement (at sea)* (2010), North Sea Force 10, 25 August 2007.

Source: Cecilie Bjørgås Jordheim.

in Microsoft Word. The code for reading *Black Walnut Grove*, the Western notational system recognized by the notational lines on a piece of paper, also plays a significant role in the piece entitled *From the West Fjord* (2011).[2]

Drawn symbols such as letters, numbers, notational scores for music, typography, choreographic notations, and cartography all show humanity's seeming need to systematize and describe the world around us. My project *Horizon. No Horizon. (The poetry of) Repetitive Vertical Movement (at sea)* (2010)[3] sought to understand the connection between the visual and auditory—both synesthetic and graphical. The drawings were produced from YouTube clips, all taken from the same category—documenting voyages in rough seas—and were always crafted from the bridge's point of view. These videos all revealed how the sea made the ship rise and fall as it confronted large waves. The horizon disappeared and reappeared. My aim was to reproduce the movement of the sea in the videos as a blind-contour rendering. The 20 postcard-sized scores, first presented in an installation entitled *NAUSEA*, each hold the name and date of the YouTube video source. The drawings of the boat's repetitive vertical movement become a score of movement, carrying the possibility of sonic reproduction.

I used the letters *A* and *e* as the base of the score on account of the ornamental resemblance between a capital A and a mountain peak (horizon), as well as between a minuscule e and a cresting wave (no horizon). I was inspired in part by the German-American abstract animator, Oskar Fischinger, who wrote in an article for the *Deutsche Algemeine Zeitung*, "Between ornament and music persist direct connections, which means that ornaments are music" (1932). The translation of the sea into music is based on the text as an ornament, rather than as a bearer of meaning; the text thus becomes a descriptive and expressive means of communication. Regarding the project's form, my intention was not to unravel the character of the sea on that particular day. Rather, I wanted my translation to exemplify more generally how movement can be materialized and systematized on paper.

We Built This City (On Rock and Roll) (2012)[4] is a piece that can be considered both isomorphic and visually intuitive. The project is based on a piece of text, lyrics by the band Starship, which I translated into a graphic architectural shape.

In 1985 Starship released the song *We Built This City (on rock and roll)*, which quickly became a hit. The lyrics express the youthful idealism of a city built on rock and roll, freed from the corporate world. (That Starship was itself a commercial endeavor adds light and delicious irony to this defiant piece.) In my project, I translated the lyrics from *We Built This City* into all available languages on Google Translate—from English into Afrikaans, from Afrikaans into Albanian, and so on, all the way through Zulu, which I then translated back into English. I then took this English version and began the process again—and then again. At this point, I had the original lyrics and three deconstructed versions. An unintended result of this was that the translation tool reduced the amount of text throughout the process; the more rounds in translation, the fewer words were left. I then redacted the text's three stages of deconstruction and the original lyrics in Microsoft Word, and the black-lining was layered horizontally and merged into one image.

What was left was a graphic architectural shape of a city skyline. The deconstruction of language led to a reconstruction of a city. In this way, *We Built This City* built a city.

While my project *We Built This City* was focused on the urban and sought to be an architectural structure based on pop music, *Black Walnut Grove* (2016–present) is more focused on the rural.

Between a maple syrup harvesting field and the outbuildings of a closed agricultural college in small-town Kemptville, Canada, there is an abandoned field of black walnut trees planted in rows. In February 2017, I placed ten blank scores (with hand-drawn horizontal lines in ink on unlaminated paper) on metal placards in that grove near Kemptville. Over time, the ten postcard-sized scores were marked by wind, rain, sun, melting snow, chew, spores, and other traces of wildlife.

Together with Chris Turnbull, the initiator of the project, I continued placing scores in the grove, one for each season. The variations in markings as a result of

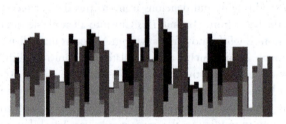

FIGURE 4.2 *We Built This City (on rock and roll)* (2012), prints 1, 3, and 5.

Source: Cecilie Bjørgås Jordheim.

the scores' exposure to the elements at various times in the year is significant. The winter scores have traces of rain, ice, and melting snow, which have washed away most of the notational lines on the paper, leaving it bleak and pale. By contrast, the late summer/early fall movements resulted in scores marked by multiple spores, seeds, and small species that have left them speckled with color.

As Turnbull remarks:

> Scoring, or marking, can be considered a seasonal practice within geographies— a way of recording and sparking ecological or atmospheric activities, of identifying "here" as well as ensuring future presences and interactions.
>
> (Turnbull 2017)

The blank scores offer a way to record the workings of the black walnut grove, to register the movements and change of the Eastern Ontario climate throughout the year. They conserve each season as ten short musical movements and become both a visual recording and a preservation of the grove's history.

Black Walnut Grove can be seen as a nod to the musical legacy and as a conceptualized version of *The Four Seasons* by Antonio Vivaldi (1678–1741), which included representational imitated sounds of buzzing insects, drunken dancers, a barking dog, and frozen landscapes, amongst others. This way of composing, known as program music, was revolutionary for the concept of music during Vivaldi's lifetime (Mellor 2018).

FIGURE 4.3   *From the West Fjord* (2011), early notation.

Source: Cecilie Bjørgås Jordheim.

The melody in the sound project *From the West Fjord* is derived from the highest top (Stetind) of a mountain range in Steigen, Vesterålen, in the Northern regions of Norway. *From the West Fjord* is an isomorphic translation connecting two graphical systems: the drawing of a horizontal line and the notational system. It is a verbalization of geography and the architecture of landscape. Using a photo, a perspective seen from the West Fjord, I manually plotted reference points into the shape of the mountains, similar to a dot-to-dot drawing. I then transferred the points to staff paper, on which the shape of the mountain indicated pitch. The distance between the dots determined the note value. In this way, the reading of geographic information set the premises for the production of sound.

In trying to translate an organic form into music, whether it is the shape of a mountain in Steigen or the movement of the sea, the challenges between and within systems become apparent. The mountain is civilized into a Western notational system, based on constant intervals, i.e., the equal tempered scale in music, which is inaccurate compared to the natural scale through which humans generally perceive sound in their daily lives. The tempered scale contains exact intervals, which are recognized in the black and white keys on a piano—from low C to high C, a stark contrast to the microtonality of the real world.

The end product in *From the West Fjord* is the music. It is an organic form made possible by one of Western society's arguably highest cultural symbols: A string quartet. My motivation for using a string quartet in this specific project lay in the possibilities of the instruments: Unlike the piano, the strings were not bound to a tempered scale. The Western notational system kept the strings from playing the true visual shape of the mountain, despite their ability to do so. By limiting the instrument to a melody consisting purely of whole tones, the instruments, and thereby the mountains, were affected by the system. The graphical notational system had, with its limitations, visually created a mountain similar to a knitting pattern.

Transforming a mountain into sound is about transforming something constant and solid into something ephemeral and linear. It is about compressing something eternal into a short interval of time. The act of conserving Stetind and its surrounding mountains through sound can be considered as a systematization equal to drawings, paintings, maps, topography, and archeological and geological research. The granite of Stetind is of Precambrian ground, making it one of the oldest rocks in Norway—about 1.8 billion years old (Andresen and Tull 1986). From an artistic perspective, a conservation of mountains through music turns into something purely abstract, an act of vanity, once the mountains' geological age is taken into perspective.

I found myself asking several questions while working on *From the West Fjord*, not only about how to translate from the visual shape of a mountain into music, but also about the context of the mountain I selected and the scenery in which it is placed. As a result of the late-nineteenth century paintings from this area depicting majestic landscapes (as discussed earlier), visitors to this region, or those who think on it, will always perceive the mountains of Northern Norway romantically. Although the tools that decided my process in *From the West Fjord* were conceptual (i.e., ideas, not concrete instruments), these mountains are charged with an art history impossible to ignore.

Geography made alive

Throughout his career, the American composer and philosopher John Cage (1912–1992) dealt with the fluctuant concept of "nature"—with nature as an uncalculated idea but at the same time as a fully constructed system that is familiar and known, omnipresent and potentially dangerous (Robinson 2010, 67). After failing to communicate the idea behind the expressive and emotional performance of *The Perilous Night* in the mid-1940s, Cage became aware of art in relation to nature:

> Cage gleaned a new mantra, which would become the banner statement for his aim from here onward: rather than self-expression, the purpose of art was "to imitate Nature in her manner of operation."
>
> *(Robinson 2010, 67)*

For Cage, the Romantic expressive thought of nature was shattered. This mantra echoed in Cage's practice: Instead of art being a construction and a conclusion of creative subjectivity, he maintained it should strive to be an action and an intervention in a larger perceptual field.

My art practice, in general, seems to have followed Cage's mantra from the 1940s, with some modifications: Nature is imitated, but as in *From the West Fjord*, the art is colored by the thought of expressivity—an expressivity possibly inherent and unavoidable in the translation from one system (mountains) to another (music). The project is affected by me, as an artist, having my own mind, logical sense, and choice of methods as parameters in the process. Empirically, one is true to a system, but one's own interpretations will always shape it.

Does my work have the ability to say anything about nature and geography? My work strives to be an interdisciplinary interpretation of reality, making it sensorially approachable; it aims to see translations and connections in an isomorphic, intuitive manner. Art, at its best, can be larger than language. Through poetry, sound, and image, geography is made alive, intimate, and available.

Notes

1 Isomorphic means "same/equal shape" and is an expression used in mathematics to indicate the similitude of two different structures. From a visual perspective, two seemingly different structures from different sources can be called isomorphic.

2 First exhibited and performed by a traditional string quartet at Stenersenmuseet, Oslo (2011) and later performed by a quartet of local musicians (consisting of viola, cello, Indian reed organ, and glockenspiel) at Friisgården in Ramberg, Lofoten under the direction of Jürg Leutert (2011). Sound available here: https://soundcloud.com/cbjordheim/sets/fra-vestfjorden-from-the-west

3 First published at UBUweb, this project was later performed a number of times by: students and professors from KHiO during the deefakt/Ultima festival (2011), an amateur

choir at Produzentengalerie Luzern (Switzerland) (2013), and a professional singer Stine Janvin Motland at Alte Schmiede in Vienna (Austria) (2014). The project was also performed on two occasions during the exhibition *All the World's a Stave* at Art Museum Nord-Trøndelag, 18 April and 6 June 2015 by members of three local choirs, all under the direction of Elina Karpinska: Namsos Kammerkor, Cygnus, and Salt.

4   First published in Matrix Magazine's *Conceptualism Dossier* (2012) and later in *The New Concrete: Visual Poetry in the 21st Century,* Hayward Publishing (2014). Large prints of the piece were exhibited at *All the World's a Stave* at Art Museum Nord-Trøndelag (2014).

References

Andresen, Arild, and James F. Tull. 1986. "Age and Tectonic Setting of the Tysfjord Gneiss Granite, Efjord, North Norway." *Norsk Geologisk Tidsskrift* 66: 69–80.

Bello, Monica. 2017. "Co-Creation at the World's Largest Particle Physics Laboratory—CERN." Paper Presented at Technology and Emotions, Oslo, November 7.

Biletzki, Anat, and Anat Matar. "Ludwig Wittgenstein." In *The Stanford Encyclopedia of Philosophy* (Summer 2018 Edition), edited by Edward N. Zalta. Accessed January 19, 2019. https://plato.stanford.edu/archives/sum2018/entries/wittgenstein/.

Blom, Gustav Peter. (1830) 1832. *Bemærkninger paa en Reise i Nordlandene og igjennem Lapland til Stockholm i Aaret 1827* [Remarks on a voyage in the high north and through Lapland to Stockholm the year of 1827]. Christiania: Tryckt i det Wulfsbergske Bogtrykkerie af R. Hviid og på hans Forlag.

"De første noder?" [The first notes?] 2019. *National Museum of Denmark.* Accessed January 7, 2019. https://natmus.dk/historisk-viden/verden/middelhavslandene/graekenland/de-foerste-noder/.

Gestalt Psychology. 2018. "Encyclopedia Britannica." Accessed May 18, 2018. www.britannica.com/science/Gestalt-psychology.

Fischinger, Oskar. 1932. "Sounding Ornaments." *Deutsche Allgemeine Zeitung,* July 8.

Foucault, Michel. (1966) 1970. *The Order of Things.* London: Tavistock/Routledge.

Heim, Steven. 2008. *The Resonant Interface. HCI Foundations for Interaction Design.* New York: Long Island University.

Jakobsen, Kjell A. 2011. "IDE1103: European Thinking from Romanticism to Present-day." Lectures. Oslo: University of Oslo.

Janson, Horst Woldemar, and Anthony F. Janson. 2004. *History of Art: The Western Tradition.* 6th ed. Upper Saddle River, NJ: Pearson, Prentice Hall.

Johannessen, Kjell S. 1989. "Analogien mellom begrep og stil" [The Analogy Between Term and Style]. *Nordic Journal of Aesthetics* 2 (4): 15–32. http://dx.doi.org/10.7146/nja.v2i4.3273.

Kristeller, Paul Oskar. (1951) 1996. *Konstarternas moderna system* [The Modern Systems of the Art Disciplines]. Stockholm: Raster.

Mellor, Andrew. 2018. "Discover Vivaldi: The Four Seasons—An Introduction." Accessed April 23, 2018. www.deutschegrammophon.com/gb/album/discover/masterworks/vivaldis-four-seasons.html.

Proust, Dominique. 2011. "The Harmony of the Spheres from Pythagoras to Voyager." *The Role of Astronomy in Society and Culture, Proceedings of the International Astronomical Union, IAU Symposium* 260: 358–367. doi:10.1017/S1743921311002535.

Ressem, Astrid Nora, ed. 2016. *Norske middelalderballader. Melodier. Skriflige kilder. Bind 4. Biografer, registre og tillegg.* [Norwegian Medieval Ballads. Melodies. Written sources. Volume 4. Biographies, registers and supplements.] "Lindeman, Ludvig Mathias," 200–202. Oslo: Spartacus.

Robinson, Julia. 2010. "John Cage and the Investiture: Unmanning the System." In *The Anarchy of Silence: John Cage and Experimental Art*, edited by Julia Robinson. Barcelona: MACBA.

Rogers, Richard, and Oscar Hammerstein II. 1959. *Sound of Music*. New York: Lyrics.

Sohma, Marina. 2016. "Song of Seikilos: Oldest Known Musical Composition Lay Hidden on a Flower Stand in Turkish Garden." *Ancient Origins*, December 5. Accessed May 9, 2018. www.ancient-origins.net/artifacts-ancient-writings/song-seikilos-oldest-known-musical-composition-lay-hidden-flower-stand-02197.

Turnbull, Christine. 2017. "Delinquent, Deliquecent: Rout/e." *Footpress: Poetry Found in Place*, April 28. Accessed January 11, 2018. https://etuor.wordpress.com/2017/04/28/delinquent-deliquescent/.

Wittgenstein, Ludwig. (1922) 1999. *Tractatus Logico-Philosophicus*. Translated by Terje Øde-gaard. Trondheim: Gyldendal.

5

KARANKAWA CARANCAHUA CARANCAGUA KARANKAWAY

Centering Indigenous presence in Southeast Texas

John Pluecker

An opening

In 2016, Mexican artist Nuria Montiel and I collaborated to build a visual and sound poem called *Karankawa Carancahua Carancagua Karankaway* at Project Row Houses in Houston, Texas.[1] The poem installation was a meditation on language and memory in the context of colonial structures of power that have historically worked to eradicate and erase one of the Indigenous cultures that inhabits Southeast Texas: the Karankawa.

The seeds of this project emerged from my investigation into archival documentation of Karankawa language assembled by colonial settlers, explorers, and scholars in Southeast Texas. That early investigation resulted in an initial, ephemeral poetic form, an artist book I titled *Ioyaiene. Ioyaiene* featured a section titled "Anti-Glossary" in which I reprinted all the words from nineteenth-century botanist, geologist, and colonial agent Juan Luis (Jean Louis) Berlandier's glossary of Karankawa. In my artist book, however, I used a mixture of paint and sand from Brays Bayou in Houston to paint over the words in Spanish and English, in order to erase the corresponding translations in these colonizing languages. I was interested in erasing settler vocabularies, gesturing toward the sound and resonance of these Karankawa words. The "Anti-Glossary" eventually appeared as a section of my book, *Ford Over* (2016).

In 2016, a collaborative process with Nuria led me out of the archive and into dialogue and conversation with contemporary Indigenous peoples of the Texas Gulf Coast.[2] The dominant historical narrative in Texas contends that the Karankawa people were driven to extinction; however, members of the Texas Carrizo/Comecrudo tribe, along with other Native Texans, refute these accounts. Elders in the Carrizo/Comecrudo tribe state that in the mid-nineteenth century, bands of Karankawa fled the coast and colonial oppression, seeking and finding refuge with the Carrizo/Comecrudo in Central Texas. Based on these narratives,

FIGURE 5.1   Installation shot of *Karankawa Carancahua Carancagua Karankaway*.

Source: Photograph by Alex Barber.

members of the Carrizo/Comecrudo tribe say that, "We are them, and they are us." In this way, tribal members reclaim their own Karankawa ancestors, grounding their identities in the stories and lifeways of their own families. Through this sound and visual poem (with audio work by Lucas Gorham, a sound artist and musician of Karankawa descent), we ended up asking: Can languages be completely eradicated, or might they survive in ways that are both perceptible and imperceptible? How might they be heard? How might inhabiting the nexus between poetry and geography provide tools for disassembling or reimagining the colonial archive?

How might artistic research and creation become methods for combating ongoing colonization and erasure? How can listening and requesting permission become modalities for a different kind of conversation?

Now, years after the installation, I want to look back at the process we initiated to realize this project. What we made was not a product, but instead a series of connections with local Indigenous peoples that led us to create an installation that made Karankawa language visible and gestured toward ongoing lifeways, despite the official historical record of eradication and elimination. What is more, our creation was not simply an installation: It was a process of conversation, learning, and listening. We ended up forming shapes out of earth, and we made something ephemeral, not final—just one concretion in the midst of a process. What we suggest through our project is that a poem is not a container or a collection of words. Poetry is, instead, a process of "being in relation" to the earth and all its inhabitants.[3]

After de-installing the poem from the space of the row house, I knew I wanted to write an essay to think through what Nuria and I had made and the process by which we made it. Now this essay has found a home as a chapter in this very book on geopoetics. So I ask myself: How might this project be understood as a geopoetical project? As Eric Magrane writes, "Geopoetics might be mulch or compost or the building of earthworks to collect stormwater runoff and plant the rain in the desert" (2018, 39). Magrane is gesturing toward a larger sense of poetry as object and material, made out of earth or organic materials and not necessarily out of simply words. Though Nuria and I were not aware of geopoetics as a term as we were doing our work, reading geopoetical work now makes it clear that our project is in dialogue with this field, especially insofar as it relates to the creation of a "speculative more-than-human geopoetics: a reflective and refractive earth-making that imagines and speculates on alter-subjectivities" (Magrane 2018, 40). The term geopoetics is derived from the Greek for "earth-making" ("geo" meaning earth and "poesis" meaning "making"). Our process for this project was not so much about making something out of the earth; rather, it was about seeking to allow the earth and our relation to it to remake us, if only for a fleeting moment.

A definitionless glossary of dirt

In a shotgun house in the Third Ward, a historic African-American neighborhood in Houston, the words of a colonial glossary were arrayed on the floor. The words were formed using laser-cut stencils made of soil and sand that Nuria and I had collected from the bayous of Houston and from Carrizo/Comecrudo ancestral lands in Central Texas. With her training in typography and printing, Nuria Montiel created the stencils that utilized the same font found in the mid-twentieth-century books through which we accessed the glossary. We built a physical space around these terms within the row house—a space for a glossary that had crystallized a spoken language into a temporary and error-filled orthography. We erased the English and the Spanish to allow the Karankawa words to exist as independent elements.

The text, that is, the glossary, became very literally a social space: One where people met and interacted, one where visitors navigated along narrow pathways through the words of the glossary, one that people wanted to inhabit for a moment. As Stefano Harney and Fred Moten write, "To say that [a text] is a social space is to say that stuff is going on: people, things, are meeting there and interacting, rubbing off one another, brushing up against one another—and you enter into that social space, to try to be part of it" (2013, 108). What is it to inhabit a social space of Karankawa language in contemporary Texas, to invite visitors to immerse themselves in a different kind of language space? What are the possibilities that emerge from a space like that? What are the potential failures?

To find the Karankawa words, we drew on multiple glossaries found in colonial records. We kept in mind the Indigenous people who gifted the colonizers with these vocabularies. We read the work of Ohlone/Costanoan-Esselen poet Deborah Miranda (2013) and noted how she valorizes the contributions of Indigenous people, particularly women, to the colonial archive of Indigenous languages. These individuals made time to dialogue with the colonial linguists, perhaps thinking that this would be a way to transmit these words into the future, despite the death and destruction brought on by colonization. We rummaged through the historical records and found mention of the Tonkawas Old Simon and Sally Washington, both of whom gifted words to the linguist AS Gatschet during the 1880s (Swanton 1940, 6). We sifted through all of the different glossaries that we could find and created an Excel spreadsheet, into which we entered all those Karankawa words.

Later, after much dialogue with local Indigenous residents of the Texas Gulf Coast, we decided to stencil the words on the ground, creating precarious structures of sand and dirt, fixed together with the slime of the nopal cactus. These letters were always crumbling, continually breaking apart and cracking. They were sometimes even stepped on and crushed by inattentive visitors. We decided early on not to rope off the words or to place any kind of barrier between the visitor and the glossary; visitors decided what attention to give or not to give to these precarious sculptures, thus mirroring a larger awareness or lack of awareness of indigeneity in this city and wider region. Every step on the wooden floor of the tiny pier-and-beam house shook the words, loosening some grains of sand and disfiguring the anti-glossary.

In the space, in addition to the words on the ground, an audio track by Lucas Gorham featured recordings of the words of the glossary, which we had asked Indigenous residents of Southeast Texas to pronounce. We conceived of this as a way of circulating the glossary, gifting it to people who largely had never seen these words in print before. Lucas is Mexican-American, with stories in his family of Karankawa ancestry. As I've been friends with him and his family for many years, I'd first heard those stories more than 15 years prior to our installation at a family barbecue. Hearing those accounts was the first experience that led me to question the dominant historiography in Texas that insisted on the long-ago extinction of the Karankawa.

Dominant historiography and other accounts

As I explained earlier, my work with the Karankawa glossaries and language was, initially, purely archival. In working on my first book of words and images, *Ford Over*, I spent years with the diaries and documents amassed by a series of men who traveled through what is now the state of Texas: Fray Morfi, Alvár Nuñez Cabeza de Vaca, Juan Luis (Jean Louis) Berlandier, and Frederick Law Olmsted. Typically, these men are called "explorers," but I question the use of this apparently objective, even adulatory, term. I didn't want to become another "explorer," as if this term was neutral or even flattering. For if "exploration" happens by settlers in the context of conquest, this is not a disinterested adventure. Colonization has rendered its agents "explorers," and "exploration" is a term that is still repeated uncritically in numerous settings, whether museums or educational institutions. Scientific progress and research—in both the social and hard sciences—are still orchestrated around this same idea of the impartial benefit wrought by "exploration." In response to this problematic perspective of history, I tried to conceive of these men as "colonial agents," men who were doing the work of empire-building, albeit slowly through the collection of materials, the recording of travel, and the documentation of language. I also tried to think of ways to explode that archive from the inside, which I subsequently aimed to achieve through a de-ordering of colonial language, that is, a process of remixing and shape-shifting.

Despite all the work I did in the archive of Karankawa language during the making of *Ford Over*, I had not engaged in dialogue about this archive with Indigenous people in Texas. I was conflicted about leaving the archive and becoming an "explorer" of unknown things or Indigenous spaces. This timidity, or unwillingness to occupy or invade spaces, was a double-edged sword. Over 15 years working as an organizer and Spanish-language interpreter, and just by living and sharing space with people in progressive circles in Houston, I made friends with many people of Indigenous descent. However, as I reached out to these friends for this project, I realized I knew nothing about the narratives of the Carrizo/Comecrudo that turn the dominant history of Texas on its head. I began to hear these Indigenous narratives only after working more closely with Nuria and leaving the quiet of the archive. I saw more clearly that ignorance of Indigenous lifeways is the default in a settler culture that breeds a lack of consciousness among its own.

Dominant historiography—in textbooks and mandated middle school Texas history classes—affirm that the Karankawa are now "extinct." The Texas State Historical Association Texas Handbook narrates the slow process of extermination of the tribe, hounded in Texas and driven south into Tamaulipas in Northern Mexico. The Handbook explains:

> By the late 1850s the Karankawas had been pushed back into Texas, where they settled in the vicinity of Rio Grande City. Local residents did not welcome the tribe, and in 1858 a Texan force, led by Juan Nepomuceno Cortina,

attacked and annihilated that small remaining band of Karankawas. After that last defeat, the coastal Texas tribe was considered extinct.

(Lipscomb 2010)

This narrative of extermination was, for a long time, the only one I had ever heard. It is surely the only one the vast majority of residents on the Gulf Coast have heard, whatever their racial or ethnic background.

Sarah de Leeuw and Sarah Hunt write:

> To not acknowledge who we are, or to leave unspecified our authorial position in relation to this paper and events unfolding all around us, is to risk perpetuating the idea that writing and knowledge is not produced by people who occupy specific temporal and sociocultural positions, positions often bound to or by colonialism.
>
> *(2018, 3)*

Following from this, it seems critical here to locate myself. My ancestors colonized Karankawa and Carrizo/Comecrudo lands. They were mainly Germans—but also Irish, Swedes, and others—who came to these lands in Texas, some as far back as the early-nineteenth century when these lands still belonged to Mexico. On the oldest branches, my family has been here for seven generations. I was not ignorant of the almost two centuries of abuse, colonization, exploitation, and profit visited upon Indigenous peoples, both indirectly and directly. In fact, my own work has focused on this history and on continuing forms of colonization and oppression in Texas; this work cannot be separated from my settler identity in these lands. While I think silence is an important strategic decision, especially for settler colonists, I don't think permanent silence or non-engagement is an ethical option; if we were to be silent, this work would be necessarily assigned to those Indigenous folks who are already most burdened by the ongoing colonial project. Though of course Indigenous peoples should lead this effort, there is a concomitant necessity for self-interrogation and participation by settlers as well.

When Nuria and I began to think about what we would do, we started from our own bodies, myself as a white-settler and, in Nuria's case, as a mestiza from Mexico City. We did not think to ask for permission to do this project. But we did think to be in conversation with Indigenous people in the Houston region, to ground our work in dialogue with these individuals of diverse backgrounds and identities. We worked to consciously undercut the assumption that we could guide the project, instead aiming to ground our work in listening and in sitting together with Indigenous folks in the region. Nuria and I designed a process of *convivio* (i.e., living together with) to find our way along the path of making this installation.

Initially, we only reached out to friends of mine who come from a variety of Indigenous backgrounds and communities, not all descended from the Karankawa. Some of those friends recommended we speak with *their* friends. Nuria and I had some initial ideas about how to orchestrate these very open-ended conversations.

We had some general ideas about topics for conversation, but we did not want to do a survey. Instead, we landed on some general and very open-ended sparks for conversation. We told my friends and theirs about our vague idea of doing a project—about our work on a glossary. We talked broadly about who we were and what we were doing. We thought to ask people where their own memory and experience of Indigenous lifeways and/or languages came from, as well as to ask about people's particular connections to Indigenous peoples and history in Texas and in other places. We thought of other questions: How much of Indigenous lifeways and languages have been eradicated? How much remains? What can be done to reclaim Indigenous lifeways and languages that have been eradicated? In the future, what place do you think Indigenous lifeways and languages should have in Texas?

Nuria and I reached out to these friends of mine to listen and to converse, to be clear who we were as human-animals and to learn more about these friends and to share our own thinking. In the end, we reached out to listen to these interlocutors and to attempt to unlearn after so many generations of learning how to abuse, colonize, exploit, and profit.

At some point, we thought to ask each person to read out loud the words they would like to read from the glossary of Karankawa. We thought to ask what it felt like to pronounce the Karankawa words. We thought to ask them if they would like to share other things, such as songs or music or poetry. It could be whatever they wanted, or nothing at all: We tried to exert no pressure or to assume anyone would want to share.

Soon after beginning the project, the curator for the round of installations at Project Row Houses, Raquel deAnda, recommended we speak with Bryan Parras, who had spoken with her about the Carrizo/Comecrudo tribe and their connections to the Karankawa. Bryan has been a friend of mine for years, but we had never really talked about the Karankawa, for the same reasons of reticence I mentioned previously. Part of what Bryan emphasized was that progressive white folks and other outsiders are often interested in Indigenous lifeways and want to engage in research; however, such a process often fails to take into account all the work already done by Indigenous people to document and un-document their own lifeways. He emphasized how many Indigenous people had been protecting sites and traditions to try to keep them from being used and abused. He highlighted how a lot of non-Indigenous people had stolen teachings from Elders, writing books—for profit—based on these teachings. He reminded us it was important to allow space for some narratives and some knowledge to remain with Indigenous folks, that it was crucial that we not set out to take possession of everything or to present everything we heard for public consumption. He mentioned that he learned the most when people corrected him, when people told him what he was doing wrong. He emphasized the importance of lifting up the words we had found and to present those words in Indigenous communities where people could reflect on them.

Bryan's words and recommendations helped to guide the project. He also told us to speak with the Carrizo/Comecrudo, and specifically with Eddie Garcia and Juan Mancías from that tribe. Bryan's words would radically reorient our process.

When Nuria and I reached out to the Carrizo/Comecrudo tribe in Central Texas, we learned that the dominant historical narrative on the elimination of the Karankawa was simply not the story that the tribe accepted as history. Eddie Garcia told us that tribal Elders had passed down the history that a band of Karankawa was driven off the Texas Coast in the middle of the nineteenth century. This group ended up in Central Texas, living with the Carrizo/Comecrudo. So according to tribal knowledge, the Karankawa went to live with the Carrizo/Comecrudo in the mid-nineteenth century, at the exact moment when dominant historiography insists the Karankawa were extinguished. The Carrizo/Comecrudo have their own understandings, namely that, as Eddie said and later Juan Mancías would repeat, "The Karankawa are us and we are them."

Eddie recommended that we begin our work by asking for permission from his own tribal Elders and other Indigenous people native to the lands of Texas. He gently suggested that we begin with, "I am on your land. I am here for you. What can I offer you? What do you need?" The emphasis was on asking for the right to be in dialogue, not assuming that there was any obligation from the colonized to talk with either a colonizer (in my case) or a mestiza from lands from the south (in Nuria's case). The emphasis was always on the fact that words are not just words, but indicative and productive of lifeways. As Eddie and Juan spoke of life-ways, they were gesturing toward an expansion beyond a sense of words as static elements on a page, into the lived experiences and cultural practices contained within those words.

Can a culture die or become extinct? Can a language die or become extinct, be extinguished?

Perhaps it hides or it intermixes. Perhaps death is not the proper metaphor.

Because a language or a culture is not exactly the same as a life.

To understand these stories, we can turn to contemporary Indigenous scholarship and activism that gives credence and asserts the veracity of claims and oral histories within Indigenous communities. As the Sisseton-Wahpeton Oyate scholar Kim TallBear writes:

> Thus, the fight for indigenous peoples—and for communities more broadly who are regularly subject to the scientific gaze—is to debate which meanings and whose meanings inform law and policy. That is where we should be working. To make sure that science, and the state, are more democratic, that our stories are heard as clearly as those of anthropologists and geneticists when the state acts to influence our lives. Or rather, that our stories should be heard more loudly than theirs when we have more at stake.
>
> *(2007, 423)*

One life is one life, and a life can end or be ended. But a language or a culture is collectively held and collectively nurtured. How does a collectivity end? Or does a collectivity end? Or are collectivities perpetually changing and morphing to create new collectivities always connected to those previous to it?

If there are people who still identify with the Karankawa, does that culture somehow still exist? It's obviously altered, but isn't there something there nevertheless? Isn't it a colonizing move to insist that a nation has been eradicated when there are living, breathing people asserting their connections to that same nation?

In what ways might a poetic gesture—whether in a row house or in an essay form—serve to generate a wider collective awareness of Karankawa presence?

There are reasons to hide one's existence or one's presence. There are reasons why one looks to protect oneself or to preserve oneself in the face of a dominant culture that is consistently hostile, negating, and attacking. And while a project like Nuria's and mine seeks to undermine colonial structures that marginalize and invalidate Indigenous thinking and work, it is important to remember that this does not effectuate decolonization. As Eve Tuck and K. Wayne Yang remind us, "decolonization specifically requires the repatriation of Indigenous land and life. Decolonization is not a metonym for social justice" (2012, 21). For this reason, I am not even sure if this project was a decolonial one, but it definitely attempted to work against and in opposition to the ongoing project of erasure and eradication of Indigenous peoples.

Study

Since engaging in this project, I've spent a significant amount of time trying to think through what we did and trying to think about the kinds of methodologies we used in our own work. What we did was not a series of interviews or a research project oriented toward "Human Subjects." Rather what we envisioned was to have intentional conversations with people whom I knew or people whom I had met over the course of 17 years living in Houston. We were not reaching out blindly into organizations or tribes, looking for people to speak with. Rather, we attempted to have more in-depth conversations with people I already knew. In this sense, what we were doing had much to do with what Stefano Harney and Fred Moten refer to as "study" that happens outside of the university: An undercommons of contact, play, and struggle that exists outside of formal academic channels. Harney and Moten write:

> Study is what you do with other people. It's talking and walking around with other people, working, dancing, suffering, some irreducible convergence of all three, held under the name of speculative practice. The notion of a rehearsal— being in a kind of workshop, playing in a band, in a jam session, or old men sitting on a porch, or people working together in a factory—there are these various modes of activity. The point of calling it "study" is to mark that the incessant and irreversible intellectuality of these activities is already present.
>
> *(2013, 110).*

The project began years before, when it was not even a project. And perhaps "projects" don't actually ever begin or end. As I mentioned earlier, it began at a barbecue in the home of old friend from a Chicano movement family in Houston. It began at a house party *tamalada* raising money for a sick relative. It began in a domestic workers' organizing group that invited me to work with women in the group to write their own stories. It began at the Indigenous-grounded marriage ceremony in San Antonio between two old friends, one an immigrant from Northern Mexico and another a descendant of generations of Mexican-American Texans. It began in a million different ways—tiny strands of connection and friendship nurtured over 17 years. All of these "various modes of activity" gave rise to this project. Without the "incessant and irreversible intellectuality of these activities," the project would never have existed. I value these kinds of long-term connections and contacts perhaps more than anything else. And because we are enmeshed in long-term relations, I am extraordinarily careful about how I nurture these ties.

In their introduction to an anthology of Lakota/Dakota/Nakota writing about the Mnisose (Missouri River) called *This Stretch of the River*, Craig Howe and Kim TallBear think through the particular predicaments faced by Indigenous people writing individually authored texts. As Kim TallBear writes, "Conscious of being accountable to our communities and families in the things that we write about and, perhaps more importantly, in the things that we do not write about, interesting conflicts can arise as we produce texts under our individual names" (2006, x–xi). Related to this, I have been exceedingly slow in writing this piece because I've been navigating all of these questions. As it is individually authored, I have been thinking endlessly about what to write about, and most importantly, what not to write about. These "interesting conflicts" continue to reappear as I set out to write and rewrite this contribution, as I consider what is mine to share and what is better left to others to tell on their own. There are great potentials to be found in this kind of slow, decolonial "study," but also great risks and a million opportunities for failure.

In response to an individualistic way of writing, *This Stretch of the River* doesn't only compile texts by individuals but also includes an edited transcript of a communal conversation between a number of writers from the anthology. In the transcript of this multi-party conversation in the book, writer Kathryn Akipa mentions a story about a particular moment of trauma and pain that affected the tribe as a result of the damming of the Mnisose (Missouri River). Akipa mentions the story in the dialogue but not in her individual writing. A footnote explains that she was uncomfortable mentioning the story of trauma and pain in her individual essay because she did not want to capitalize on painful memories in her individual piece, but that in the dialogue it seemed appropriate because it represented "an exchange or mutual remembering rather than an individual promulgation" (101–02, 2n).

There are some stories I won't share here, some dynamics that I—a seventh-generation settler in these lands—have no business discussing on my own. No one told me not to expose these stories, but I can feel in my gut that to do so would be a mistake. Or, more than a mistake, it would be a taking, an appropriation, of something shared with me but that nonetheless is not mine to tell.

Other people's stories, their autonomy

Nuria and I didn't do this work to occupy a space or to explain Indigenous lifeways still present in Texas. We did it to attempt to learn how to occupy space differently in a time of rapaciousness, greed, and profit.

Of the various Indigenous interlocutors, there was a mix of different ethnicities and tribal affiliations.[4]

I am not going to define the identities of all of these people here because I don't think it is my business to do that. Identity and affiliation are delicate formations, and I don't think it is my place to represent them. They each represent themselves every day.

Nuria and I consciously did not build a platform for Indigenous folks to speak because those platforms and organizations have already been made by Indigenous people themselves. These Indigenous spaces—whether protests or ceremonies or ritual spaces or barbecues—are there for people to assemble and engage in dialogue. There are particular dynamics and ongoing tensions and conversations in those spaces, and it is not my place to amplify those dynamics or attempt to explicate them.

Settlers have an eye for conflicts and divisions within Indigenous communities. It is a way to regain power for settler bodies and lifeways, a way of pointing to disagreement and dissension and thus avoiding power imbalances between Indigenous and settler. I am wary of settler colonists who—after generations of abuse, colonization, exploitation, and profit—now arrive to extract the stories of peoples who have been so thoroughly marginalized and erased. These are not my stories to tell.

At a certain point, we asked for permission to build an art installation from the Elders of the Carrizo/Comecrudo and other Indigenous folks with whom we were in conversation. We asked for permission to make a visual and sound poem. We asked for help from our interlocutors, whom we eventually identified as "consejerxs" (i.e., counselors). But we also asked if our own project was in alignment with their goals, or their thinking, or their feelings in their guts. We believed in feelings in guts. We asked what their work was, how they thought of their own work and their own identity. What we had to offer was very little—certainly not enough. We were *humbled* by the meagerness of what we had to offer.

And just because I received permission at one point to sit and listen to these friends talk does not mean that I have permission to repeat them. I don't actually have much of a desire to go back to them to ask them for permission to tell their stories. They own their own stories, and I do not envision encroaching on that narrative sovereignty. Their stories are their own to tell, and they are not mine.

I remember how the process of conversation with these friends unfolded: stories told around my dining table in my house in the East End in Houston, how the light filtered in as these friends told stories. I remember tears. I remember the quiet pauses and sighing. The difficulties of telling certain stories of loss and yearning. What has been torn away is more than a language or a culture, they reminded us. What have been torn away are lifeways.

The stories are grounded also in particular places. The stories of a place are formed by that place and take the form of that place. And this piece is formed by all the places and spaces where the stories were birthed—though it is also removed from those places where these stories were told: It was written on my computer, in my office, in my home in Houston's East End.

If you want to hear those stories or to be in relation to the people we spoke with, then you can ask for permission to do so. If you want to hear those stories, then you need to be in relation to the people where the stories were told. I can't tell you those stories unless you are in relation to those people, those places, the species in that place. Does that make sense?

The installation and this composition are mere gestures toward the existence of something else. Another kind of relation. It seems important that my relationship with most of these interlocutors was as "friend" or "friend of friend," that is, as someone who continues to have a relationship, someone who continues to be in relationship with. In some ways, it is a return to the root word for "Texas," which is *taysha*, the Indigenous Caddo word for friendship.

Whenever we had conversations with Indigenous people throughout the course of the project, we emphasized that we were not scientists or experts or even academics. One of us is an artist, we would say, and the other is a poet. We were honest about not knowing what we were planning to do.

As an essay—an act of trial and error—I've had trouble knowing where to begin this chapter or where to end it. This makes sense, because this process of working in relation with Indigenous groups in what is now colonized as Southeast Texas has not been linear. It has also not been simple.

There are stories that are not to be mined—mined in the sense of "made mine," and mined in the sense of extraction, removal, unearthing, probing. As colonialism extracted numerous glossaries from Indigenous peoples, they created these collections of words and phrases in isolation from their communities. The glossaries had no relationality with the community, in the sense of "being in relation" that Kim Tallbear has written about extensively. Words and narrative do not exist in isolation; as the Carrizo/Comecrudo *consejeros*, Eddie and Juan, talked about, these words and narrative are part of the collective lifeways of the tribe, so it would be impossible to make them individual belongings ("made mine") or to extract them from the community.

As Eddie Garcia and Juan Mancías from the Carrizo/Comecrudo tribe both emphasized: Who asked you to be on this land? Who have you asked permission from? Who do you continue to ask permission from? When we say *who*, we can think of humans, but also of the earth and the trees, the insects and the grasses, the doves and the detritus. We can enlarge our sense of personhood to include the dirt below us, the dirt that carries the bones of ancestors within it. This land is not separate from the Indigenous people who have walked it for millennia. Eddie and Juan emphasized that every day in Texas, people—settlers and descendants of enslaved peoples and Indigenous alike—are walking on their ancestors. This is because the bones of Karankawa and other Indigenous nations are literally in the dirt. There is no separation between the dirt or the trees or the human or non-human animals.

We are all made of the same primordial substance. Settlers are not just on *Indigenous land*: This land is *literally composed* of Indigenous peoples' ancestors. This was the initial impetus for making letters out of earth: However temporarily, to place the glossary back into relation with the land, and with the people who inhabit it. Though there are many stories that do not feel mine to tell, this is one that bears repeating, and it is a story that came to form the basis for our aesthetic and poetic decisions about the form of the poem/installation.

Kim TallBear always emphasizes in her speaking and writing that this land is not *sacred*, because *sacredness* is a Western concept that sections off some things as higher than others that are mundane or *not-sacred*. So the earth we used to make the Karankawa glossary is no more or less special than any other earth. And this different relationship to earth remade us and remade our art and our poetry.

The limits and the garage

The sand that we used to make the letters of the glossary on the floor of the row house is still stored in my garage. Nuria and I have had many plans or ideas over the years since the installation about what to do with it, but we still haven't landed on something definite. We are taking our time thinking about it. About what to do with the earth we used for the installation. We are still not sure. We have thought to bottle it and give it to the people of Indigenous descent who were generous enough to assist us with the project. We have thought about returning it to the bayous or to the tribal lands of the Carrizo/Comecrudo. Perhaps what we really need to do is just ask the dirt. Or ask the Carrizo/Comecrudo members who gave us permission to use it.

Since doing this project, I've continued to live in relation to the land and to the Indigenous friends who were kind enough to participate in the project. These conversations have led to other invitations. In 2017, Bryan Parras invited me to participate in a large-scale Indigenous-led protest in Eagle Pass, Texas, on the Mexican border to demand the closure of an open-pit coal mine. More recently, Bryan asked me to help to welcome a delegation of Lummi tribal members from the Northwest part of the US on a pilgrimage they were making to Miami to demand the return of a blackfish to its ancestral waters (what settler culture tellingly calls "killer whales"). In the process of writing this piece, the Indigenous consejerxs that we worked with continue to be engaged in social justice struggles for workers' rights, immigrant rights, against the border wall, and a number of other struggles. I shared this piece with a number of them, initially to get permission to publish this. However, they did not have the time to read it through in the midst of the intensity of attacks here in Texas coming from both the Trump administration and the right-wing politicians who control state government. I am committed to bringing this piece to them in the future, to make sure that the words in this piece—especially those that are theirs—are returned to them. The installation was just one point in this process; this piece is another point in an ongoing relationship.

A poem became a reason to converse more deeply, to engage each other in dialogue. It became a ground upon which we might continue to engage. The poem in that row house and this chapter do not seek to retell a story: A story of colonization and dispossession, broken treaties and virulent racism in an endless settler loop. The poem and this chapter attempt to breach the dominant historiography and the hegemonic lifeways exploiting and ravaging these lands, even now, in this petrochemical landscape of refineries, pipelines, ship channels, and tankers.

There is something poetic and important about how we decide to handle the earth housed in my garage. That earth stands in for the earth as whole, or for these lands I inhabit—these unceded Karankawa territories. What do I do with these lands? How do I live in relation to them and to all its human and non-human inhabitants? What is to be done?

Notes

1 More information about and related visuals for this project can be found here: www.johnpluecker.com/kcck-w-nuria-montiel-2016.
2 In this piece, I use first names to refer to friends, that is, to people with whom I have a relationship of friendship. I think it is important to use that level of informality that corresponds to closeness and intimacy, since closeness and intimacy are key parts of this writing. I refer to people who I do not know personally by their last names.
3 I use the term "being in relation" as Sisseton-Wahpeton Oyate scholar Kim TallBear uses it in her own work. For example, she wrote on her blog (accessed January 15, 2019), *The Critical Polyamorist*: "Being in relation requires doing and asking. This is because we cannot do everything for ourselves, or for others." The thinking in this paper is very much in debt to all of her work in a variety of academic and public forums.
4 I would like to thank those consejerxs (counselors) here by name: Ana Reyes Bonar, Julia De León, Laura Floyd, Eddie García, Liana López, Juan Mancías, Bryan Parras, Rainflowa, Angela Sánchez, and Monica Villarreal.

References

Craig, Phillip Howe, and Kim TallBear. 2006. *This Stretch of the River*. South Dakota: Oak Lake Writers' Society.

de Leeuw, Sarah, and Sarah Hunt. 2018. "Unsettling Decolonizing Geographies." *Geography Compass* 12 (7): 1–14. https://doi.org/10.1111/gec3.12376.

Harney, Stefano, and Fred Moten. 2013. *The Undercommons: Fugitive Planning & Black Study*. New York: Minor Compositions.

Lipscomb, Carol A. 2010. "Karankawa Indians." *Handbook of Texas Online*, June 15. www.tshaonline.org/handbook/online/articles/bmk05.

Magrane, Eric. 2018. "Geopoetics." In *Counter-Desecration: A Glossary for Writing Within the Anthropocene*. Middletown: Wesleyan University Press.

Miranda, Deborah. 2013. *Bad Indians: A Tribal Memoir*. Berkeley: Heyday.

Pluecker, John. 2016. *Ford Over*. Las Cruces: Noemi Press.

Swanton, John. 1940. *Linguistic Material from the Tribes of Southern Texas and Northeastern Mexico*. Washington, DC: Smithsonian Institution.

TallBear, Kim. 2007. "Narratives of Race and Indigeneity in the Genographic Project." *The Journal of Law, Medicine & Ethics* 35 (3): 412–424. doi:10.1111/j.1748-720X.2007.00164.x.

Tuck, E., and K. Wayne Yang. 2012. "Decolonization Is Not a Metaphor." *Decolonization: Indigeneity, Education & Society* 1 (1): 1–40.

6

GEOPOETICS OF *INTIME* AND (*SUND*)

Performing geochronology in the North Atlantic

Angela Rawlings

Foreshores

The counterclockwise collusion of the North Atlantic Drift, the Irminger Current, and the Labrador Current impacts subarctic, arctic, and temperate foreshores around the North Atlantic. While planetary rotation is counterclockwise (from east to west), the Coriolis Effect deflects air and water currents to the right in the northern hemisphere, causing a tendency for clockwise flow in the North Atlantic Ocean (Harper 2004, 22). The North Atlantic Drift, through ocean and air currents, extends northeast from the Gulf Stream to continue its path across the many vulnerable shorelines of Scotland's western seaboard. The Drift presses its storms farther north-northeast, touching Denmark and Norway, its climate impact felt as far inland as Sweden. Arcing as an extension from the North Atlantic Drift, air and water currents of the Irminger Current deviate from the Coriolis Effect to circulate counterclockwise near Iceland's south shore, connecting with the Labrador Current by the southern tip of Greenland. The Labrador Current cuts a path south to reconvene with the Gulf Stream and North Atlantic Current along North America's Eastern Seaboard (Robinson 2006, 124). The upthrust of warm air from the North Atlantic Drift keeps sea ice away from these North Atlantic shorelines while modulating the climate of impacted land (Petersen, Sack, and Gabler 2015, 69).

The volume of English-language media reportage from North America and the United Kingdom promulgates a larger worldwide awareness of and emphasis on North Atlantic shorelines. Media focus during the twentieth and twenty-first centuries has alternated between military actions in and on the Atlantic Ocean, news of trade routes, and the development of submarine natural resource extraction (oil, fishing, and energy harvesting). During the Cold War, extensive hydrophone networks were laid on the Atlantic seabed so that American and

UK governments could monitor and anticipate the movement of Russian submarines (Benson and Rehbock 2002; Duke and Stockholm International Peace Research Institute 1989). Ships and submarines from these countries continue regional patrol of the GIUK Gap, making landfall first in littoral zones (Nicolson 2015). An antagonistic interdependence of Norway and the United Kingdom's governments and corporations plays out within North Sea oil extraction (Noreng 2016, 21). These geopolitical subjects continue to preside in world news as climate change and global heating place foreshores as central players impacted by storminess, glacial melt, rising sea levels, and ocean acidification, among other topics.

It is on the slow and rapid refiguring of North Atlantic foreshores that I research—through artistic practice—how to perform geochronology at the introduction of the Anthropocene. I use the same methodology—that is, artistic practice-as-research—to examine the Anthropocene's entanglement with climate change. The result of these studies is the production of the geopoetic works Intime and *(SUND)*. Geochronology determines the ages of sediment, fossils, and rocks to assemble a geologic planetary history. Researching how to perform geochronology through the use of embodied practice situates creative work within a materialist paradigm. The surficial geochronology of immediate contact is therefore currently named the Holocene, with the proposed Anthropocene an immediate lens for considering humans as geomorphologic agents. The Anthropocene's proposed start dates have produced wide discussion on topics including nuclear detonation, colonization, and industrialization; each proposal emphasizes site engagement, with climate change a very felt indicator of all within the context of the foreshore and a "catastrophic now."

Geochronologists who are part of a working group aiming to give the Anthropocene its formal designation note that "the expression of the Anthropocene in the environmentally sensitive coastal systems [including beaches, tidal flats, and deltas] . . . represents a diverse patchwork of deposits and lacunae that reflect local interplays of natural and anthropogenic forces" (Zalasiewicz, Williams, and Waters 2014, 46). As an intrinsic component of coastal systems, foreshores are delineated by how tides draw, erase, and redraw the relationship between land and water. The tide at full flood and emptied ebb marks the two surficial borders of the foreshore. As geographic phenomena, shores are in movement, becoming or going. They are vulnerable to shifts in temperature, inhabitation, climate, and storminess—and house within their sediment and any neighboring sea cliffs the collective stories of deep time. Climate change spotlights the precariousness of shorelines with the prospect of rising sea levels. Contacted and redrawn by wind and air currents, foreshores obviate erosion and deposition—two primary markers of geomorphic narrative construction. Shorelines are not only settings but also actors in urgencies[1] coming to a future near you. No longer storied as tranquil or Romantic, shorelines are a space from which bodies may need to flee in a not-too-distant future. As a site for performing geochronology, the foreshore becomes both endangered and endangering within a temporal whorl.

Intime

In time

Pronunciation: /ɪn/ /taɪm/, ĭn tīm

Etymology: in tīma (Old English), in tīmô (Proto-Germanic), i time (Danish), í tíma (Icelandic). See also *tide*.

Prepositional phrase

1 When or before due
2 Immediately
3 Eventually
4 Successive continuum of past, present, future
5 Rhythmically synchronous

Intime

Pronunciation: /ɛ̃.tim/

Etymology: intimus (Latin)

Adjective

1 Intimate
2 Inner (Wiktionary 2019)

The term "intime" implies the French word for "intimate" and is a conjunction of the English-language phrase "in time." As a series of coastal/tidal actions, *Intime* is geopoetics writ large—a visual and proprioceptive performance poem inscribing "O" onto foreshores at low tide (see Figure 6.1). I propose geopoetics as planetary-aware artistic praxis that interrogates (in)comprehension of planetary crisis through a refiguring of human–non-human relations. In *Intime*, this refiguring is staged through the performer's enacted proprioception, understood via American poet Charles Olson's definition as "the data of depth sensibility/ the 'body' of us as object which . . . produces experiences of . . . SENSIBILITY WITHIN THE ORGANISM BY MOVEMENT OF ITS OWN TISSUES" (Olson 1997, 181). Proprioceptive refiguring may increase awareness of abiotic entities (physical and chemical components of ecosystems) as communicative agents via a geosemiosis, loosely defined as "the action of signs . . . that leads investigators on a fruitful course of hypothesis generation" (Baker 1999, 12) in geology.

As sites prone to the geomorphic acts of deposition, erosion, and intrusion, foreshores provide an impermanent surface through which to interrogate the deep time and climate change affiliated with the Anthropocene's inaugural narrative. What is foreshore is not what was foreshore, or what will be foreshore. By extrapolating the results of geochronological dating, one might reveal a narrative of *be*foreshores, and predict fore*thcoming*shores. The foreshore acts as a powerhouse of "betweenings," a wobbly or warbling signature of time changes. Foreshores, in this sense, may be keystone geographical features. Foreshores are formed through deposition, erosion, intrusion, extrusion—with examples being submarine eruptions as extrusive events,

FIGURE 6.1   Still from edited video in which Laureen Burlat and Angela Rawlings perform *Intime* on Kinghorn, Scotland's foreshore in February 2016.

Source: Copyright Angela Rawlings.

while marine flora and benthic infauna communities mimic intrusive contributions. The multiple narratives of geochronology, constructed through stratigraphic compilations of geomorphic data, play out in the deep prehuman time and speculative futures of the foreshore.

Intime proposes embodied, site-respondent possibilities of how to relate to large-scale, temporally distinct geologic events through attunement with geomorphic processes and geophysical forces. The point of departure is the North Atlantic Drift in collusion with the Irminger Current and Labrador Current; counterclockwise circulation of surface air and ocean currents rove from the British Isles to west-coast Scandinavia and southern Iceland. *Intime* is performed on foreshores impacted by such air and ocean currents. Each performance lasts the length of the participants' available time and energy, averaging an hour. *Intime* performances have occurred in: Loch Long, Scotland (2016); Lomma Bay, Sweden (2017); Herøya Industripark, Norway (2017); Hjörseyjarsandur, Iceland (2016/7); Nidarø, Norway (2017); and Kinghorn, Scotland (2016/7). In this chapter, I will describe the performances at Loch Long and Nidarø, which are representative of all the performances on account of their number of participants and foreshore materiality.

To prepare for an *Intime* performance, I check the tide chart that corresponds with the foreshore where the group will perform. The variable distances between ebb (low tide) and flood (high tide), influenced by spring (corresponding with full or new moon, when tide pulls are at their strongest) and neap (quarter moons, weak tide pulls) tides, mean that the littoral zone is exposed for different periods of time on any given day. At each site, a group of two or more participants commence the performance; this takes place when the low tide begins its flood so that the full foreshore is exposed and—if sandy—will provide a surface for footprints to create a temporary tattoo of our path on the ground. As the tide floods the tattooed sand, it "erases" the mark we have made—though our physiological interaction endures through video documentation, within our memories, and by its impact on the multiple species within the site's benthic and ornithological assemblages. I also attempt to choose a date closest to the spring or neap tide so as to maximize the depth of the littoral zone; this provides the largest stage on which *Intime* may be performed. Finally, I select a diurnal hour so as to perform the interaction in light conditions suitable for documenting with video (viewable at https://vimeo.com/233804882).

I choose the familiar activity of walking in circles in order to estrange habitual engagement with my body, the sites, and my collaborators. How does the quality of the ground guide one's movements? How do wind speeds, weather, temperature, and more-than-human inhabitants inform the movements we make? Walking in a counterclockwise circle for more than an hour provides a platform to open to the materiality of the site and for proprioceptive observation and attunement. I observe my movements, sense-abilities, interactions, and reactions with the sites' human and more-than-human co-constituents. Walking in a counterclockwise circle is a form of structured improvisation. I notice when my collaborators indicate their own attunements through their physiological engagements with and in the site.

Practice-as-research documented in *Intime* captures an in situ approach to how performance unfolds in the flux of the Anthropocene's introduction. *Intime* is documented with video, which I then edit into a split-screen archive where the videos function as a cosmopolitan assemblage of sites and temporalities. My intention is to create an aesthetic presentation of the documented geopoetics, incorporating distance and close shots of endurance performance alongside non-human activities on or near foreshores. The results of this assemblage exemplify knowledge through practice via the enactment of concepts that other practitioners may use to process, reflect, and respond both to *Intime* as performance and to *Intime* as video installation. North Atlantic air and water currents are conjured through the counterclockwise movement of humans as they inscribe their presence on a foreshore. Watching the video may provide a tuning-in moment where a viewer reflects on embodied activation through previous or future engagements with foreshores, thereby evoking tacit comprehension of geomorphic agents and foreshore co-constituents.

Intimate with Loch Long, Scotland

Researching how to perform geochronology in the Anthropocene through an embodied practice invites an engagement with the materials entangled within this proposed epoch. Those materials include plastic deposition, nuclear (including military) presence, industrial production and waste, and the geomorphologic processes affiliated with climate change. Gender and cultural studies professors Astrida Nei-manis and Rachel Loewen Walker argue for

> reconfiguring our spatial and temporal relations to the weather-world and cultivating an imaginary where our bodies are makers, transfer points, and sensors of the 'climate change' from which we might otherwise feel too dis-tant, or that may seem to us too abstract to get a bodily grip on.
>
> *(2014, 559)*

To reconfigure my own spatio-temporal relationships to and with foreshores and air and water currents, I began a practice of tracking[2] North Atlantic storms whose paths in 2015 would contact the shoreline of Loch Long, Scotland, where I lived at the time. The first storm I tracked had recently moved south along the coast of the Eastern Seaboard of the United States and had then swept east along ocean currents to bevy it toward the United Kingdom. After it impacted the west coast of Scotland, the storm swept north-northeast and then looped north-northwest and west to bring rain and wind to Iceland's south coast. This remarkable counterclock-wise movement of a large collusion of ocean and air currents impressed me. The storm would touch land in two continents and multiple countries, affecting the daily movements of humans and more-than-humans dwelling there. I wondered at the lifespan of a storm, and its larger counterclockwise motion. (How) Could I develop a corporeal relationship through artistic practice with the larger body of ocean and air currents that carry a storm? If I could strategize a way to embody a similar motion to this unusual counterclockwise path of a storm, it might assist me in devising transformative action induced through experiential, embodied knowl-edge acquisition.

The idea of North Atlantic counter(clockwise) movement increased when I saw my first half-submerged submarine drift down Loch Long from Royal Naval Armaments Depot (RNAD) Coulport armaments facility. Vanguard-class subma-rines carrying Trident ballistic nuclear weapons are constantly on patrol within the North Atlantic, departing from west-coast Scotland to move through the GIUK Gap for 90 days. Submarines are material indicators of nuclear warfare, and their presence at Loch Long reminded me of one of the Anthropocene's proposed start dates: The first nuclear bomb detonation by the United States Army in 1945. This hidden movement in the ocean's depths further prompted my research into embod-iment of large-scale movements implicated within related or accessible materials of a foreshore site.

A few months into my practice as research, the French visual arts master's student Laureen Burlat joins me through an ERASMUS placement, positioning her proximal to a working artist of her selection. We walk Loch Long's rocky foreshore, searching for plastic bags washed ashore—a geosynchronous indicator of the new Anthropo-cene circulated within oceans—and we move in counterclockwise circles. For our *Intime* performance on 26 February 2016, we commence at a contemplative pace, which at times finds us synchronized but more often finds us meandering around the circle at our own paces. The slick, rocky foreshore ensures we step with care. As we gain familiarity with the quality of the rocks, we test our speeds by sprinting, but this is short-lived for the danger of turning an ankle; we maintain a slower pace that allows us to pause for impromptu beachcombing. After some time, we synchronize our meditative movement, stepping equidistantly and in the same movement across the circle as a gradual balancing act. Laureen pauses to beachcomb or to stare out at the ocean; our speeds are in opposition at one point: She stands still as I jog.

The structural clarity of walking a circle is complicated through the material quality of the foreshore we encounter. Slick rocks impact ease of passage and constrain the possibilities of the movement vocabulary we co-construct. We learn how to adjust foot placement as we walk, and we primarily pace both heart rate and breathing to accommodate the durational work. The circulation brings me into polyrhythm with the lapping waves as I attempt to walk with, or counterpoint to, each audible crash. Site attunement becomes evident through proprioceptive attention to how constant our care is when maintaining balance on the rocky and slick foreshore. Our circulation acts as geosemiotic communication to more-than-human species in our presence through the impact of each footstep, the displacement of rocks by walking or beachcombing, and the chemical signals of the pheromones we emanate as we circle. The hallmarks of our initial movement experiments with *Intime*—including walking, running, beachcombing, synchronicity and polyrhythm, and equidistant placement—provide a blueprint for the adaptation of the practice to other sites.

By aligning our bodies with the counterclockwise movements of ocean and air currents, and learning tidal rhythms in conjunction with moon cycles and local foreshore constituents, we open ourselves to experiential knowledge acquisition that emplaces more-than-human entities as focal to site attunement. The spatio-temporal refiguration acquired through site activation and proprioceptive awareness has altered what I notice within the foreshore. My initial attention to how I move, see, and displace, as well as to how I may be seen shifts through extended circulation to foreground the site's materials (e.g., how waves and wind move, the material qualities of rocks and their placements) and eventually the development of a tacit understanding of interconnected movements (physiological, ecosystemic, geopolitical) co-authoring the narrative. As Donna Haraway writes, "the order is reknitted: human beings are with and of the earth, and the biotic and abiotic powers of this earth are the main story (2016, 55)."

In time with Nidarø, Norway

My process includes public performances, and I invite audiences to participate in this work-in-progress. A public invitation to participate in *Intime* in Norway was issued through Kunsthall Trondheim as part of a group exhibition co-curated by the "multispecies think tank" (Bencke, Antonsen, and Ortíz Lundquist 2017), Laboratory for Aesthetics and Ecology (LABAE). Prior to our *Intime* circulation in Norway, LABAE invited exhibition contributors to alter their curatorial statement; I left the statement on a foreshore overnight, inviting an Icelandic spring tide's revision (see Figure 6.2)

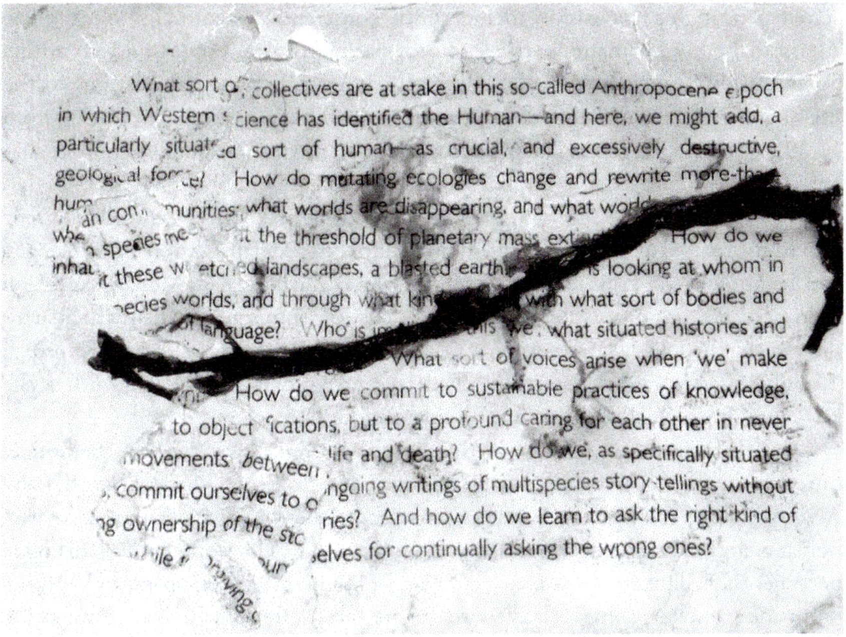

What sort of collectives are at stake in this so-called Anthropocene epoch in which Western science has identified the Human—and here, we might add, a particularly situated sort of human—as crucial, and excessively destructive, geological force? How do mutating ecologies change and rewrite more-than-human communities: what worlds are disappearing, and what worlds when species meet at the threshold of planetary mass extinction. How do we inhabit these wretched landscapes, a blasted earth. Who is looking at whom in species worlds, and through what kinds with what sort of bodies and language? Who is implicated, what situated histories and What sort of voices arise when 'we' make How do we commit to sustainable practices of knowledge, to objectifications, but to a profound caring for each other in never movements between life and death? How do we, as specifically situated commit ourselves to ongoing writings of multispecies story-tellings without ownership of the stories? And how do we learn to ask the right kind of selves for continually asking the wrong ones?

FIGURE 6.2 An Icelandic spring tide revised LABAE's curatorial statement for the exhibition "A New We" at Kunsthall Trondheim.

Source: Photograph by Angela Rawlings.

through water saturation, seaweed and sand redactions, and textual displacement and loss caused by material exposure to biotic and abiotic forces.

On Trondheim's Nidarø foreshore, we catch the tide a few days after the full moon. The foreshore is thin along the Nidelva as it meets the North Atlantic, and its thick gray mud attracts feeding shorebirds including gulls, geese, magpies, mallards, and an imported Asian duck. Prior to commencing the *Intime* circulation, local visual artist Yngve Zakarias offers an oral history of our site, including settlement over 1,000 years ago, the presence of World War II bunkers, and the current contentious municipal plans for urban development that threaten an endemic fungus growing solely on hawthorn trees proximal to the shoreline. The potential presence of an eleventh-century medieval monastery on Nidarø adds to the city's tension (Sørbø and Langseth 2017). Because we are on a contested site, our counterclockwise circulation resonates through local politics via movement common to circular picketing in protests. When Zakarias's schedule precludes his participation in our performance, a female-centered circulation of *Intime* becomes apparent to, and exciting for, several participants.

As with any larger-group *Intime* performance, I encourage the participants prior to starting to attend to their own bodily inclinations; if they need to rest, drink, or leave, they should feel free to come and go without worry of breaking the group activity. Through participation, including the conscientious removal of one's self to attend to bodily inclinations, *Intime* affords participants an opportunity to attune to the site by noticing their physiological actions, reactions, and relations to the site's materials (including human and more-than-human bodies). Circulating from 10:30 to 11:45 a.m., the participants enact this attendance to bodily requirements, entering and exiting the circle with some fluidity. Laboratory for Aesthetics and Ecology curators Dea Antonsen, Ida Bencke with her baby Alvin, and Elena Ortíz Lundquist marvel at the thick, squelching mud. Norwegian University of Science and Technology (NTNU) researchers Libe García Zarranz, Kim Ménage, and Heli Aaltonen join the procession, enriching discussion through remarks particular to their research fields. Danish-American visual artist Rosemary Lee and Kunsthall Trondheim co-curator Katrine Elise Pedersen take a break from their work at the art gallery for a quick circulation on Nidarø's foreshore.

Dea sets herself the task of collecting human-produced garbage that she notices mired in the mud and forms a small cairn near the circle with her collection. Baby Alvin is carried the entire procession, passed between the arms of most circle participants and spurring discussion about motherhood. The constancy of dialogue becomes the hallmark of this *Intime* circulation. While most participants opt at some point for a quieter, contemplative circulation, more often the collective walks see participants moving side-by-side, in twos or threes, and sharing in active conversation on life and work experiences.

After the circulation, I spend a few days prior to the exhibition editing two videos of *Intime* circulations. The first is an expansion of the composite *Intime* video, which is later projected onto a pillar in the gallery space. The second is a Nidarø-specific video where the circulation is overlapped at a few junctures, producing a ghost-like effect as participants double, fading in and out over minutes (see Figure 6.3). This

FIGURE 6.3 Still from edited video, showing a ghost-like overlapping of participant circulation in *Intime*.

Source: Copyright Angela Rawlings.

video is projected onto Kunsthall Trondheim's front window, where it loops as an interface with outdoor passersby and the indoor attendees.

Beyond the moment of participation, durational performance instigates conversations among participants—an indicator of the transformative action embedded within *Intime*'s experiential knowledge acquisition that is demonstrated through self-reflective discourse. I witness this in conversation with NTNU professor Libe García Zarranz, who has described her experience of *Intime*, emphasizing the counter of counterclockwise as complementary to her pedagogic research into counternarratives (García Zarranz 2019). LABAE curatorial intern Andrea Pontoppidan has a second-hand engagement with *Intime*: She has viewed the video installation and conversed with Nidarø participants. In a reflection for the Norwegian Writers' Climate Campaign, she writes, "We are inextricably linked to materials that are part of a completely different premise than ours. Can we find a new intimacy here?" (Pontoppidan 2017) Her focus on the potential to find intimacy through estrangement, thereby resituating our sense of interconnection within a geosemiosis, underscores the aspiration of this practice's knowledge production. Drawing on the reflections of García Zarranz and Pontoppidan, I reposition my understanding of *Intime*'s impact on fellow researchers: I enfold their *Intime*-positioned narratives into future tellings of how *Intime* performs.

Mapping Intime as *(SUND)*

(SUND) is a gallery-floor installation that plots a meteorological mapping of wind currents over the North Atlantic. The installation is constructed of gaffer tape, and wind currents are depicted using the meteorological symbol of wind barbs (which indicate wind speed and direction). On the date of each *Intime* performance, wind currents roved from the British Isles to west-coast Scandinavia and southern Iceland. In the installation of *(SUND)* within the gallery space, I remap the wind currents with fidelity to cardinal direction. *(SUND)* was installed in the Inter Arts Center's Black Room for a February 2017 exhibition, as well as in Kunsthall Trondheim for the "A New We" exhibition running September to December 2017.

Gaffer tape is a much-used tool for a theater technician, used to mark both the position of set pieces and blocking for performers. This tape emplaces; it bookmarks place for future return. The tape denotes position and viewing. For *(SUND)*, I use gaffer tape, with its theater referents, to design wind barbs on a floor. The marking system of a theater house converges with the meteorologist's air current symbol, both offering straight-line designs of two or more lines. Short or long lines on the wind barb's vertical-line ascender indicate wind speed in knots. Speed-lines are positioned to be drawn from the cardinal direction from which the winds blow. The re-creation of wind currents mapped onto the floor proposes blocking or choreography through which exhibition attendees might navigate and activate the gallery space in relation to North Atlantic wind paths. Their counterclockwise, roving movement grafts the circulatory processes and fluctuating tempos of geomorphic entities onto the gallery as tangible site.

Any (*SUND*) installation is directly related to a performance of *Intime*, so I take a snapshot of the wind speed and direction over the North Atlantic that aligns with the dates and times when an *Intime* performance occurs. Kunsthall Trondheim's floor is gray, and black gaffer tape is the most legible on it. A single wind barb in Kunsthall Trondheim provides a measurement schematic; all wind barbs are cut by hand.

At Kunsthall Trondheim, Laboratory for Aesthetics and Ecology curators Dea Antonsen and Elena Ortíz Lundquist assist me as I finalize my installation near midnight. Their curatorial decisions on where to situate the wind barbs include referencing my recommended placement for indicators of wind direction and speed. They also consider how the barbs will choreograph the space and indicate standing-points to some gallery attendees. They think through the relationship of the barbs to the other artwork in the gallery and strategize the physical spacing and visual impact of the barbs' placement. Dea and Elena layer proprioceptive, experiential knowledge acquired through their *Intime* circulation onto their cura-torial training. Revisiting their own counterclockwise circulation on Nidarø (see Figure 6.3), they consider how their placement of wind barbs—as the meteorologi-cal representation of geosemiosis brought into the gallery space—may simultane-ously communicate a blocking (where to position one's body) and a choreography (direction or speed through which to navigate the space) to exhibition attendees. Their curatorial choices envision a path through the space that reflects elements of *Intime*'s encoded practice.

In her essay for the Norwegian Writers' Climate Campaign, Andrea Pontop-pidan describes a tour of Kunsthall Trondheim's "A New We" exhibition, led by co-curators Dea Antonsen and Ida Bencke:

> We went from work to work and there was a flow and self-confidence in our movements. Control in the transitions. But suddenly Dea stops and says, "It's as if what we are doing now is to suggest that there is only one way to go through the exhibition. But that does not have to be. There are other choreographies that are suggested in the room." She points down to the floor, where little black characters spread over the entire floor area of the room.
>
> *(Pontoppidan 2017)*

The installation work cannot exist without direct engagement with tangible sites. (*SUND*) becomes a cosmopolitan assemblage, where multiple sites commingle, and where responses to sites enable response-abilities. The potential choreographies afforded through (*SUND*)'s placement within a gallery enables audience response to the referential sites inscribed through the gaffer tape, as well as to the immediate site of the gallery.

Conclusion

What began as attempted embodiment of counterclockwise ocean and air currents, the circulation of submarines, and the whorl of periwinkle shells or cochlea has grown to a ritual entangling of human bodies with the liminal space of the fore-shore. *Intime* performances approach the question of what it means to make perfor-mance at the crux of climate change and at the introduction of the Anthropocene as a geological delineation of time. *Intime* is an invitation to put one's self in the position of sedimentary particles, exposed to wind and weather. Interconnected-ness, attunement, and collaboration are integral to *Intime's* practice as research. By investigating the sites, movements, and collaborative practices, *Intime* is a platform for experiential knowledge acquisition that holds the possibility for pedagogical and creative application. Practiced on foreshores at low tide, *Intime* explores mul-tiple temporalities and intimacy both through the durational performances and within edited video documentation.

Moon cycles pull tides. Days dip darker past the autumn equinox. *Système pré-paratoire infrarouge pour l'alerte* (SPIRALE) tracks ballistic missiles using infrared satel-lite imaging (Deriu 2010). The cyclic motion of tectonic plates. The sinistral whorls of cochlea, of periwinkle shells. In the new drowned Wonderland, Alice runs the Caucus-race with Mouse, Eaglet, Dodo, Lory, and Duck (Carroll 1865, 29–33). Sea-birds circle overhead. Wiccans close the sacred circle by walking widdershins (Gri-massi 2000, 445). Eelgrass washes ashore as entangled loops. The rising tide erases sand spirals. We hold our writing hands midair, trace the letter O with our index fingers.

Notes

1 "I name these things urgencies rather than emergencies because the latter word connotes something approaching apocalypse and its mythologies. Urgencies have other temporali-ties" (Haraway 2016, 37).
2 I tracked storms through meteorological data visualization software, including Windyty.

References

Baker, Victor R. 1999. "Geosemiosis." *Geological Society of America Bulletin* 111 (5): 633–645. doi:10.1130/0016-7606(1999)111.

Bencke, Ida, Dea Antonsen, and Elena Ortíz Lundquist. 2017. "A New We: A Multispecies Think Tank." *Laboratory for Aesthetics and Ecology*. www.labae.org/past/#/a-new-we/.

Benson, Keith Rodney, and Philip F. Rehbock. 2002. *Oceanographic History: The Pacific and Beyond*. London: University of Washington Press.

Carroll, Lewis. 1865. *Alice's Adventures in Wonderland*. London: Palgrave Macmillan.

Deriu, Floriandre. 2010. "SPIRALE, Premier Pas Vers l'alerte Avancée." *Ministère Des Armées*. www.defense.gouv.fr/actualites/articles/spirale-premier-pas-vers-l-alerte-avancee.

Duke, Simon, and Stockholm International Peace Research Institute. 1989. *United States Military Forces and Installations in Europe*. Oxford: Oxford University Press.

García Zarranz, Libe. 2019. "Thresholds of Sustainability: Cassils and Emma Donoghue's Counter Narratives." In *The Other Side of 150: Untold Stories and Critical Approaches to Canadian History, Literature and Identity*, edited by Linda Morra, Louis-Georges Harvey, and Sarah Henzi. Waterloo: Wilfrid Laurier University Press.

Grimassi, Raven. 2000. *Encyclopedia of Wicca and Witchcraft*. St. Paul, MN: Llewellyn Publications.

Haraway, Donna J. 2016. *Staying with the Trouble: Making Kin in the Chthulucene*. London: Duke University Press.

Harper, Kris. 2004. *A Student's Guide to Earth Science*. London: Greenwood Press.

Neimanis, Astrida, and Rachel Loewen Walker. 2014. "Weathering: Climate Change and the 'Thick Time' of Transcorporeality." *Hypatia* 29 (3): 558–575. doi:10.1111/hypa.12064.

Nicolson, Stuart. 2015. "What Do We Know About Faslane, the Home of Trident Nuclear Weapons?" *BBC News Scotland*, August 31.

Noreng, Øystein. 2016. *The Oil Industry and Government Strategy in the North Sea*. New York: Routledge.

Olson, Charles. 1997. "Proprioception." In *Collected Prose*, edited by Donald Allen and Benjamin Friedlander, 181. Berkeley: University of California Press.

Petersen, James, Dorothy Sack, and Robert E. Gabler. 2015. *Fundamentals of Physical Geography*. 2nd ed. Stamford: Cengage Learning.

Pontoppidan, Andrea. 2017. "At Bevæge Sig Med Vindstrømme." *Forfatternes Klimaaksjon*. https://forfatternesklimaaksjon.no/2017/12/02/at-bevaege-sig-med-vindstromme-essay-af-andrea-pontoppidan/.

Robinson, Allan R. 2006. *The Global Coastal Ocean: Interdisciplinary Regional Studies and Syntheses*. Cambridge: Harvard University Press.

Sørbø, Kari, and Marit Langseth. 2017. "Fortsatt mye uklart tre måneder før anleggsstart på nye Trondheim spektrum." *NRK Trøndelag*, June 22.

Wiktionary the Free Dictionary, s.v. "Intime." Accessed January 6, 2019. https://en.wiktionary.org/wiki/intime.

Zalasiewicz, Jan, Mark Williams, and Colin N. Waters. 2014. "Can an Anthropocene Series Be Defined and Recognized?" *Geological Society, London, Special Publications* 395 (1): 39–53. Geological Society of London. doi:10.1144/SP395.16.

7

SEISMIC, OR TOPOGORGICAL, POETRY

John Charles Ryan

> The different points of this deep glen seem as if they would fit into the opposite
> fissures which form the smaller glens alternately on either side. The whole is
> indeed a grand natural spectacle, and is an indubitable mark of the vast convul-
> sions which this country must at one period have undergone.
>
> (Explorer and surveyor John Oxley [(1820) 1964, 300])

The Northern Tableland bioregion of New South Wales, Australia, is a plateau dis-
tinctive for its geological and botanical diversity (Dunn and Sahukar 2003). Intensively altered since European colonization in the early-nineteenth century (Butzer
and Helgren 2005), the bioregion comprises a network of sharply incised gorges—
described in corporeal terms by Oxley as convulsive glens—around which a conservation system, including the Oxley Wild Rivers National Park, has developed in
recent decades. Known commonly as the New England of Australia, the Tableland
is "a strange, almost inverted landscape" of gentle, undulating plains aside deep,
tortuous chasms (Haworth 2006, 23). In 1818, after crossing the lowlands of the
Liverpool Plains and traversing the upwarped Great Escarpment marking the western edge of the bioregion (Oilier 1982), Oxley and his companions became the
first Anglo-Europeans to set foot on the Tableland. At once attracted to and repelled
by the unsettling contrasts of the fissured landscape, the explorer-surveyor evoked
the Tableland as "exceedingly grand and picturesque," yet "divided longitudinally
by deep and apparently impassable glens" (1964, 293–294). To be certain, Oxley's
journal reveals a contrapuntal phyto-geological consciousness of the inverted terrain. Alternating perceptually between vegetation and land forms, he recalls that
"the rocks were covered with climbing plants, and the glens abounded with new
and beautiful ones" (294).

Ideals of the picturesque and scenic—inherited from European landscape aes-thetics—were confounded by the grotesquerie of the botanical and the seismicity of the geological Oxley faced. The chasm walls, for instance, were "clothed with stately trees [and] creepers" growing "so extremely thick that we found it impossi-ble to penetrate through them" (Oxley 1964, 303). With a taxonomic eye, however, Oxley noted rocks "covered with epidendra [orchids], bignoniae [bignonias], or trumpet-flowers, and clematides [clematis], or virgin's bower" (294). In the pre-sent era, these "grand natural" spectacles—synonymized as glens, gorges, ravines, chasms, and valleys in Oxley's journal—have become increasingly vital sanctuaries for Tableland flora. Shielded from the overwhelming impacts of colonization on the plateau, endemic plants—such as those observed by the explorer-surveyor—populate the rim and inner flexures of glens. Some species are exceedingly site-dependent. The golden-flowering Ingram's wattle (*Acacia ingramii*), as an example, is endemic to the Apsley-Macleay ravines near the town of Armidale (Wright 1991, 97). The vulnerable gorge hakea or corkwood oak (*Hakea fraseri*), moreover, occurs at a few locations in the Tableland, but its actual distribution remains difficult to ascertain (Commonwealth of Australia 2018). As Oxley's account suggests, devising language for the "extremely deep, entrenched gorge system that cuts cross-wise through the Tableland to the eastern sea" (Haworth 2006, 25) demands phyto-geopoetic awareness oscillating between vegetal and geological phenomena.

Through examples from my own phyto-geopoetics in the Tableland, this chap-ter develops a postcolonial, interspecies, and collaborative mode of writing Aus-tralian gorge country. The title, "Seismic, or Topogorgical, Poetry," riffs on literary critic Stephen Burt's analysis of the topographical genre—or *mode*, as he prefers to call it—in his essay entitled "Scenic, or Topographical, Poetry" (2012). Diverging slightly from the topographical tradition—with its prevailing emphasis on scenic views and local specificities—topogorgical poetry attends to the seismic scales, per-plexing prospects, pastoral inversions, deep temporalities, exacting transcorporeali-ties, and biopolitical legacies of chasmed landscapes. Rather than strictly denoting earthquake activity, however, *seismic* in a phyto-geopoetic formulation signifies the immense spatio-temporal proportions of the Tableland. What's more, topogorgical poetry responds to the multisensorial errata, perceptual abnormalities, and bodily convulsions that have fallen outside the topographical purview since John Den-ham's poem "Cooper's Hill" ([1668] 1709). To this effect, I contextualize phyto-geopoetics in relation to radical landscape poetry (Tarlo 2011, 2013), interspecies collaborative art (Broglio 2011; Jevbratt 2010; Ryan 2015), and field-based crea-tive praxis that centralizes walking (Tarlo and Tucker 2017). Foregrounding the imperialist walking of Oxley early in the chapter allows me to argue that phyto-geopoetics proffers a linguistic counterforce to the historical transfiguration of the Australian environment and, more specifically, the (neo)colonial violence enacted on its plants and their habitats.

In my Tableland "earth writing" (Springer 2017), I have developed three approaches interweaving the botanical and geological sciences, experimental art, and concrete poetry. Following the theorization of topogorgical poetry in the next section, the elaboration of these techniques supplies the structure of my discussion. The first compositional tactic, gorge-walking, is predicated on sensory openness to the more-than-human world and a willingness to negotiate—physically and metaphysically—the "frightful precipices" and "deep and apparently impassable glens" of the plateau (Oxley 1964, 307, 293–294). The second strategy is concrete composition that inscribes the morphological articulations of plants within chasms. This approach encompasses the reading, performance, and exhibition of installation-sized poems, in collaboration with an Australian botanical artist, as a mechanism for enhancing community awareness of threatened Tableland plant(scape)s. The third technique comprises the planting, composting, and seeding of sonnets as a de(re) generative and de(re)compositional phyto-geopoetics. Focused on dialogically and materially writing *with*—rather than monologically and immaterially *about*—flora, land forms, microbes, and the elements, this technique intersects with the use of deconstruction and decay in filmmaking (Ramey 2016, 5–32), process-based environmental art (Wallis and Kastner 1998), and vegetal subjectivity in experimental poetry (Burk 2015). Sonnetic composting also makes generous use of first (plant-) person address to enliven the more-than-human subjects of phyto-geopoetics.

Toward topogorgical poetry: from the scenic to the seismic

> Here should my Wonder dwell, and here my Praise,
> But my fixt Thoughts my wandring Eye betrays;
> Viewing a Neighbouring Hill, whose Top of late
> A Chappel crown'd, till in the common Fate,
> The adjoyning Abbey fell.
>
> (Denham 1709, 9, ll, 3–7)

The topographical mode of poetry emerged in 1642 with Denham's poem, "Cooper's Hill." It was advanced by WH Auden in a poem he wrote in 1948, "In Praise of Limestone" (Auden 1951) and reinterpreted widely by contemporary poets (Tarlo 2011). The mode is synonymous with scenic, prospect, local, and loco-descriptive poetry (Aubin 1936; Burt 2012, 598; Foster 1970). In *Topographical Poetry in Eighteenth-Century England* (1936), Robert Aubin characterizes the mode as "describing *specifically named actual localities*" (vii, emphasis original). For Stephen Burt (2012), the formal markers of the topographical consist of: titular place names; the deictics of *here*, *there*, and *now*; sensory mimesis through form, enjambment, syntax, and linguistic density; and a structure approximating the tracking of the eyes or movement of the body across a site (602). Focusing on the multi-scalar imbrications between site and region, topographical poetry calls attention to "the particular ways of seeing, hearing, walking, remembering that constitute the poem, and that are as likely to occur in a built-up environment as in one that looks wild" (Burt 2012, 611). John Foster (1970), moreover, delineates five characteristics of the topographical, namely, extensive description, time-projection, space as patterning device, extended metaphor, and an overarching moral position (403). The topographical outlook is concerned with the effect of three-dimensional space on poetic composition and, as a result, employs "*extensive* description" (396, emphasis original) in which qualifiers bear prominent spatial resonances. Time-projections, additionally, engage the present, historical past, mythological past, and future as the poet apprehends the *landscape*—site, locality, place, region, environment, terrain, topography—through historical and moral optics (399; 403).

Notwithstanding its moralistic fixation on prospects and its pretense of detachment from the landscape, topographical poetry presages recent developments in geopoetics as "geographical creativity" (Cresswell 2014, 142), "creative geography" and "geophilosophy" (Magrane 2015, 97), and "earth writing" (Springer 2017). Recognizing the potential value of transdisciplinary synthesis, geographer-poet Tim Cresswell (2014) argues that "geography and poetry can inform each other in the practice of writing creatively" (141). Through narrative, lyrical, and experimental orientations, geopoetics maintains a receptivity to "the philosophical-creative possibilities of collaborative processes with other-than-humans" (Magrane 2015, 94, 96). Embracing interspecies possibilities, geopoetics reconfigures the "telescopic" stance of the historically dominant topographical mode as "kaleidoscopic" human–non-human interchange "not concerned with distancing but with comprehensiveness, a circular all-at-onceness" (Brownlow 1978, 25, on the poetry of John Clare). Phyto-geopoetics, then, considers "the philosophical-creative possibilities" of vegetal life, collaborative meaning-making with the botanical world, and the embeddedness of plant agency, semiosis, sentience, and intelligence in places. On cognition in plants, for instance, biologist Anthony Trewavas (2016b, 543) asserts that intelligence—including that of plants—is connected intrinsically to the place in which it is expressed. To be certain, in Oxley's journal, the "botanic supplies" (1964, 314) of the imposing gorges presented a material interface with the Tableland as the men crawled over decaying logs and became tangled in creepers. From a phyto-geopoetic stance, however, plants are more than scenic objects or aesthetic elements; they are minded beings, co-creative agents, collaborative instigators, and geopoetic participants in "a practice of radical experimentation in making new worlds" (Magrane 2015, 94).

With this phyto-geopoetic framework in mind, I propose *topogorgical poetry* as a variant of topographical poetry. Prompting collaborative methods of earth writing appropriate to the peculiarities of chasmed landscapes, topogorgical poetry heralds a movement from the *scenic* to the *seismic*—from the visual emphasis of pleasing prospects to the corporeal intergradations of convulsive country. Whereas the scenic depends principally on sightlines across landscapes, the seismic demands looking (and sensing) downward into colossal depths—or looking (and sensing) upward from those depths as one ascends through the spatio-temporal inscriptions of geological strata. In addition, the topogorgical mode strives to imbue human discourse with the enunciations of glens—in this instance, Gara Gorge, Dangar's Gorge, Apsley-Macleay Gorges, and others I have visited regularly in the Tableland—hence disrupting the monologic language of the beautiful, picturesque, and sublime. Topogorgical poetry, as such, is a radical landscape poetics *in/of* the Australian Tableland bioregion and *of/with* its flora. As a postcolonial mode, the topogorgical responds to the European colonialist and Anglo-Australian neocolonialist eradication of plants and botanical communities in the Tableland. Through poetry as an "act of remaking and reimagining the world" (Handley 2015, 341), the historical fragmentation of vegetal nature-cultures in the region prompts opportunities for multispecies renewal. As a form of "discursive recuperation" (DeLoughrey and Handley 2011, 3)—intervening in the dominant discourses of plants as aesthetic, utilitarian, and taxonomic objects—phyto-geopoetics envisions (and *ensenses*, or, grasps through the sensory faculties) new ways of being with vegetal life. In this context, ecocritics Elizabeth DeLoughrey and George Handley (2011) maintain that "since the environment stands as a non-human witness to the violent process of colonialism, an engagement with alterity is a constitutive aspect of postcoloniality" (5). A phyto-geopoetic enactment of the topogorgical mode hinges on a twofold awareness of both plants and land forms as well as site–site and site–region transactions. The *topogorgical* neologization is prompted by the work of poets, such as Harriet Tarlo and Elisabeth Bletsoe, who manipulate the *mise-en-page*, line breaks, and other formal dimensions of their poetry, thereby incorporating the articulations of plants and sites into their work while creating an interspecies "textual ecology" (Scott 2015). These poets also collaborate in the field with visual artists and privilege walking as a generative act (Bletsoe 2010; Bletsoe and Hatch 2013; Tarlo and Tucker 2017).

Gorge-walking as embodied phyto-geopoetics

From "I turned the corner and entered the mind of the forest"

> I turned the corner and entered the mind
>   Of the beech forest. The seen was not a scene
>   But a psyche. The trees' old way of thinking
>   Coppiced from within me. I walked innerly
>   A while towards eternity. It was no ordinary
>   Overcast midday before Labour Day. Should-
>   Ered by the Great Escarpment, I gaped east over
>   Spinal ridges of the Bellinger River Valley. I heard
>   The drawled and well-treed clauses of glacial speech.
>   Through haziness beneath, prone figures of Cenozoic
> History sprawled towards the Tasman Sea: sacral
>   Curves, lumbar hollows, those vertebral foramen
>   Of time itself ever so expansive in its brevity.
>   My body dropped through basalt strata of
>   Other epochs as I rounded the elbow below
>   Point Lookout and crashed face-first into the
>   Very thought of the forest.

>   (Ryan 2017a)

Oxley's journal evokes a topogorgical disposition. His narrative of trekking also conveys acute anxiety over mobility-inhibition—the slowing down, complication, preclusion, or cessation of movement. On the one hand, his mobility-inhibition resulted from vegetative barriers: "It was a continued ascending and descending of the most frightful precipices, so covered with trees and shrubs and creeping vines" (Oxley 1964, 307). On the other hand, the fractured terrain frustrated his ambulation as the steep slopes and loose stones rendered "any attempt to descend even on foot . . . impracticable" (295). Notwithstanding its colonialist imperative, Oxley's procession across the engorged landscape can be read in terms of *moving-with* others including, if only briefly, the Traditional Owners of the Tableland when, for instance, he surprised a "solitary native . . . [mostly] deprived of the use of his limbs" (302) and hobbling slowly away. Other co-ambulators, of course, were fellow members of the expedition. At Oxley's command, botanist Charles Frazer "ascended the peaked hill on our left, and had a most extensive prospect" (319). Domesticated and wild animals, furthermore, were gorge-walkers accompanying—and encountered by—Oxley. The party's horses frequently became incapacitated by the exacting terrain. A mare "literally burst with the violent exertion which the ascent required" (310). Even marsupials succumbed. The men pursued a kangaroo to a waterfall, "down which he leapt and was dashed to pieces" (298). In addition to the presence of human and more-than-human gorge-walkers, there were omnipresent traces of ambulation, especially the tracks of emus.

Contingent on topography, weather, and seasonality, as well as on the constitution and affinities of the ambulator, the habitus of *gorge-walking* stimulates perspectives—and brings about exertions—specific to chasmed landscapes. As the first written account of walking the Tableland, Oxley's journal supplies a bioregional springboard for elaborating phyto-geopoetics as a framework and the topogorgical as a mode. The negotiation of the *seismicity* of the landscape—strenuous climbs, harrowing descents, fallen trees, slippery rocks, and unstable footing—factors into his language. At times, his diction mimics the form of the walked landscape, "broken and divided by ravines and steep precipices" (295). Moreover, hydrogeological bodies interpose agentially in the writing-walking *umwelt* as "Hunter's River will *permit* us again to turn easterly" (295, emphasis added). And, somatic metaphors with deep-time inflections provoke empathetic identification with the Tableland in statements such as "How dreadful must the convulsion have been that formed these glens!" (296). To a limited extent, Oxley's expeditionary narrative prefigures contemporary walk-texts, including poetic ones, which "convey the experience of the walk rather than merely inscribing statements about that experience" (Marland and Stenning 2017, 5). To be precise, while the tradition of the wayfaring nature writer extends at least to Thoreau and Bashō, a more recent upwelling of attention to walking as a creative praxis has emerged with Alice Oswald's *Dart* (2002), Alec Finlay and Ken Cockburn's *The Road North* (2014), and other poetic works in which the prosody embodies patterns of ambulation.

In the context of gorge-walking as creative praxis, my large-format concrete poem entitled "I Turned the Corner and Entered the Mind of the Forest" (2017a) is a phyto-geopoetic walk-text composed for A2 dimensions (16.5 x 23.4 inches), presented at community venues in rural New South Wales as a ten-minute performance and exhibited alongside the illustrations of Australian botanical artist David Mackay (Ryan and Mackay 2017; see Figure 7.1). The process of writing the topogorgical poem combined episodes of solitary ambling and collaborative walking with David through a relictual grove of Antarctic beech (*Lophozonia moorei*) at

I Turned the Corner and Entered the Mind of the Forest

I turned the corner and I entered the mind
Of the beech forest. The seen was not a scene
But a psyche. The trees' old way of thinking
Coppiced from within me. I walked inwardly
A while towards eternity. It was no ordinary
Overcast midday before Labour Day. Should-
Ered by the Great Escarpment, I gaped east over
Spinal ridges of the Bellinger River Valley. I heard
The drawled and well-treed clauses of glacial speech.
Through haziness beneath, prone figures of Cenozoic
History sprawled towards the Tasman Sea: sacral
Curves, lumbar hollows, those vertebral foramen
Of time itself ever so expansive in its brevity.
My body dropped through basalt strata of
Other epochs as I rounded the elbow below
Point Lookout and crashed face-first into the
Very thought of the forest. Away from picnicky
Clamour. Farther away from the yowl and yammer
Of randy roisterers, of backpacking boisterers to a
Lyrebird percussing in the brush downslope from
Us. When I had to rush rudely by a camera-laden
Cadre of eco-tourists, sidestepping their hanker
For communion with the wildness of New England
And so leaving my feathery ground-dwelling fellow
To his flirtatious spring swaggering. Did I mention?
It was the day before Labour Day and there were all the
Typical signs of a prodromal state: edema and irritation,
Contractions, perspiration and the vague indication of
Colostrum, for some of us, that is, and that was how
I entered the mind of the ferny *Lophozonia* forest.
A vestige species, identified first by William Carron
And W.A.B. Greaves along the Upper Clarence River.
After that, Charles Moore, in homage to Carron, called
The tree *Fagus carroni*. Then Ferdinand von Mueller, in
Homage to Moore, renamed it *Fagus moorei*, though,
Before all of them, Carl Ludwig Blume propounded
The term *Nothofagus* for "false beech" but meant
Notofagus for "Southern beech." And so it was:
Lophozonia moorei, on pre-Labour Day, with its
"H" intact nonetheless despite aspirations of genus.
Barbecuers bellylaughed at the comedies of treeness.
From the second lookout, I heard utes growl in first and
Second gears to Waterfall Way, everyone, including the
Forest, ecstatically indifferent to the accumulating "h"s.
I concur with Maiden: "I have quite satisfied myself
That the separation of *Nothofagus* from *Fagus* is
justifiable." And *Fagus*, the Northern Hemispheric
Beech, a child of the Middle Eocene, a meagre fifty-
Million years it's been. But *Nothofagus* pollen can be
Seen in Tertiary sediments eighty-million years in age. A
Gondwanan taxon with recollection of supercontinental
Drift. Its bones ground in the rift between Australia and
Antarctica in the Late Cretaceous. It witnessed the
Era of Mammals. Then the trees witnessed us.
Although we name them, we cannot know them:
Red Beech, True Beech, Colonial Beech or Mountain
Beech, Negro-head Beech "owing to the rich dark colour
Of the foliage," Maiden noted. But, for its Indigenous one, he
Knew "of none although it is probable they had a name for
So conspicuous a tree." I turned the corner and entered
The mind of the forest. The seen was not a scene
But a sensation. The trees' old way of seeing
Bore winged seeds within my being.

so *lophozonia moorei* formerly
nothofagus moorei, speaking,
twisting, glacial beeches
along eagles nest track
populating escarpment
at juncture where yellow
asters, purplish solanum
and creamy paper daisies
are beginning to fade away
where acid of oligotrophic
soil is summoning raucous
congregation of epiphytes
mosses, ferns and orchids
along footpath girded on its
downward aspect by smooth
steel handrails and tidied up,
anticipating spring arches,
where a collapsed beech,
chainsawed, is disclosing
its clotted crimson heart
in coronary rays incised on
a cross-section of memory,
evanescent opaque views
over gullies made of gums
and wallabies "where the
land is frequently covered
in mist" as young botanist
G.N Baur said in the 1950s
and when, nearing weeping
rock, knobbly beech shapes,
announced themselves as
caespitose, stunted, multi-
stemmed, tufted presentations
and clusters, gnarled limbs
burgeoning from boulders,
whole boles cloaked thickly
in lush assemblages of clingers,
generations of knotted trees
leaning in throning synchrony
bivouacked to this scarp brink
scape fluctuating with ruptures
of water dribbling from bluffs
accruing in quagmires below
slick cliffs glistening in timid
sun, sudden microcosms of
bracket polypore, undulating
undersides having the colour
of cooked salmon, and there!
yes, *dendrobium falcorostrum*
with succulent sectioned stems
tapering, though no porcelain
blossoms dangling yet from its
edge, exclusive to beech orchids,
sanctuary of banksia of platypus
valley lookout high up: could it be
lignotuberous *neoanglica*, its
leaves stippling in grey-white
feltness?

I turned the corner and entered the awareness
Of the beech forest. The seen was not a scene
But an essence. Its language was my own but
Forcefully different. It seemed a presence of
Mindful mindlessness, or a timed occasion
Of timelessness, reverberating in the cerebral
Protuberances of tree roots. Its telos was autotelic,
A complete end in itself. So was mine for a moment,
Freed from all striving apart from a meeting of selves.
There was something multiplying vegetatively from
Within me, on this refugium between tablelands
And sea. Countless bryophytic bodies: mosses
Liverworts and hornworts, greenly veneering
Burly buttresses of beeches, most vivacious hues
Of Gondwanan refuge, except, of course, for superbly
Strutting lyrebirds. I saw a universe when peering up into
The canopy, camera poised and ready to record what my ball-
Point couldn't otherwise: craggy branches festooned fully
In epiphytic masses with patches of corky grey bark
Faintly evident in places. I gasped at the inability
Of my eyes to take in the forest totally. Though
Perhaps the gasp was from pinched nerves in my
Neck. Better yet, the gasp was grasp for glacial air. I
Swear: leaves appeared miniscule from my point-of-view
But I trusted Maiden, effusing of "the very dark-green foliage
Is striking and the habit of the leaves is handsome," as well
As Blume, before him, observing the "leaves summer or
Winter green, consisting of two rows, folded along
The side-nerves or not." The canopy appeared
To me an orchestrated cacophony, as contorted
And convoluted as concepts underneath all my three
Feet circumventing moss-clad impediment in circumstance
Compelling a sense of precarious balance. Then, through ferns
Of the forest floor, a woody vine coiled like a lasso, colony
Of woolly things moving in opportunistically. Did you
Once believe the forest bore no mind? Believe me. It
Was watching me back. Was watching my back
As I blundered through its quarters, messing up its
Antiques boorishly. It was patient with me as I wrote
This poem recklessly which I thought foolishly might
Be able to express to you the essence of the mind of the
relict beech residence. As I said, it was the day before
Labour Day, prodromal spirits were trembling, as I
Perspired febrily in the cool sticky stratosphere at
Four-thousand, five-hundred feet. Ascending the
Incline to Banksia Point, beeches began to disappear,
And I strode an airy church of casuarinas and eucalypts,
Then back to the parking area before Point Lookout, where
I motored off just like that, with anxiety of boisterers trailing
Behind me. Almost fifteen kilometres to the sealed road.
Gut-wrenched by corrugations and the choking dust
From overtakers. I slowed for cows, calves, sheep,
And stockmen, rumbled over cattle grids
Passing Dutton Trout Hatchery, Yaraandoo
Educational Centre to the highway junction left
Towards Armidale. Wollomombi village on the right.
Tree-ferns diminished. Going deeper and deeper into the
Heart of the beech forest. The seen no more a scene but
Breathing memory. The trees' way of living coppicing
From within me. Still walking inwardly each day
Towards eternity. The glacial trees' old mode
Of seeing bearing winged seeds within
My breathing, thinking, being.

FIGURE 7.1   Layout of "I Turned the Corner and Entered the Mind of the Forest."

Source: Layout by John C. Ryan (2017a).

Point Lookout, a mountain prospect located in New England National Park on the eastern edge of the Tableland. The phyto-geopoetic form is mimetic of the damp and rocky microhabitat of *L. moorei*. Two helical columns of verse—approximating cliffsides but also connoting "the vast convulsions" of thought itself as a material-evolutionary inheritance—surround a Corinthian-like figure compositionally centralized to induce the imposing presence of a beech. I wrote-walked at the site at various times of the year so that each experience of the vegetated terrain differed markedly from previous ones. Although of a moderate grade by Tableland standards, the Point Lookout trail instigated bouts of mobility-inhibition. Slowing down, losing my footing, bending forward, and steadying my eyes at the humusy ground to negotiate fallen logs recalled Oxley's own multisensorial preoccupation with "vegetable mould, the produce of decayed trees for ages" (Oxley 1964, 319).

Concrete verse as community phyto-geopoetics

From "Rock orchid hyphae"

Profound gouges
found on hide-leather edge
where beetle mandibles

chewed abscesses—
charred blotches with rimes
of ash, like cigarette

burns on old mattresses.
Fitful wind shakes
organs of rock orchid—

whole stiff gorgon quakes,
transmitting shivers
along ridges of stretched

stems, those pseudobulbs,
half-clothed in membrane,
feeling of filthy paper

lantern material left outside
over many winters,
di s in t eg r a ti n g and peeling

b

a

c

k.

(Ryan 2017b)

The second phyto-geopoetic technique I have tested is concrete-visual composition mimetic of the articulations of plants in the depths—and on the peripheries—of Tableland gorges. Toward a model of community-minded phyto-geopoetics, this approach engages the public through the exhibition of installation-scale poems focused on fostering appreciation of Tableland flora and expanding awareness of broader conservation issues in the New England region. For the traveling exhibition entitled *New Perspectives on Tablelands Flora* held at booktores and galleries in New England during 2017, phyto-geo-poems such as "I Turned the Corner and Entered the Mind of the Forest" and "Rock Orchid Hyphae" were displayed along with David Mackay's variously sized botanical illustrations, which he made using chalk pastel, black ink, graphite, charcoal, acrylic paints, paper, illustration film, and other media. Synchronizing words with their structure and presentation, concrete poetry refers to isomorphic "work that has been composed with specific attention to graphic features such as typography, layout, shape, or distribution on the page" (Drucker 2012, 294). In visual and sonic registers, a concrete poem strives for the cohesion of aesthetic expression. Indeed, intermedia developments in the late-twentieth century brought concrete poetry into exchange with the visual arts, film, performance, and animation (Ibid., 295). A close relative of concrete poetry is visual poetry in which "the visual form of the text becomes an object for apprehension in its own right" (Berry 2012, 1523). To align meaning and form, both poetic traditions emphasize, among other textual elements, lineation, indentation, intra- and inter-linear space, punctuation, type size, and linguistic density. As examples, Indigenous South Australian poet Ali Cobby Eckermann's "Thunder Raining Poison" (2016) and its American antecedent, Gregory Corso's "Bomb" (1958), exemplify the biopolitical resonances made possible within the concrete-visual frame. Eckermann's performative piece—arranged as a series of recurring questions and conjuring a mushroom cloud—responds with potent effect to "the impact of atomic bomb testing on our traditional lands at Maralinga in South Australia by the British government during the 1940s–60s" (Eckermann 2016, 151).

In conjunction with concrete-visual techniques, installation poetry offers a compelling means for bringing the presence of biopolitical subjects—bombs, but also plants and gorges—to the page, exhibition space, intellect, and imagination. The ritualistic act of hanging topogorgical poetry in a gallery setting, moreover, confers tactility to the language of the seismic and frees the texts from the manacles of the "codex page" (Berry 2012, 1523). Toward intermedial synthesis—and text-image synergy—the visual-verbal interchange centered on particular Tableland plants and ensured that neither form of creative expression gained pre-eminence over the other. The upshot was a participatory ecology of artists, viewer-readers, and phyto-geopoetic forms. Whereas the larger installation poems (A1, or 23.4 x 33.1 inches, and A2, or 16.5 x 23.4) provoked visual experience at a distance—even before they were read or heard—smaller-scale compositions (A4, or 8.3 x 11.7, and A5, or 5.8 x 8.3) persuaded members of the audience to move closer and engage proximately in the dialogue between illustrations and poems. The exhibition experience approximated the oscillation between perspectives that happens when walking in natural environments: beholding from afar, walking toward, examining at close range, looking up then down, listening, pondering, and remembering (see, for example, Thompson 2013).

My hanging of topogorgical poetry was inspired by contemporary work, ranging from Ian Hamilton Finlay's poem-prints *The Blue and the Brown Poems* (1965) and garden lithopoetics (2012) to "poetry in public spaces" (Chamberlain and Gräbner 2015). As part of a Tableland phyto-geopoetics, the A3-sized (11.69 x 16.53 inches) "Rock Orchid Hyphae" is an installed concrete-visual poem emphasizing the seismic over the scenic or, more precisely, the scenic *within* the seismic. The poem constitutes an extended meditation on the haptic sense, figured as a gravitational yearning for the contact between bodies—both macroscopic (humans, orchids) and microscopic (insects, fungi)—that inspirits ecological systems. Known also as rock lily and outstanding dendrobium, the king orchid (*Dendrobium speciosum*) of the poem grows at a site called Budds Mare on a slope above the Apsley-Macleay Gorge in the Oxley Wild Rivers National Park. Accumulating a mass of debris washed over successive seasons from upslope, the large lithophytic orchid has been sliding incrementally away from its precarious hold on a rock face. Fungal symbionts populate the inner topographies of orchids, including those of the genus *Dendrobium*. As the tubular filaments forming the fungal mycelium and controlling vegetative growth, hyphae are complex structures that need to sense the presence of other hyphae in order to mediate fungi–fungi and fungi–orchid communication (Money 2011). The form and, more specifically, the lineation of "Rock Orchid Hyphae" reflect the reaching of hyphae across the internal spaces of the orchid in anticipation of encounter with one another. In macroscopic terms, the poem is mimetic, in a chiasmatic sense, of the Apsley-Macleay chasm at the rim of which the orchid grows. Deriving from the Greek *khiasma* (signifying two things positioned crosswise), and homonymic with *khasma* for a yawning gulf, *chiasma* denotes the interstitching of elements.

The form of "Rock Orchid Hyphae" generates multiple significations as the reader tracks up and down, left to right, or diagonally across the chiasmatic arrangement. Rather than gazing hypnagogically (i.e., drowsily) across a picturesque prospect, the eyes track saccadically (i.e., rapidly and in jerky movements) over multiple focal points/stanzas across the chasm or beneath the viewer/reader's position on the gorge rim/enjambed line. Another poem in the exhibition with topogorgical effects is "Pure Forest and the Deterioration of Expectation" (Ryan 2017c) with its textual protuberances and precarities above a talus slope of disintegrating words. On close inspection, the poem becomes legible, but not immediately so. The reader-viewer is left to decipher whether the enigmatic final phrase "to persuade me?" is the voice of the poet, critic, forest, or collective *me*. Through the integration of concrete-visual composition, installation, and performance, a community phyto-geopoetics emerged via these poems and others. At exhibitions, readings, and other public events, the phenomenological traces of Tableland plants—beeches, orchids, wattles—within their habitats engaged members of the community otherwise not entirely familiar, or even comfortable, with poetry as a written convention of the codex page. The interplay of listening to, apprehending, and contemplating phyto-geo-poems aimed to cultivate, in attendees, knowledge of—and empathy for—the plants and places of the Tableland.

Composted sonnets as collaborative phyto-geopoetics

"Nocturne" from "What would the trees say?"

Devoted I am to this mode, being
an ascetic in a dirtless crevice,
bivouacked to a Gondwanan terrace,
disciplined I am to disagreeing,
neither helmet nor harness guaranteeing
suction on such crumbly, precarious
chasm talus, lacking even a tarsus
for traction nor a tongue, though decreeing
I found my devotion, go find yours too.
Squat beside me, although for not too long,
for I will soon have too many chores to do.
The glacial nocturne swiftly coming on
and solitary I shall make it through—
Farewell and thanks for clambering along.

(Ryan 2017d)

Deconstruction and decay subvert an ontology of a creative object as a stable signifier propounding a coherent meaning. As a case in point, experimental techniques of "distressing" film emphasize the ways in which the image breaks down through burying, exposing to wind, hanging outside, and submerging in water (Ramey 2016, 23). To create the short film *Self Portrait Post Mortem*, for instance, Canadian filmmaker Louise Bourque buried family footage in her backyard garden, retrieving the film several years later to disclose "an unearthed time capsule . . . with nature as collaborator" (Bourque in Takahashi 2008, 58). In so doing, Bourque calls attention to de(re)composition as an intrinsically collaborative process predicated on medial, human, animal, vegetal, microbial, fungal, and elemental assemblages. Interspecies collaboration in art and music has the potential, as Lisa Jevbratt (2010) writes, to "generate unprecedented respect and understanding of other species" (1). In particular, human–non-human interchange activates limbic resonance as the internal topographies of thought-*full* organisms synchronize (Jevbratt 2010, 10). Although Jevbratt privileges mammalian minds, the principle of resonance extends as well to plant minds when understood in relation to recent studies of vegetal intelligence and cognition (Trewavas 2016a, 2016b). In reference to the Tableland, my collaborative interspecies phyto-geopoetics implicates de(re)generative processes hinging on linguistic dissolution and reconstitution. Accordingly, decay becomes a reservoir of renewal as the object—be it film, artwork, song or, in the case of "Nocturne," topogorgical sonnet narrated in the first person by a hakea (i.e., a type of Australian shrub)—acts as a material agent within the Tableland landscape. In other words, the object is neither separate to nor imposed upon the landscape.

"Nocturne" is part of the de(re)generative sequence of 24 Petrarchan sonnets that are collectively entitled *What Would the Trees Say?* and are co-authored with Tableland plants (Ryan 2017d). The sonnet's hakea subject occupies a crumbly stone plinth situated vertiginously above Wollomombi Gorge with an unimpeded sightline to a plunge waterfall once considered Australia's highest. Cast in intertextual relation to Joyce Kilmer's lyric poem "Trees" (1914), the sequence incorporates concrete-visual forms, archaic language, taxonomic allusions, and glossolalic gestalt (or spontaneous verbal associations) to elicit a sense of constrained linguistic disarray. Projected at exhibitions as a continuous digital loop, the sonnetic cycle endeavors to prompt a hypnotic response from viewer-readers as organic text-forms morph into one another. Toward its collaborative provenance, the sequence has been planted, composted, digested, and seeded since September 2017 at Wollomombi Gorge. An ethos of human–non-human reciprocity arises through the ceremonial earthing and unearthing—gorging and disgorging—of poems and fragments. An assemblage of seen and unseen chasm-dwellers works over the source-sonnets, contributing terms, connotations, inflections, syntax, marginalia, elisions, and deletions, which I then integrate into successive versions of the phyto-geo-verses. The open-ended cycle, accordingly, fluctuates with seasonal conditions, soil moisture, root activity, burrowing insects, microbial digesting, human forgetting, and other de(re)compositional processes. Weeks of heavy rain, for instance, rapidly intensified decay, which resulted in a papier-mâché clod. Agents of de(re)construction reformulated a couplet from "Trees" as: "I think that I shall never/A poem lovely as a tr," injecting hesitation and indeterminacy into Kilmer's classic as well as into the Petrarchan sequence as a whole.

As another example from the co-authored sequence, the palimpsested sonnet "Fettle" came about when two source poems fused to produce composite lines such as "thy year/of cruelest seeing/on munted shores of vernal being" (Figure 7.2). Eschewing the third-person omniscient point of view, "Fettle" and other phyto-geo-poems comprising *What Would the Trees Say?* deploy the first (plant-) person address unabashedly throughout. Endowing Tableland flora with perceptual acumen, the sonnets engender polyphony and heteroglossia to invert the prospect mode and turn the critical gaze of canny plants toward the fallible human subject. The sonnet cycle, in this way, joins contemporary poetic work, notably, Louise Glück's *The Wild Iris* (1992) and Les Murray's *Translations from the Natural World* (1992), narrated from the standpoint of vegetal life and promulgating a view of plant-subjects as percipient. In reference to narrative theory, Marco Caracciolo (2014, 110) differentiates between *consciousness-attribution*, in which a reader attributes mental states to a subject, and *consciousness-enactment*, in which a reader is granted access to the mind of a subject communicating in the first person. As hakea and other plants examine and express their interior lives, liberated as they are from the distancing function of third-person voice, the topogorgical sonnets become ever-evolving collaborative

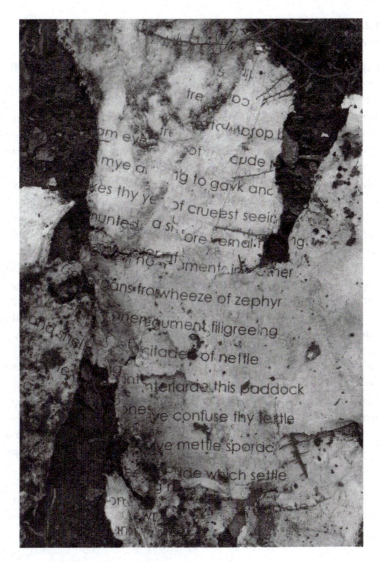

FIGURE 7.2 Phyto-geopoetic palimpsesting.

Source: Photograph by John C. Ryan.

forays into consciousness-enactment in the Tableland. *What Would the Trees Say?*, moreover, adds ecopoetic and conceptual inflections to a Tableland geopoetics. With a material emphasis that diverges from—yet complements—gorge-walking and concrete-visual versification, sonnetic composting underscores the intimate imbrications between plants, gorge habitats, broader climatic variables, and human tactility. Specific to the Tableland and micro-scalar in orientation, the poems call attention to phyto-geopoetics as an indeterminate process of world-making.

Conclusion: phyto-geopoetics and the making of new worlds

Beginning with nineteenth-century explorer-surveyor John Oxley's walk-text, I have defined phyto-geopoetics as a conceptual model and topogorgical poetry as "radical experimentation in making new worlds" (Magrane 2015, 94). In the spatio-temporally seismic Tableland of northern New South Wales, the confounding of perception that Oxley experienced necessitates a radical depriveleging of both the landscape prospect and the human subject within the topographical tradition. Toward this end, the chapter has outlined three different, but interrelated, phyto-geopoetic methods vis-à-vis a practice of Tableland writing. These approaches synergize as an integrated phyto-geopoetic whole. The multi-sensorial nature of gorge-walking signifies the writer's embodied immersion in the gorges and in vegetal worlds. The community-focused display of concrete-visual poems situates the writer within a broader human milieu delineated by the bioregion and engendering public awareness of botanical conservation urgencies. Although markedly conceptual in orientation, sonnetic composting foregrounds the potential of cross-species collaboration within phyto-geopoetics and, importantly, destabilizes individuated authorial subjectivity. As a multi-part configuration, this phyto-geopoetic model recognizes the vegetal domain as an agentic presence within—and percipient contributor to—"textual ecologies" of place (Scott 2015). Thus, in its imbrication with plant life, "geographical creativity" (Cresswell 2014, 142)—including poetry—becomes as convulsive as the vegetated land itself. In particular, prosody and form render poetry highly apposite to phyto-geopoetic experimentation. As critic Mary Jacobus (2012) observes, poetry can put us in touch with "the stilled voice of the inanimate object or the insentient standing of trees" (78). For Jacobus, poetry enables readers to appreciate how "things look back at us. They may even seem to talk back" (5). While phyto-geopoetics contests the "insentient standing of trees," the model engages with the premise that things look back at us—that things interpolate and, even, interpellate. As a being-called-by things, poetry articulates these sensations and encounters in plantscapes.

References

Aubin, Robert. 1936. *Topographical Poetry in Eighteenth-Century England*. New York: Modern Language Association.

Auden, W. H. 1951. *Nones*. London: Faber & Faber.

Berry, Eleanor. 2012. "Visual Poetry." In *The Princeton Encyclopedia of Poetry and Poetics*, edited by Roland Greene, Stephen Cushman, Clare Cavanagh, Jahan Ramazani, Paul Rouzer, Harris Feinsod, David Marno, and Alexandra Slessarev, 1522–1525. Princeton, NJ: Princeton University Press.

Bletsoe, Elisabeth. 2010. *Pharmacopoeia and Early Selected Works*. Exeter: Shearsman Books.

Bletsoe, Elisabeth, and Frances Hatch. 2013. *LPBmicros: Missal Birds*. Rhydyfelin, UK: The Literary Pocket Book.

Broglio, Ron. 2011. *Surface Encounters: Thinking with Animals and Art*. Minneapolis, MN: University of Minnesota Press.

Brownlow, Timothy. 1978. "A Molehill for Parnassus: John Clare and Prospect Poetry." *University of Toronto Quarterly* 48 (1): 23–40.

Burk, Wendy. 2015. *Tree Talks: Southern Arizona*. Fort Collins, CO: Delete Press.

Burt, Stephen. 2012. "Scenic, or Topographical, Poetry." In *A Companion to Poetic Genre*, edited by Erik Martiny, 599–613. Hoboken: Wiley-Blackwell.

Butzer, Karl, and David Helgren. 2005. "Livestock, Land Cover, and Environmental History: The Tablelands of New South Wales, Australia, 1820–1920." *The Annals of the Association of American Geographers* 95 (1): 80–111.

Caracciolo, Marco. 2014. *The Experientiality of Narrative: An Enactivist Approach*. Berlin: De Gruyter.

Chamberlain, Daniel, and Cornelia Gräbner. 2015. "Poetry, Public Spaces, and Radical Meeting Places—Invitation." *Liminalities: A Journal of Performance Studies* 11 (3): 1–5.

Commonwealth of Australia. 2018. "*Hakea fraseri*—Gnarled Corkbark, Fraser's Hakea." *Species Profile and Threats Database*. Accessed April 30, 2018. www.environment.gov.au/cgi-bin/sprat/public/publicspecies.pl?taxon_id=16093.

Corso, Gregory. 1958. *Bomb*. San Francisco, CA: City Lights Books.

Cresswell, Tim. 2014. "Geographies of Poetry/Poetries of Geography." *Cultural Geographies* 21 (1): 141–146. doi:10.1177/1474474012466117.

DeLoughrey, Elizabeth, and George B. Handley. 2011. "Introduction: Toward an Aesthetics of the Earth." In *Postcolonial Ecologies: Literatures of the Environment*, edited by Elizabeth DeLoughrey and George B. Handley, 3–41. New York: Oxford University Press.

Denham, John. (1668) 1709. *Coopers-Hill: A Poem Written by the Honourable Sir John Denham, Knight of the Bath*. Reprint. London: H. Hills.

Drucker, Johanna. 2012. "Concrete Poetry." In *The Princeton Encyclopedia of Poetry and Poetics*, edited by Roland Greene, Stephen Cushman, Clare Cavanagh, Jahan Ramazani, Paul Rouzer, Harris Feinsod, David Marno, and Alexandra Slessarev, 294–295. Princeton, NJ: Princeton University Press.

Dunn, Irina, and Roshan Sahukar. 2003. "The New England Tableland Bioregion." In *The Bioregions of New South Wales: Their Biodiversity, Conservation and History*, edited by Irina Dunn and Roshan Sahukar, 157–170. Hurstville, NSW: National Parks and Wildlife Service.

Eckermann, Ali Cobby. 2016. "Thunder Raining Poison." *Poetry* 208 (2): 150–151.

Finlay, Alec, and Ken Cockburn. 2014. *The Road North*. Bristol: Shearsman Books.

Finlay, Ian Hamilton. 1965. *The Blue and the Brown Poems*. New York: Atlantic Richfield Company & Graphic Arts Typographers.

———. 2012. *Ian Hamilton Finlay: Selections*. Edited by Alec Finlay. Berkeley, CA: University of California Press.

Foster, John W. 1970. "A Redefinition of Topographical Poetry." *The Journal of English and Germanic Philology* 69 (3): 394–406.

Glück, Louise. 1992. *The Wild Iris*. New York: Harper Collins Publishers.

Handley, George B. 2015. "Climate Change, Cosmology, and Poetry: The Case of Derek Walcott's Omeros." In *Global Ecologies and the Environmental Humanities: Postcolonial Approaches*, edited by Elizabeth DeLoughrey, Jill Didur, and Anthony Carrigan, 333–351. New York: Routledge.

Haworth, Robert. 2006. "The Rocks Beneath." In *High Lean Country: Land, People and Memory in New England*, edited by Alan Atkinson, J. S. Ryan, and Iain Davidson, 23–34. Sydney: Allen & Unwin.

Jacobus, Mary. 2012. *Romantic Things: A Tree, A Rock, A Cloud*. Chicago: University of Chicago Press.

Jevbratt, Lisa. 2010. "Interspecies Collaboration—Making Art Together with Nonhuman Animals." Accessed May 6, 2018. http://jevbratt.com/writing/jevbratt_interspecies_collaboration.pdf.

Kilmer, Joyce. 1914. *Trees and Other Poems*. New York: George H. Doran Company.

Magrane, Eric. 2015. "Situating Geopoetics." *GeoHumanities* 1 (1): 86–102. doi:10.1080/23 73566X.2015.1071674.

Marland, Pippa, and Anna Stenning. 2017. "Introduction." *Critical Survey* 29 (1): 1–11. doi:10.3167/cs.2017.290101.

Money, Nicholas. 2011. "Introduction: The 200th Anniversary of the Hypha." *Fungal Biology* 115 (6): 443–445. doi:10.1016/j.funbio.2010.09.014.

Murray, Les. 1992. *Translations from the Natural World*. Paddington, NSW: Isabella Press.

Oilier, C. D. 1982. "The Great Escarpment of Eastern Australia: Tectonic and Geomorphic Significance." *Journal of the Geological Society of Australia* 29 (1–2): 13–23. doi:10.1080/00167618208729190.

Oswald, Alice. 2002. *Dart*. London: Faber and Faber.

Oxley, John. (1820) 1964. *Journals of Two Expeditions into the Interior of New South Wales*. Reprint. Adelaide: Libraries Board of South Australia.

Ramey, Kathryn. 2016. *Experimental Filmmaking: Break the Machine*. New York: Focal Press.

Ryan, John. 2015. "Plant-Art: The Virtual and the Vegetal in Contemporary Performance and Installation Art." *Resilience: A Journal of the Environmental Humanities* 2 (3): 40–57. doi:10.5250/resilience.2.3.0040.

———. 2017a. "I Turned the Corner and Entered the Mind of the Forest." Unpublished Poem.

———. 2017b. "Rock Orchid Hyphae." Unpublished Poem.

———. 2017c. "Pure Forest and the Deterioration of Expectation." Unpublished Poem.

———. 2017d. "What Would the Trees Say?" Unpublished Poem.

Ryan, John, and David Mackay. 2017. *Between Art and Poetry: New Perspectives on Tablelands Flora*. Mount Lawley, WA: International Centre for Landscape and Language.

Scott, Clive. 2015. "Translating the Nineteenth Century: A Poetics of Eco-Translation." *Dix-Neuf* 19 (3): 285–302. doi:10.1179/1478731815Z.00000000083.

Springer, Simon. 2017. "Earth Writing." *GeoHumanities* 3 (1): 1–19. doi:10.1080/23735 66X.2016.1272431.

Takahashi, Tess. 2008. "After the Death of Film: Writing the Natural World in the Digital Age." *Visible Language* 42 (1): 44–69.

Tarlo, Harriet. 2011. "Introduction." In *The Ground Aslant: An Anthology of Radical Landscape Poetry*, edited by Harriet Tarlo, 7–18. Exeter, UK: Shearsman Books.

———. 2013. "Open Field: Reading Field as Place and Poetics." In *Placing Poetry*, edited by Ian Davidson and Zoë Skoulding, 113–148. Amsterdam: Rodopi.

Tarlo, Harriet, and Judith Tucker. 2017. "'Off Path, Counter Path': Contemporary Walking Collaborations in Landscape, Art and Poetry." *Critical Survey* 29 (1): 105–132. doi:10.3167/cs.2017.290107.

Thompson, Catharine Ward. 2013. "Landscape Perception and Environmental Psychology." In *The Routledge Companion to Landscape Studies*, edited by Peter Howard, Ian Thompson and Emma Waterton, 25–42. London: Routledge.

Trewavas, Anthony. 2016a. "Intelligence, Cognition, and Language of Green Plants." *Frontiers in Psychology* 7: 1–9. doi:10.3389/fpsyg.2016.00588.

———. 2016b. "Plant Intelligence: An Overview." *BioScience* 66 (7): 542–551.

Wallis, Brian, and Jeffrey Kastner. 1998. *Land and Environmental Art*. London: Phaidon.

Wright, Peter. 1991. *The NPA Guide to the National Parks of Northern New South Wales: Including State Recreation Areas*. Woolloomooloo, NSW: National Parks Association of NSW.

8

ROUT/E

Chris Turnbull

In walking/hiking, which is a central element of my poetry and visual/installation work, I am interested in historical and ecological variables, as well as in the material/built objects that synthesize or emerge (conceptually and physically) from the territories and topographies through which I move. In a writerly and artistic way, movement includes a constancy of tracking/reading within landscapes and ecologies that are non-static; there is repetitive or emergent evidence of prior and current movements, settlements/habitations, knowledges, and species interactions.

Movement and emergent shifts of meaning inform, in part, my interest in a poetics of trajectory. I may place (or engage) visual and written projects across geographies, in large or small forms. These projects may offer concurrent, possibly disruptive, and multiple views of land/surfaces by focusing on text—or on the articulation of text as visual, built, and sonic—as a feature of landscapes. An example of this is my serial, multi-textual landscape installation entitled "land/mark," initially installed at Fieldwork (curated by ceramicist Susie Osler in Maberly, Ontario) and then taken across geographies and maps. One textual fragment from "land/mark" that read "just before the hour" was inscribed, slightly revised as "just before the new," on a tidal erratic along the Gullane Sands (between Aberlady and Gullane, Scotland). The wording was revised yet again, this time to "just before the," for the title of a wooden, multi-view, interactive cube installed on a small rise in a fringe of red pine forest as part of an exhibit curated by artist/writer Sophie Edwards for the 4ElementsLiving Arts Festival in Kagawong, Manitoulin Island, Ontario. Interested participants could look into and out of the cube simultaneously or share the cube with other individuals, who could look through one of the other sides. The cube itself could be moved. Viewers were exposed to multiple reflections: one's eye, a concrete poem, a portion of the landscape, or the inner workings of the cube itself. The movement of the viewers changed their exposure to different features. Eventually, subject to weather and other elements, the cube fell apart and the concrete poem inside it scattered into discrete alphabet pieces among leaves and pine needles along the trail.

My poetics includes text as physical form; projects may contain layered text and image within the design features of a page or be performed polyvocally, connected by pagination and textual/visual references. Much of the work I have written and many of the projects I've devised have a strong focus on obvious movement (e.g., the body, the eye, the transitory elements of sound) combined with "slow" change in localized environments, rendering the form and the content of the text conditional. In my writing or installation projects, I often try to include elements that are "off" the page; I aim to achieve this through use of space, textual or visual emergence, and indicators of voice/dialogue (performance).

Since 2010, I have curated *rout/e*, an outdoor installation project through which poems are placed alongside (or slightly off) trails in Eastern Ontario, mostly in its rural areas. I invite poets from various countries and locales to contribute a published or unpublished poem to be "planted" alongside a trail or like area. I construct a stand for each poem and plant/install a few poems each year on different trails (within varying landscapes). I monitor them in various seasons and over years until they vanish. I use a WordPress site, named *rout/e* (Wordpress, 2019), to post informal narratives and images that focus on the poems and particular aspects of the spaces they are inhabiting—on the poems, for example, in relation to encounters and effects of environment. The narratives do not follow a chronological order; rather, they document the project and various pieces over time. The poems themselves remain to be found or encountered by any species that use the trail; on the *rout/e* website, I deliberately do not relate exact locations of the poems. Instead, locations are usually generalized by area or entrance point, such as "Marlborough Forest" or "Baxter Conservation Area" or "Heaphy Road." Updates on the website are sporadic because my own returns to the poems are largely unscheduled or inconsistent.

Trail-users tend to be familiar with the possibility of encountering a wide range of tracks and species within landscapes as they move along marked spaces (trails) to get from one point to another on a map or from one experience to another within understood habitats (e.g., "forest"). Poems are unusual within trail networks in rural areas—and there is always the possibility that the poem will not be encountered each time a trail is used. The trails themselves are used sporadically and seasonally by hikers, ATV riders, snowmobilers, and mountain bike enthusiasts. For each of these user groups, speed of travel will influence what is noticed and encountered.

rout/e has a simple format: Poems are printed in black and white on 20lb bond, 8.5 x 11-inch paper (portrait or landscape), which is placed on a plywood square roughly 8.5 x 11 inches and sealed under plexiglass. This facing is screwed to a 2 x 2 x 48-inch stake, and the piece as a whole is inserted into the ground. Now and again, I have planted *rout/e* poems that do not follow this format—for example, postcard-sized pieces, or a large solar panel style frame—usually for particular projects that highlight specific components or elements of a landscape. An example of this deviation is derek beaulieu's visual poem, *Translating Translating Apollinaire*, which was installed in 2014 at Baxter Conservation Area in Kars, Ontario to fill in

FIGURE 8.1 Cecilie Bjørgås Jordheim, movements 7–10 in a row of black walnut trees from *Black Walnut Grove,* late winter 2017.

Source: Photograph by Chris Turnbull.

the space of a missing panel in a solar array. Another example is a series of poems planted and rotating yearly between 2014 and 2018 in an abandoned black walnut grove at the former Kemptville Campus of the University of Guelph (Kemptville, Ontario). The most recent of these plantings has been a seasonally complete collaboration called *Black Walnut Grove* (winter 2017–winter 2018) with artist/poet Cecilie Bjørgås Jordheim. Each season, she sent me ten blank scores or "movements" on 5 x 7-inch cards, which I then placed in the placards; I posted seasonal narratives and visual records online at *rout/e*'s website, documenting the interactions and markings that occurred each season.

Trails, used by multiple species to move through and along various terrains, are indicators of movement, recorded cartographically and/or through print signatures (tracks) on land-based surfaces. They reveal travel and habitat networks, enable community–community or neighbor–neighbor accesses, enhance development of local economies and connections, and fashion and possibly redefine the interactions and movements of diverse species. In the emergence and use of trails and habitable human spaces, however, the ecosystems that trails mark, cross, and can provide access to often also become critical elements of human systems of economy and production. The intricacies of ecosystem processes that ultimately define these spaces "as spaces" are difficult to imagine and express outside human experiences or valuations. These experiences and valuations are embedded in the language used to describe both place and access to place.

Where I live, landscapes and ecologies are directly tied to unceded territories, agriculturally based and settler cultures, riverine provincial or national borders, and natural resource uses. Abstractly—and yet demonstrated in concretely political and economic ways—languages, lands/ecosystems, and accesses/movements among and between species are reciprocal and proximal elements of co-habitation. In many ways, coming upon a poem causes sensory redirection from landscape/habitat enjoyment or exploration: On the one hand, the visual and sonic elements of the poem itself foster recognition of familiar relationships between language and its structures (the form of the poem on the page, the rhythms of speech and shape); on the other hand, the poem is an obvious discrepancy that raises questions: What, in *this* place and space, is relevant or immediate; why is the poem *here?*; what is *here?*

Distance, geology, transportation, seasonal conditions (snow/ice, flood, access barred by foliage), material forms on/near the trails, and my own consideration of where to place the poems affect where I plant them. What sort of landscape, what sort of viewability (by others), what sort of exposure to humans, other species, atmospheric conditions, and obvious reference points will impact engagement with the poem or impact the poem itself? Do I need permission? Who will access the area, the language, the poem, regardless of boundary, page dimension, and textual changes incurred by external factors? What does "monitoring" involve—especially given that *rout/e* poems are usually planted at distances far apart?

If I am planting a poem or poems in a conservation area, I request permission, such as from managers of the Baxter Conservation Area. However, I usually place poems either in verge areas between obvious properties or in areas of abandoned settlement. When trails run between properties (as marked by trail signs or postings), I do not cross No Trespassing signs—even though these are largely intended to prevent hunters from crossing property boundaries. Conservation managers have been very supportive of the project, despite the slow, unruly disintegration of the pieces themselves within the conservation zones. I've never come across a property owner while I have been posting or monitoring the poems, and only once have I encountered another trail user when I have been planting poems.

Mobility, access to transportation, and geographical distance affects whether poets who contribute to *rout/e* can physically get to the trail to find the poem in its location. Social media and email assist with geographical or physical constraints: I email contributor-collaborators a short description of the poem in its environment, and I sometimes send them an update. Each poem has a link or qr code to the poet's website (if they have one). There is a qr code for the *rout/e* blog, although wireless connection to access the qr codes from the trail isn't always available or consistent.

The act or impetus of reading a poem in *rout/e*—encountering the poem and engaging with it—is relational; the poem changes with multiple exposures and contact with the outdoors, readers, and species interactions. The poet's initial, printed use of language is not singular but multidimensional and emergent. The poem's text—that is, the physical print—alters as it is affected by atmospheric effects of seasons or changes in the texture of the page (e.g., wrinkles, rips, fading). Furthermore, coming across a poem on a trail can be dislocating; a hike, for example, may be temporarily suspended to include a literary encounter, in the way that one might otherwise pause to look at the effects of ecological dispersals and developments or make note of habitat, an ephemeral pond, an insect, or a flower. It is possible (and okay) for one to not "know" about the diverse workings of a poem, a pond, a flower, or an insect (or be able to name these things). It is possible to pause and not dwell on literary analysis, data collection, or natural history knowledge—instead to just wonder about the origins of the structure itself in the place the person has come across. While poetry is a known form, it also comes out of a literary culture. Not everyone who accesses these trails would willingly attend a poetry reading or purchase a poetry book: It is simply not relevant to the everyday, nor is it a method they would use to deepen meaning. Yet, a poem a person may encounter in this context may also compel them to consider what they do know in these places they move through or toward: What sorts of knowledges might become dislocated or synthesized, or what additional things might be observed? In addition, literary forms are unnecessary for species who are intent on survival; the non-human species who use these trails and ecosystems also encounter the poems. The plywood that the poem sits on may be an attractant for an entirely different set of reasons than the visual or spatial; the plexiglass covering the poem offers a surface for

FIGURE 8.2   Hiromi Suzuki, "A Swimmer." Planted 2016 in Limerick Forest North at the edge of a flooded track that cuts through a marsh. The track is seasonally accessible. Only when the water freezes or when there is a season of drought. At other times, it is flooded by marsh spilloff.

Source: Photograph (2018) by Chris Turnbull.

landing on. The dynamic of geographies offers any number of potential changes of states between/across vast or microscopic distances, linguistic relationships, and spatial expressions and territories.

As is the case in almost any act of reading, *rout/e* compels readers to engage in intermittent encounters in diverse locations, occurring at unknown times and within diverse habitats. Individuals and other species walking a trail may see/have seen a poem just off-trail: on the outskirts of a track, tucked behind cattails at the edge of a marsh (Amanda Earl); at the liminal edge of a bridge and between an ephemeral stream or marsh (George Bowering; rob mclennan); under red pines next to a squirrel midden in a managed forest (Nico Vassilakis); ringing a nut grove or tucked into the hollow of a windfall (Eric Magrane); replacing a solar panel or between marshes at a federal wetland in a managed forest (derek beaulieu); wrapped in sheep wire at the edge of a local creek (jw curry); occupying the abandoned placards (postcard-sized) of a forgotten black walnut grove (Sandra Ridley; Monty Reid; derek beaulieu; Cecilie Bjørgås Jordheim); on the verge of a fen, along a track that once led to a (now absent) train station (Christine LeClerc); caught in the rock bar of a local river accessible by fishermen and adventurous kids (Monty Reid); laid out along a trail among cedar, lichen, and field, the delicate soil around juniper bushes and under a pine pelleted by sapsucker holes (Linda Russo); overlooking a swale and at the edge of a railbed (now a truncated trail that leads to a historic bascule bridge in Smiths Falls) (Jordan Abel); at a point of an industrial heritage site where a river diverges (Jason Christie); overlooking a river used for milling and toward a red telephone box (Bruno Neiva); along a section of the Rideau Trail piled with discrete waste piles, and moved (Angela Rawlings); at a marsh edge, the track an aqueous border between sections of draining marsh or a dry flat bed in times of drought (Hiromi Suzuki); or at a settlement fence line facing toward old-style telephone poles that cross a marsh (Paul Hawkins), to name a few of the locations and poets who have participated in *rout/e*.

With the exception of a few occasions, I have been mostly unaware of what others observed (or did) when they came across a poem that was part of *rout/e*. Once, I observed members of a youth organization walk past rob mclennan's poem, "Situated along various trails in a similar fashion," as they traveled along the Cedar Loop trail in Marlborough Forest. The poem was at the end of a wooden bridge, and they walked by it without noticing it. It had been planted at the edge of a creek beside a small footbridge, but since then has been moved three times by someone/ some people unknown to me: from beside the bridge to its end; then again so that it lay flat on the bridge rail; and then again, stake gone, the poem hammered into the cedar tree at bridge end. On a monitoring trip, along the Rideau Trail at Wood Road (which also served as a hiking outing with my then-6-year-old son and his friend), we came upon Angela Rawlings's "I will not ruin the environment." They were unaware that I had planted the poem there; I watched as they conducted, impromptu, a hip hop rendition of her poem, complete with sticks, bug nets, and dance, as they read it out loud, the sticks beating on the garbage, the poem, the ground. In another instance, I watched as a hiking companion encountered a

FIGURE. 8.3 derek beaulieu, visual poem, 2011. Poem planted in Marlborough Forest at marsh edge. The poem was later paintballed; the poem on paper was found intact, months later. The plexiglass was shattered, the stake missing, the plywood floating in the marsh.

Source: Photograph by Chris Turnbull.

selection of five poems by Linda Russo—part of a larger piece called "Daynotes on Fields and Forms (Flittings)," which appears in her collection, *Meaning to Go to the Origin in Some Way* (2015). The poems themselves are directly expressive of interspecies dynamics and multiple (linguistic, cultural, ecological) implications and developments that evolve from settled, co-inhabited, and co-opted locales and regions. I had planted the poems a good distance from each other along the Earth

Star Loop in the Marlborough Forest. Each time my companion encountered a poem, he took note of the biographical details and took pictures of each poem, commenting on elements of each. Noting a birch leaf on the surface of one of the poems, however, he was about to brush it off and then paused, noting that the leaf was now part of the poem and should be left where it was. He was unaware that I'd planted the poems.

Of surprise to me—perhaps it should not be so surprising—unknown individuals sometimes monitor and look after the *rout/e* poems they encounter. This is indicative of a form of stewardship that is not unusual, exactly, but which I didn't expect to extend to poems found in landscapes. For several years, for example, Angela Rawlings's "I will not ruin the environment" was "caretaken" by some unknown individual(s): Her poem was moved from one side of the trail to the other, where it was situated in a new place and maintained over time. The trail itself had piles of garbage along it for the first two kilometers, subject to ATV (all-terrain vehicle) drops. Someone placed the poem within a window frame that also contained a pile of garbage. The poem faced the trail and had unmistakably impacted the individual enough that it served also as a statement as much as a poem. It lasted there for seven years; each time I hiked to it, it was clear someone else was still ensuring it lasted because, while the text and the stand were slowly changing, the place where the poem had been planted had become subject to spring flooding as a result of ATV tracks changing the track structure. Someone ensured that the poem did not become submerged in the deeply flooded ATV track, and thus crushed by ATV wheels. Another example is Amanda Earl's poem, "until even now," which I planted among wild strawberries at the edge of a beaver pond. Someone attentively mowed around it each year; after it lost its stake, someone carefully placed it in the branches of a tree. In another example, a cyclist encountered Bob Hogg's poem and commented about it on the *rout/e* website. Finally, Jason Christie's poem, "Trailing," was moved three times at the Merrickville Industrial Heritage Site; someone eventually nailed it to a tree, using the same screw I'd used to fasten the poem to the stake. The plexi cracked, it seems, across the language: A crack on the surface of the page.

Each *rout/e* piece is similar in design to types of conservation or ecological signage that one might see in parks and other protected areas, yet *rout/e* has some of the formalities of a small-press publication. In contemporary times, small-press print magazines are distributed by various methods: by hand, often at localized poetry or artistic "gatherings" that emphasize core communities of individuals interested in both the writing work and the community itself, or between individuals who live within a fairly close distance or in places that are accessible by various means of transport; and by mail, to individuals who are too far apart geographically to permit the publications to be passed around by hand. Various small-press items can be endlessly distributed, over time, between members of the community or in other ways; this style of circulation acts as a form of moving archive and keeps the ephemera live. Key components of the small-press magazine are its form/design, its content, and methods of issue, publication, and distribution.

Virtual small-press publications follow the same pattern, for the most part, as their printed equivalents, but the communities who access them can be more greatly dispersed (depending on technological access) and greater in number as a result. In addition, communities can be more extensive—whereas small press printed publications may include a literary community, limited to particular geographies, a virtual publication can include multiple communities across varied disciplines (e.g., literary, scientific, historical). *rout/e* shares elements of design and process (solicitation, printed page), but is not directed at any one community and does not bother with traditional distribution; the blog opens possibilities of addressing a wider community, but the blog itself is at a remove from *rout/e* plantings because it is virtual; the blog offers a limited view of the poem (images and text) and, because of natural processes and ecological changes, the environment to which the poem contributes.

Each year in which poems are planted constitutes a sort of issue, even if this is not explicit. While poems are more often planted singularly within a year, and not as a group effort per se, *rout/e*'s directions sometimes change to include other possibilities, such as planting poems in urban centers or across greater distances, or engaging in a chapbook project. For example, to trace the evolution of poet/artist Angela Rawlings's poem, "I will not ruin the environment," Rawlings and I collaborated with Michael Flatt of Low Frequency Press (Buffalo, NY) on *The Great Canadian*, a chapbook that highlighted Rawlings's poem as part of a work-in-progress. *The Great Canadian* as a chapbook refers to the development of a manuscript over time (by the artist), the elements that impact "I will not ruin the environment" at the level of language and recorded by image, and the act of the poem planted, then re-placed, on the trail. I contributed photos from the trail and site of "I will not ruin the environment" to the chapbook.

To some extent, the possibility of a fluid "form" for *rout/e* poems has also enabled greater collaborations between me, poets, designers, paper-makers, publishers, and a "public." An example of this would be poems planted as part of a literature class at Memorial University in Corner Brook, Newfoundland; the course instructor solicited the poems from the literary community in Corner Brook; students took the lead in constructing versions like mine for planting outdoors; students and poets then participated in a planting "walk," which featured impromptu readings during the walk. Another example would be a group planting and reading at Petrie Island in Ottawa, Ontario: The poets and I planted the poems with the assistance of volunteers at Petrie Island, who made the stands for us and supplied tools and other equipment for planting the poems. More recently, *rout/e* was part of the CSArt (Community Supported Art) subscription program in Ottawa. I invited Ottawa-based poets to contribute poems to be planted, collaborated with local printer/papermaker and poet Grant Wilkins on making a chapbook of these poems for subscribers to CSArt, planted the poems across Ottawa, and organized a poetry walk and readings by the poets for the subscribers. Locations for the poems were determined with the assistance of Katherine Forster, who runs *wild.here*, a business that highlights the use of naturalized niches in urban centers. Forster uses a blog to identify areas within Ottawa's urban spaces that are forgotten, reinhabited, or largely dismissed. Both Forster and Wilkins helped to plant the poems at the various locations in Ottawa.

The grove project is another sort of side path of *rout/e*: As mentioned earlier in this chapter, I planted poems in small, metal 7 x 4-inch placards that I discovered at the base of black walnut trees while exploring an abandoned nut grove at the local agricultural college in my community. Since then, I have learned that decades ago, ECSONG (Eastern Chapter of the Society of Nut Growers) planted this nut grove. This landscape design offered possibilities of placing row-based poems, which opened the opportunity for a loose "conversation" between the poems, as well as a linear, grid-like presentation of the poems within a transforming and mostly human-forgotten/abandoned ecological space dominated by nut trees, cedar hedges, and various field species. At one point in time, the nut grove would have been monitored by ECSONG or individuals at the college. Now, it has been abandoned and left for squirrels and other species, much like the poems.

The first group of poems that I installed in the black walnut grove were by Monty Reid, derek beaulieu, and Sandra Ridley. Over time, these pieces moved or vanished into different locations within the grove, as a result of wind and other elements; sometimes one or two became buried under leaf litter between seasons, only to reemerge in a different season. The line of the trees, the plot of the land, and the larger area interact with each other in unique ways, both visually and ecologically. The poetry by Ridley, beaulieu, and Reid replaced the textual information that had once defined the trees by location, species, and ecology.

Between the late 1990s and early 2000, *rout/e* had been a printed pamphlet, in five issues. These issues featured poems that were mainly by Vancouver poets, but the pamphlets also included some solicited pieces from other poets I knew of elsewhere. There was a hiatus between *rout/e*'s strictly print form and its current "footpress" format as a result of other projects and life in general. The hiatus was also a result of my own sensitivity to what seemed to be a disconnect between academics and community members outside of academia (not necessarily non-academics). For the former group, poetry/language was an accessible part of their culture, while for individuals I knew in the latter group, poetry and other literary forms were at a significant remove from their interests; at times when they had been exposed to poetry, there was an implicit expectation of learning it by rote or of being tested on what poems "meant." In general, what these individuals avoided was the language tied to trying to "explain" poetry; they simply wanted to enjoy the language of poetry.

Moving *rout/e* from a print-based, small-press publication to a form that occupies landscapes was a way to address some questions I have encountered as a result of my observations of individuals in rural communities encountering poetry or poetic forms. I was curious about how poetic forms, the conveyance of meanings through language-presented-as-poetry, or the conveyance of sound and performance through visual and textual combinations, entered the lives of individuals in my community. For all intents and purposes, it doesn't, for the most part. While the majority of the individuals I encountered in this context did not object to poetry as an endeavor, neither did they seek it out regularly once they had no longer been required to memorize it or explain it as a component of a final mark at school or elsewhere.

I was curious to see what would happen if I combined relationships (with land, with species, interpersonal, and community-based) that local geography fosters with the sorts of tenuous meanings and spatial navigations that language invokes. *rout/e* functions within forms of design that impact the construction/habitability of multiple communities and their cohesions: the shaping of trails and locales; off-road gathering and hunting niches; species monitoring zones; property boundaries; archaeological, historical, and ecological areas; the shaping of landscape and terri-tory and markings; the shaping of the poetic "stand"; the poem itself; and the visual structure of letters and their sonic syncopations or visual frames and extensions. These forms of design have embedded structures and trajectories that are experi-enced both in places and of places. I wondered if a person might read a poem if the poem was *on the way* (to, from, between, over, under) as opposed to if they were sim-ply carrying the poem "with" them, via a book or electronic device—memorized poetry aside. If language can be considered as having geographical underpinnings, then linguistic structures (such as prepositions) offer ways of tracing movements as well as of allocating coordinates for staying in place. I wondered what sorts of other interactions might affect the readability and the presence of the poem; how the poem—on its stand—might emerge/merge over seasons and in the place it was planted.

How and where pieces are located (their territorialities) and their distribution/dispersal includes their topographical effects, means of access, community/species interactions, landscape design, and the aesthetics of page design. Structural for-malities within landscape (settlement, trail, hills, stone fences, tree plantings) are mimicked in my own work, but *rout/e* contributions are often placed among these elements and reflect them back. That is, a person can read the poem but also observe the interaction of light on the plexiglass covering the page, which works to blank out the poem and obscure it, as well as to mirror the landscape components imme-diately above or nearby it. Over time, there is textual and material degradation, scatter, or loss (by theft or random acts of destruction). For example, derek beaul-ieu's work in Marlborough Forest was paintballed and the structure destroyed. The page remained intact, as did the plywood on which it rested. Sandra Ridley's grove poem was chewed and torn in the winter months. Pieces of it were found within the grove, scattered, in a different season. Poems by Christine LeClerc, Amanda Earl, Angela Rawlings, bruno neiva, jw curry, Bob Hogg, Jason Christie, Jordan Abel, and many others, have vanished. This vanishing is part of *rout/e*.

Movement and motion involve disruption and displacement; the chronologies and truncated narratives that underlie, and at times suspend, lived experiences—that emerge as processes—accompany a variety of linguistic, acoustic, and visual forms. My poetic and installation work incorporate these elements partially as design; my work is often instigated outdoors as much as on the page as a way to engage linked transitory relations, interactions, and assumptions. A geopoetic approach—which would also integrate the language and spatialization of the page and the surfaces/ topographies of engagement, of movement, of choice regarding which landscapes to move through—considers the various features and processes that affect interactions with the poems and the habitats/ecologies through which they are encountered. To read the poem, one has to enter its physical and printed spaces, themselves conditional and transitory—move toward it, and make contact.

References

Rawlings, Angela, and Chris Turnbull. 2015. *The Great Canadian*. Buffalo: Low Frequency Press.
"*ROUT/E:* Footpress. Poetry Found in Place." *Wordpress,* January 20, 2019. Accessed December 22, 2018. https://etuor.wordpress.com/.
Russo, Linda. 2015. *Meaning to Go to the Origin in Some Way*. Bristol: Shearsman Books.

PART II
Reading

9

LYRIC GEOGRAPHY

Maleea Acker

What are the geopoetic possibilities for unfolding place? How might creative process and product (specifically, poetry and poetics) contribute to geographical thought? How can I use my training as an artist to find a way to talk about poems and place within the social sciences?

This piece is a geopoetics in practice. It is a response to a call issued by geographers for a more situated, creative, entangled way of writing about the world. The form is a reflection of the content. Dialogue—that essential interaction between self and other, between disciplines, and within and outside the academy—should be an essential method in geography to enable humans to ethically engage with the world. Thus, the left-hand text is mine; the right-hand text features other voices. The right-hand quotes have been selected as a kind of curatorial exercise—a bringing in of authors from both within and outside of geopoetics. The quotes enter into conversation with one another; they also converse with my own work, on the left-hand side. The form itself owes a debt to Jan Zwicky, who, in *Wisdom and Metaphor* (2003) and *Lyric Philosophy* (2011), provides an example of how academic and creative work might each speak more fully when presented in tandem.

Some of the left-hand text occurs in short prose vignettes, some takes place in more aphoristic prose, and some is poetry itself. The vignettes of daily life (in Ajijic, Mexico) are meant to support the poems written in Ajijic and to illustrate the concepts raised in the prose. The prose work sets the stage; the vignettes are my observations; the poems are the creative culmination of the study. I argue that, together, they offer a new way of doing geography within geopoetics.

How does one write about the place in which a poem occurs? Or the emotions that occur when one is in a resonant place? Like Wylie's benches in Mullion Cove (2005), the place a poem occurs does not remain static. It changes under different light. It changes depending on the person who has apprehended it or attended it.

The moments of apprehending a place with the intention of practicing poem-making, as well as the writing itself, are moments when the ghostly hovering of a more resonant understanding of the world comes close, when one is able to catch its coattail for a moment before it slips out of reach.

As a poet, I find Massey's (2005) idea of place as a *throwntogetherness* to be especially powerful when examining the act of writing a poem. *Throwntogetherness* brings in the geography, as Massey points out, of a living, changing landscape that is and is not home. The places where poems are written are records of the constantly changing effects of culture, environment, and weather on a writer.

In my own case, I wrote poems in Ajijic about connection, loss, home, inner emotional landscapes, and outer cultural landscapes. They are a geography of lyric experience in that they chart physical locations that serve as interior exteriors—safe zones—where I can connect with the emotional and intellectual sphere where a poem's beginning lies. I write to chart what these places mean to me in the world. Place impels me; it is the actor. Place also unfurls around me; it becomes the subject.

Doreen Massey

What is special about place is not some romance of a pre-given collective identity or of the eternity of the hills. Rather, what is special about place is precisely that throwntogetherness, the unavoidable challenge of negotiating a here-and-now (itself drawing on a history and a geography of thens and theres); and a negotiation which must take place within and between both human and nonhuman. This in no way denies a sense of wonder: what could be more striking than walking in the high fells in the knowledge of the history and the geography that has made them here today? This is the event of place (2005, 140).

Don McKay

A mystic who is not a poet can answer the inappellable with silence, but a poet is in the paradoxical, unenviable position of simultaneously recognizing that it can't be said and saying something. . . . Language is not finally adequate to experience and yet is the medium which we—the linguistic animals—must use. What to do? The poem's own soundplay holds the clue: "we must answer *in chime*," [Scott 1926] a term suggesting both rhymed resonance and one that harmonizes compatibly with the appeal (2011, 31).

Place is integral to most poets. I argue that the places where we create are ground zero for a connection with physical location and with lyric experience, which is itself a kind of bodiless embodiment of being. Though I anticipate the suspicion that this claim might draw and its openness to critique from feminist and critical geographies concerned with politics and power, this piece is a way of testing the waters.

Inclusion of the creative voice means acknowledging the strange familiarity of an individual voice—irrespective of class, cultural background, or ethnicity—a voice in a poem that emerges as a leveling call. One reads the particular story but experiences what I argue can be a common understanding—universal feelings of connection, loneliness, love, loss, as in Zwicky's poem "Glen Gould . . . ." Whether a reader has been north of Lake Superior or not, the physical details provide specificity that anchors the emotion. They also allow for a reader to substitute in their own details, specific to their own place (e.g., hearing music, seeing light on bare rock, driving while thinking of someone) while still accessing the poem's "precise and nameless" joy. It is this universal aspect of poetry (and indeed, of all the arts) that I fear is being (purposefully? naïvely?) neglected when examining the contributions geography and humanities can make to one another's disciplines while still speaking politically to the human experience.[1]

If geographers are serious about petitioning for inclusion of the arts in geographical conversations, we must be ready to acknowledge this aspect of art. Art as the ultimate freedom. Art as a voice that can speak pre-politically, even as it engages in the political, because art's engagement with politics *is* its dealings in emotion and fundamental human experience.

Jan Zwicky

Glenn Gould: Bach's Italian Concerto, BWV 971

North of Superior, November,
bad weather behind, more
coming in from the west, the car windows furred
with salt, the genius of his fingers
bright, incongruous, cresting a ridge
and without warning the sky
has been swept clear: the shaved face
of the granite, the unleafed aspens
gleaming in the low heraldic light, the friend
I had once who hoped he might die
listening to this music, the way
love finds us in our bodies
even when we're lost. I've known very little,
but what I have known
feels like this: compassion without mercy,
the distances still distances
but effortless, as though for just a moment
I'd stepped into my real life, the one
that's always here, right here,
but outside history: joy
precise and nameless as that river
scattering itself among
the frost and rocks.

                                    (2004, 84)

The following pages contain three parts: a description of daily life in the place where I wrote these poems, an exploration of how this place affected what I created, and the poems themselves. The description and exploration are observations of place; the poems are the culmination of a study of place. All, however, should be seen as creative work that contributes to a geographical understanding of a place: In this case, the highlands of central Mexico.

Miranda Ward

Any line between "creative-critical" writing and "traditional academic writing" or "overtly creative writing" is not always going to be clear, and nor should it be (2014, 757).

Emily Orley

I found it impossible to take up the two positions of critic and practitioner simultaneously: analyzing my artistic practice made me too self-conscious and disturbed my creative process. I found, however, that I could inhabit the "site" between the distinct positions by concentrating on my writing as practice, juxtaposing creative modes alongside more traditional academic forms (2009, 159).

Owen Sheers

Landscape and poetry share the same grammar and semantics of association and suggestion, if not the same vocabulary (2008, 173).

I lived in Ajijic, Jalisco, Mexico for six months in 2015. I was filled again and again with music, food, language, birdsong, the shouts of the truck driver selling his wares up and down every street, until the shell of my skin overflowed and I walked every evening, spilling out the edges of my body. Until the ground itself bore the record of my perambulations. Until my own body expanded into the landscape. Until I was the smell of *guayaba*, garbage, dog, tortilla, smoke, meat, fat, brick, dust.

Sarah de Leeuw

Your skin membrane   a watery system
veins like tributaries   thin splinters and blue
sparks. I brush up against
        the estuary of your heart   aortas
draining into the salty rush
oxygen uptake   breathing from the side
    of your mouth   red blood cells
like salmon roe. Riparian ribs. Swim me
swim me   your body seal-slick
you take me in your mouth
                        (2015, 80)

Tom Bristow

Our sense of place is related to our sense of affective being, which is
constituted by the many selves expressed by a person during the course
of one life. These selves pass by us while we are in transit. . . . Conversely,
the lyrical self is rarely divided like this; however, it offers a singleness of
eye from which we witness the calamitous division of the world and the
emergence of protean selves (2015, 92).

Ajijic is a village on the shore of Lake Chapala. Surrounded by volcanoes (including the regularly erupting Colima), Mexico's largest lake, and 8,000-foot mountains, the village nestles into eight blocks of sloped land between the mountains' precipitous rise and the lake's shallow expanse, and extends for three kilometers along the lake edge.

The village streets are cobblestone. Murals by local artists dot the walls of buildings. On one, the circle of life: A man catches fish swimming down from the air, while a woman standing in water gives life to fish with her hands.

Along the lake's edge runs a walkway, approximately two kilometers long, called the Malecón, where I walked each evening. During each day, I worked on a shaded rooftop that I took to calling the *nave espacial*, or spaceship, for its resemblance to a ship's cockpit and its limitless view.

Emily Orley

The first stage [of engaging with place] involves a self-reflexive awareness of ourselves in place. This begins with a close, hushed and stilled observation of the place's details. If the visitor in a place is still and quiet enough, and pays close enough attention, she can become aware of what the place is "doing" around her, as well as her own response. This cannot be rushed (2009, 190).

Tim Lilburn

I want to go back. It seems ludicrous to want this. Want to bed down beside things. . . . I want to be married there, home, quiet, looking around. . . . Maybe it's just that I want to be heard as matter, heard as animal, want to be heard in this broad a way, and want the rest that such recognition would spread along the whole muscle of self (1999, 78).

Lynn Hejinian

Knowledge is always situated (2000, 229).

Each afternoon, I completed a daily perambulation that took me to the liminal zone of lakeshore and then back to the house, where I watched months of midnight thunderstorms, rain pelting down and the windows lighting up the interior like a cave.

Each place I passed through contributed to my writing process: streets I became familiar with, corners where certain foods such as *tamales* or *comida casera* or drinks such as *ponche* were sold; the corner where Gabriel worked washing and parking cars; playgrounds that were full or empty depending on the time of day; the alcove where Max's aunt sold *chicharrón*; sidewalks; groups of trees at the corner of Marlene's house with its green, shaded patio; views of the mountains from Marika's rooftop or the roof of my own house. Passing them was akin to passing touchstones. It was as if I was sketching myself onto the body of the town. I sought to become part of the village's story as it unfolded itself in its routines, socializing, and trade. I wanted the connection not just to the natural phenomena—birds, trees, mountains—but also to the people. Using Lilburn's courteous gaze was my tactic when writing purely about landscape. But to dwell in Ajijic was, as in Heidegger's understanding of dwelling, a closing of distance between people and things.

Rebecca Solnit

Walking wasn't only a subject for Wordsworth. It was his means of
composition. . . . In *The Prelude* he describes a dog he used to walk with
who would, when a stranger drew near, cue him to shut up and avoid being
taken for a lunatic (2001, 112).

Harriet Hawkins

Creative geographies therefore not only offer chances to meet
representational challenges of contemporary ways of thinking place,
but they potentially become the means to intervene in the processes of
place-making.

. . . To draw is to discover, to be led to see, to be drawn into an intimate
relationship with the object. So it was with my drawing in the field, this was
a practice of slow careful looking that was not so much about recording the
place but discovering it, coming . . . to know (2015, 7).

Martin Heidegger

Dwelling involves a lack of distance between people and things, a lack of
casual curiosity, an engagement which is neither conceptual nor articulated,
and which arises through *using* the world rather than through scrutiny
(1993, 28).

Tripe and cake

The body is an animal
out of which longing springs.

The hands, spiders. Eyes,
mustached bus drivers. Forehead,

a corrugated tin roof
over spangled Guadalajara girls.

I am a saltshaker of happiness.
My thirst knows this plastic table

like the back of your hand.
Why won't you cave to the unachievable pleasure?

My shaker breaks loose and cartwheels the main street,
through dust heaps, potholes, past the nicked

knuckles of the stone worker. It settles
in a foundation crack. It wicks back and

forth in the evening's drumroll. It dries
and shims the leftover heat. It wants things.

It's almost three and dressed as a gypsy.
It's eating tripe and cake.

It's not speaking because it can't
eat it and want it, too.

Brenda Hillman

Four survival tools for contemporary culture that poetry is especially
good at providing . . . : the sense of who we are in our environments; the
understanding that every word and phrase matters and can be of interest;
the idea that meaning circulates on many levels; and the conviction that the
strange mystery of our existence can be represented (2006, para. 15).

Rebecca Solnit

Poetry is a philosophical and descriptive foray into the world, and it has
some permission that I want to give myself sometimes, to make associative
leaps, to ask the reader to work a little, to evoke as well as define. Somebody
yesterday was using that phrase of Paul Klee's, to take a line for a walk:
I want to take the language on a walk (quoted in Elkin 2013, para. 14).

My daily destination for the months I lived in Ajijic was a small stone wall on the Malecón, with the mountains at my back and the lake before me. I sat there each early evening as the light faded. Waiting for a poem to happen felt like attending to the rich tapestry of a complete but unhinged life. I was peeled open. As families passed, their toddlers on tricycles, or couples sauntered by, their conversations barely audible, I felt the connection I had with the world both expanding and contracting; what I wrote took in the world around me, and both was and was not a bodily representation of what I heard and felt.

The wall on the lakeside edge has a lip on the stone to allow one's feet to rest when sitting. The land, the lake, and the trees curve forward in one's peripheral vision. A group of willows frames the view of the mountains to the west; the mountains rise steeply, putting a kind of pressure on the back of the neck when one faces the lake. The lakeshore itself is flat, grassy, sandy. Depending on the whims of the adjacent state of Michoacán, which releases water from bordering dams, an invasive green water lily called *lirio* may blanket the foreshore and the shallows, rising and falling like a second stretch of greened earth.

John Ashbery

Any day now you must start to dwell in it,
the poetry, and for this, grave preparations must be made, the walks of sand
raked, the rubble wall picked clean of dead vine stems, but what
if poetry were something else entirely, not this purple weather
with the eye of a god attached, that sees
inward and outward? What if it were only a small, other way of living
like being in the wind? or letting the various settling sounds we hear now
rest and record the effort any creature has to put forth to summon its
spirits for a moment and then
fall silent, hoping that enough has happened? Sometimes we do perceive it
this way, like animals that will get up and move somewhere and then drop
down in place again . . .

(2014, 145)

In summer, the sun falls behind the mountains to the west of Ajijic at approximately 8 p.m.; I arrived to something extraordinary. A man stood beside a red car on the sandy lakeshore. The car, a hatchback, sat with its trunk and doors open. On the roof were two portable speakers. In the back lay an alto saxophone and a tenor. The man, in his 60s, variously stood looking at the lake, arranging things inside the car, pouring a drink, or adjusting his speakers—out of which music drifted. John Coltrane, Miles Davis, Gilberto Gil, Keith Jarrett, Michel Guglioni, JS Bach, Lionel Richie, Carlos Jobim. Occasionally, he would join in with a saxophone, letting the notes hang in the falling light, over the strolling couples, the horses, the birds, trees, and moving water.

People gravitated to him. They would stop and ask him about a song, ask him to play at weddings, let their dogs play with his dog, or just stop and listen to his soundtrack, the evening's landscape made audible.

We became friends; his name is Antonio, but I called him the Dictator, because he chose for us. It was an affectionate term. His musical choices were always apt. I would sit on the Malecón's wall, with families passing behind me, his music filling the air between us. I wrote to a soundtrack that was a man trying to teach a town about music other than *banda* and *ranchero*. Even after we met and became friends, we had a tacit agreement: He created his work out on the sand; I created mine while sitting on the wall. If he saw me arrive, which he usually did, he would lift his chin in the self-sufficient and quietly euphoric acknowledgement of an artist working near another artist. I think we both felt, like Bachelard, that we had been "promoted to the dignity of the admiring being." His music made the place we came to each night a studio for our respective work. Most of my poems were written at the feet of his choices.

Gaston Bachelard

In analyzing images of immensity, we should realize within ourselves the pure being of pure imagination. It then becomes clear that works of art are the *by-products* of this existentialism of the imagining being. In this direction of daydreams of immensity, the real *product* is consciousness of enlargement. We feel that we have been promoted to the dignity of the admiring being (1969, 184).

Hayden Lorimer

In short, acoustics matter as much to the poetics of place as they do to poetry. When read aloud, and when experienced in person, inventive geographical writing is carried along by its rhythmic qualities. Place condenses when it is formed of breath, intonation, emotion and stress (2008, 182).

Robert MacFarlane

Language does not just record experience stenographically, it produces it. Language's structures and colours are inseparable from the feelings we create in relation to situations, to others and to places. Language carries a formative as well as an informative impulse (2010, 118).

The dictator

The Dictator puts on Music for 18 Musicians,
Section IV. It is the last hours
of the year, before Maximino
gives his children to the Americans.

They still live in two rooms
beside his carpenter's
workshop illuminated by stars
by night, by dusk, by day.

Section V begins.
The iron worker has finished
his calla lily doors. He's doing
something pedestrian now, maybe

the framing for a toilet. The carpenters
drag their new saw inside, up
the furred cement stairs. In two rooms
two beds sleep five, on the roof

the cistern, the tree that hides
bathing, the neighbours from which
there is no invitation.
In an hour, the bats will emerge

from their little ceramic shells.
As the last of the sun stumbles
into the lake the Dictator,
parked at the edge with his extraordinary

car battery and his ridiculous heart,
turns up the speaker on the roof
and keys up some Lionel Richie,
then a little Brahms. It's early

Lionel Richie so it's okay.
Notice the stomped,
broad-leafed grass, the gelatine air
at its inconsequential fulcrum,

JD Dewsbury

This practice of writing drives something out of you and, like a mirror, becomes a screen for presenting your self to yourself; and the future calls as always, here materialized in whatever medium your activity is becoming archived in, but not in these moments within any precise destination but with the hope and munificence of open lines of flight that someday, someone, somewhere will share the ecstatic point that the writing is bringing into the world . . .

. . .We want so much from words, to hear some words: do we fail to realize that they say more than we know; that there are silences, gaps, which communicate so much, too much. How do you write that (2014, 148)?

RP Blackmur

Words, like sensations, are blind facts which, put together, produce a feeling no part of which was in the data (quoted in Hillman 2006, 158).

Mary Ruefle

You might say a poem *is* a semicolon, a living semicolon, what connects the first line to the last, the act of keeping together that whose nature is to fly apart (2012, 4).

Sarah de Leeuw

Literary practices and visual arts are more than sites from which to draw geographic information (Salter and Lloyd 1977; Bunkse 1990) but are rather geographic practices unto themselves (2003, 19).

the last cloudbank spitting
out a star. Is it possible
he is playing out our life?
Baker, Sosa, Ibañez, Bach,

pulled in all directions,
all incomputable, we are
little racing dogs just bathed.
Someone lights a bonfire

in the field of the gypsies.
Maxi's four children
fix themselves at the edge.
The oars of the fisherman

flake the mother of pearl lake
as the Dictator pulls
out all the stops, slides
on his black gloves, lifts

his wine from the car's
roof. Cities fall. He takes
a drink, they bubble up again
as the dust of the unremarkable

end meets its maker, the one
with the skin of damp gold
and the mind of a dog
and the hearts of a child.

Ali Blythe

I am placing the future you
in my perfectly unappointed
room with the numerals
on the door by typing

minuscule code into a glowing
handheld campfire with my thumbs.
Soon it will rapidly become
impossible to separate

what is happening
to my body from what
you are doing to it.

(2015, 62)

Being in Ajijic saved me through an intertwining of myself with people and landscape. This is a risky thing to say in geography. I can only qualify this by pointing to other artists, such as Lilburn, who built a listening cave in the side of the Saskatchewan hills and slept in it for a summer, hoping to connect himself bodily to the land. Or McKay (2001), who calls a porch a threshold between humans and wilderness, where communication between the two is aided.

My geographical practice, similarly, acknowledges walking through places and charting my feelings and the ways in which I interact with people as both a creative practice and a turn toward dwelling at home, toward choosing to engage *with* place rather than to distance myself from it.

Charting the journey to where the creation of a poem takes place (the walk through the village), as well as the place itself (the last curve of the Malecón), shows how a writer arrives at the act of creation; it demonstrates the intertwining that can happen between place and geographer; it validates artistic process as well as product. The place where a poem happens is a world unto itself. It is its own geography, its own emotional landscape. Attending to that place is also a form of courtesy. This is the gift that the place receives. The gift that the poet/geographer receives is not just the poem, it is the arrival at a version of home. Geography could learn this form of exchange from poetry. When I wrote what is now my third book of poetry in Ajijic, the mountains, the lake, the Dictator, the children riding behind me on their bicycles took what was a tired, sad, singular person and linked her back into the world. Poetry is, in the end, about charting an attachment to place, and this becomes a reciprocal relationship. Geography should be no different.

Matthew Zapruder

I have lived in the black crater

of feeling every moment
is the moment just after
one has chosen forever

to live in the black crater
of having chosen to live in the black crater.
                                                    (2010, 100)

Tim Lilburn

Sorrow is the alteration of self before extreme dissimilarity; it is admission
of the unlikeness of what one cranes toward and one's exclusion from
its beauty, its community, and thus is what knowledge of the thing's
uncontainability feels like; sorrow is what fashions courtesy, work of
reverence, toward what one would know utterly (1999, 65).

Jan Zwicky

It is the meanings of certain experiences that are ineffable.
This would explain why we seem to be able to say so much about
the experiences themselves, while continuing to insist that they are
indescribable.

But what is meaning? Meaning, I would like to say with Robert Bringhurst,
is what keeps going (2012, 199).

Toward the end of my time in Ajijic I took a small trip to the other side
of Lake Chapala for a Sunday meal. I traveled with friends: Max, his sister
Hortencia, her husband Jorge. We would meet their daughters, and their
daughters' husbands and children, in the village of San Pedro, where we
would assemble at a chicken shack, stop for cups and napkins, again for fresh
tortillas, then for gas, for juice, for tequila, before finally arriving at a small
park that lay between the village and the lakeshore—another Malecón,
where Mexican families gathered with music and food for extended
lunches, lifting folding chairs out of the truck box, laying out an enormous
meal, turning up the stereo, enjoying the light through the shading willows,
leaning against the warm fender of the truck's nose while watching the
children careen back and forth.

I didn't know any of this when the morning began. But I could imagine
it, having been to enough family gatherings during the previous months.
I could imagine the languidness, the laughter, the easy pour as the men
filled their plastic cups with ice and drink again and again. When we started
out, there was just the cobbled street where we lived, the 1980s truck with
its cracked windshield and metal dashboard, the open beer we held, our
gaze forward into the bright noon, wedged into the front bench seat. Jorge
turned the engine over. Hortencia was teasing Max. She squeezed my left
hand. On my other side, Max leaned against the door and threw his head
back in laughter. The houses drifted by and we turned onto the *carretera*,
climbed the hill to the outskirts of the village. We passed the spot on the
lake where so many of my poems had taken place. I hadn't been in a vehicle
for perhaps two months. I loved the people I was with like family. My heart
broke open, then, I think, in that village, *as we left it.* My chest swung open
like two doors, and a profound release spread from it to the rest of my
body. As if some inner muscle I had not even known I was clenching had
suddenly and completely relaxed. I had never felt the sensation before in my
life. I have not felt it since. It was like arriving to a place as I was leaving it.
It was like dwelling at home as I was carried from that home.

Rishma Dunlop

Today, as a poet, as a [geographer], I reject my former professors who
criticized "mixed" metaphors, who told me in Honours English that if
one was to write literary criticism as a scholar, one could not be a creative
writer, one would have to choose. I refuse to choose (2002, 33).

Tim Cresswell

The challenge for cultural geographers of landscape is to produce
geographies that are lived, embodied, practiced; landscapes which are never
finished or complete, not easily framed or read. These geographies should
be as much about the everyday and unexceptional as they are about the
grand and distinguished (quoted in Kate Anderson 2003, 280).

John Wylie

The geographies of love, in other words, would speak against any solipsism
or narcissism, and equally against any sublimation of self and other, any
abolition of the spaces between us, any coincidence of self and landscape
(2009, 286).

Don McKay

Have you heard this?—in the hush
of invisible feathers as they urge the dark,
stroking it toward articulation? Or the moment
When you know it's over and the nothing which you
have to say is falling all around you, lavishly,
pouring its heart out.

                                      (2000, 59)

Hesitating once to feel glory

Sometimes I think we can see
the world before it began,
and that's what makes us
so sad. Before the world began

there were swallows flying
across a lakeside field
as the sun allowed the trees to shade it.
There were leaves fallen

during dry seasons that made
a golden road. And there was
silver and stone and clover,
and a man on horseback

with a dog with no tail
that loped across the field
in a lazy semi-crescent as though
drawing the orbit of a small moon.

There was a burro
on a ten-foot length of rope
stomping a dust patch in the earth.
And there were pelicans

with injured wings handfed
by a waiter and so many willows—
so many! growing by the water's edge.
There was the clink of bottles

before the world began
and so its sound still
makes us melancholy
the way ice can, booming

on a river in spring
or tilling a glass in a woman's hand.
Stones, too, uncovered from earth
pockmarked with clam houses,

and also clams. Pianos, there were
pianos too, their cascade made us
restless, they could not offer
more nuance than the half note.

Rishma Dunlop

An erotics of place is necessary. An erotics of place engages us in thinking about how we know the world in sensual, primal ways. In academia, it is fear of the open heart that controls academic institutions and modes of writing research. To combat this fear, we need eros, a deep, loving connection to ourselves, to others, and to place. And within this eros, narrative scholarship becomes a way of loving ourselves, others, and the world more deeply. Scholarship becomes *florecimiento*, a flowering, or opening of the heart, a tawny grammar (2002, 37).

Tim Robinson

Among the echoes of all these steps—rash or wary, ritualistic or whimsical, processional or jiggish, trespassory or proprietorial—it is impossible to isolate the particular resonance I had hoped to amplify further, that of the good step, the one equal to the ground it covers. . . . Having now acted out to the best of my capacity the impossibility of interweaving more than two or three at a time of the millions of modes of relating to a place, I can feel in the tiredness of my feet what any sensible thinker would have gathered from a moment's exercise of the brain, that the good step is inconceivable. And this book in its oblique and evasive way had undertaken the conceiving of what I knew to be inconceivable (1986, 363).

Things kept coming
before the world began, and stacked
and tumbled over themselves
in drifts like snow,

insensible. The world
before the world was annotated,
expansive, all the stones
the boys could throw

never hesitating once to feel
glory, to feel jealousy,
boredom, and the nostalgia
the grass feels as it clambers

above itself, and loses
its former lives in the clean,
disintegrating thatch
and dust and clay.

The sadness of the alternate
armed rower, who walked his boat
to shore! The sadness of the far shore
and the thud of a foot against a ball,

the bent hook of wire hanging
from a tree's lost branch stub,
the question in the ibis' voice
the sudden flash of a red bird

like a compass of ink in the brush.
Before the world began
there were bells that never
rang the correct time, and wings

and spheres of sad eggs in water.
The burrow walked his circle
and the carpenter never saw
his children further

than 6th grade. He never
painted his room yellow or cooked
on anything but a burner
on a board. And the neighbour,

after the party, she never
gave the plate back though
she said she would,
she always said she would.

Note

1 For examples of the reticence geographers have displayed toward depictions of universal experience or emotion, to which poetry contributes, see Tolia-Kelly's work on affect (2006) or Cresswell's investigation of more-than-representational geography (2012).

References

Anderson, Kay. 2003. *Handbook of Cultural Geography*. London: Sage.

Ashbery, John. 1991. *Flow Chart: A Poem*. Manchester: Carcanet Press.

Blythe, Ali. 2015. *Twoism*. Toronto: Ice House.

Bristow, Tom. 2015. *The Anthropocene Lyric: An Affective Geography of Poetry, Person, Place*. Palgrave Pivot. Basingstoke: Palgrave Macmillan.

Bunkse, Edmunds V. March 1990. "Saint-Exupery's Geography Lesson: Art and Science in the Creation and Cultivation of Landscape Values." *Annals of the Association of American Geographers* 8: 96–108.

Cresswell, Tim. 2012. "Nonrepresentational Theory and Me: Notes of an Interested Sceptic." *Environment and Planning D: Society and Space* 30 (1): 96–105.

de Leeuw, Sarah. 2003. "Poetic Place: Knowing a Small British Columbia Community Through the Production of Creative Geographic Knowledge." *Western Geography* 13 (14): 19–38.

———. 2015. *Skeena*. Halfmoon Bay: Caitlin Press.

Dewsbury, J. D. 2014. "Inscribing Thoughts: The Animation of an Adventure." *Cultural Geographies* 21 (1): 147–152.

Dunlop, Rishma. 2002. "In Search of Tawny Grammar: Poetics, Landscape and Embodied Ways of Knowing." *Canadian Journal of Environmental Education (CJEE)* 7 (2): 23–37.

Elkin, Lauren. 2013. "The Collector: Rebecca Solnit on Textual Pleasure, Punk, and More." *The Daily Beast*, July 2.

Gaston Bachelard. 1969. *Poetics of Space*. Boston: Beacon Press.

Hawkins, Harriet. 2015. "Creative Geographic Methods: Knowing, Representing, Intervening. On Composing Place and Page." *Cultural Geographies* 22 (2): 247–268.

Heidegger, Martin. 1993. *Basic Writings*. Edited by David Farrell Krell. New York: Routledge.

Hejinian, Lyn. 2000. *The Language of Inquiry*. Berkeley: University of California Press.

Hillman, Brenda. 2006a. "Cracks in the Oracle Bone: Teaching Certain Contemporary Poems." *Poetry Foundation*. www.poetryfoundation.org/learning/essay/239762.

Lilburn, Tim. 1999. *Living in the World as If It Were Home*. Toronto: Cormorant.

Lorimer, Hayden. 2008. "Poetry and Place: The Shape of Words." *Geography* 93 (3): 181–182.

Massey, Doreen B. 2005. *For Space*. London and Thousand Oaks, CA: Sage.

MacFarlane, Robert. 2010. *The Wild Places*. New York: Penguin Books.

McKay, Don. 2000. *Another Gravity*. Toronto: McClelland & Stewart.

———. 2001. *Vis à Vis*. Wolfville, NS: Gaspereau.

———. 2011. *The Shell of the Tortoise*. Wolfville, NS: Gaspereau.

Orley, Emily. 2009. "Getting at and into Place: Writing as Practice and Research." *Journal of Writing in Creative Practice* 2 (2): 159–171. doi:10.1386/jwcp.2.2.159/1.

Robinson, Tim. 1986. *Stones of Aran: Pilgrimage*. Mullingar: Lilliput.

Ruefle, Mary. 2012. *Madness, Rack, and Honey*. Seattle: Wave Books.

Salter, Christopher L. and William J. Lloyd. 1977. *Landscape in Literature*. Washington: Association of American Geographers.

Scott, Duncan Campbell. 1926. *The Poems of Duncan Campbell Scott*. Toronto: McClelland & Stewart.

Sheers, Owen. 2008. "Poetry and Place: Some Personal Reflections." *Geography* 93 (3): 172–175.

Solnit, Rebecca. 2001. *Wanderlust: A History of Walking.* New York: Penguin.

Tolia-Kelly. 2006. "Affect—An Ethnocentric Encounter? Exploring the 'Universalist' Imperative of Emotional/Affectual Geographies." *Area* 38 (2): 213–217. doi:10.1111/j.1475-4762.2006.00682.x.

Ward, Miranda. 2014. "The Art of Writing Place." *Geography Compass* 8 (10): 755–766.

Wylie, John. 2005. "A Single Day's Walking: Narrating Self and Landscape on the South West Coast Path." *Transactions of the Institute of British Geographers* 30 (2): 234–247.

———. 2009. "Landscape, Absence and the Geographies of Love." *Transactions of the Institute of British Geographers* 34 (3): 275–289.

Zapruder, Matthew. 2010. *Come on All You Ghosts.* Port Townsend, WA: Copper Canyon.

Zwicky, Jan. 2003. *Wisdom and Metaphor.* Wolfville, NS: Gaspereau.

———. 2004. *Robinson's Crossing.* Boler, ON: Brick.

———. 2011. *Lyric Philosophy.* Wolfville, NS: Gaspereau.

———. 2012. "What Is Ineffable?" *International Studies in the Philosophy of Science* 26 (2): 197–217.

10

EKPHRASTIC POETRY AS METHOD

Candice P. Boyd

Introduction

This chapter presents ekphrastic poetry as a research method for interrogating non-representational geographies.[1] Based on my own geopoetics practice, the chapter treats everyday practices of life as saturated phenomena, imbued with affect, that are afforded limited capture by ekphrasis. The concept of saturated phenomena was first articulated by Jean-Luc Marion, a post-phenomenologist, and has been defined by a scholar of Marion as objects that are "so 'saturated' with 'intention' that [they] exceed any concepts or limiting horizons that a constituting subject could impose on them" (Mackinlay 2010, 57). As such, a saturated phenomenon, e.g., a painting, can be constantly reinterpreted via the singularity of encounter and seen each time anew (Marion 2002). This notion has resonance with non-representational methodologies that seek to reveal the ways in which the world unfolds through mundane instances, fleeting potentialities, and temporally unstable relations between multiple actants and actors that similarly "take place" as events or happenings (Dewsbury 2003).

The chapter begins by presenting how ekphrasis is traditionally understood as a process of reinterpreting art. From there, I consider how it might be possible to reimagine ekphrasis as a method of presenting (or re-presenting) geographical fieldwork findings where embodied, sensuous, affective, and material conditions and concerns are at the heart of the research. I argue that this necessitates a very different approach to the conduct of fieldwork than what is prescribed by classical ethnographic research methods. These arguments are grounded in my own geo-poetics through which I explored the non-representational geographies of thera-peutic art making (Boyd 2015; Boyd 2017a). On that basis, I then offer an *ars poetica* that considers the processes of creating ekphrastic research poetry. To achieve this, I outline my aesthetic and methodological choices. The chapter concludes with some brief thoughts about the value of research poetry.

Understanding ekphrasis

Put simply, ekphrasis is a process of translating one art form into another. In terms of poetry, it usually involves poetic responses to an image or a sculpture (i.e., the ekphrastic object). Ekphrasis has a definition based in its etymological roots as well as one that reflects its more contemporary manifestations. As a word, its origins are via Latin, from Greek. Ekphrasis means "description," and comes from *ekphrazein: ek* means "out," and *phrazein* means "tell" (*OED* n.d.). The roots of the word are important as ekphrasis is not just about interpreting but also about *revealing* the ekphrastic object in some way. As Moorman (2006) explains, there are many different approaches to traditional ekphrastic poetry. The ekphrast might be simply trying to describe the scene that they see, in a painting for example, or trying to figure out what the artwork is about. Or, they might be taking on the voice of the artist, speaking to the subject of a painting, or speaking as a character in a painting. By doing so, the ekphrast is attempting two things: "To make sense of what they see and make meaning on the page" (Ehrenworth 2003, 7). According to Ehrenworth (2003), this requires the poet to engage in empathic imagination, which often goes beyond simply narrativizing an image.

In the contemporary context, ekphrasis is less about verbal representation of a material phenomenon than it is about rewriting, re-presenting, translating, or transforming it in some way (Clüver 1997). According to Kennedy (2012), contemporary ekphrastic poetry exploits the tension between likeness and unlikeness of visual and textual forms in ways that let the ekphrastic object signify something new or "represent something else" than what the original artist may have intended. Lund (1982) suggests that contemporary ekphrasis does this through the relational processes of combination, integration, and transformation: Combination refers to a kind of bi-medial communication where ekphrastic object and poem "talk to one another"; integration is where the poem takes on features of the object in its composition; and transformation refers to the capacity of ekphrastic poetry to extend the ekphrastic object—or to see "more than" or "beyond" what is represented. For these reasons, Kennedy (2012) argues that contemporary ekphrastic poetry is both critique and inquiry.

While ekphrasis has traditionally taken the form of poetry, it is not limited to a dialogue between image and text. Sometimes referred to as "reverse ekphrasis," it is possible to pair two art forms across a range of sign systems from verbal to pictorial, sonic, and even kinetic (Bruhn 2000). Therefore, using an ekphrastic process, a dancer may respond to a visual work through movement, a composer may similarly respond to poetry or paintings, a painter may respond to dance or music. In each case, the ekphrastic object (or original artwork) is transmuted into a new work that reflects something of the ekphrast's encounter with it. This metamorphosis from one art form into another is simultaneously aesthetic, affective, and sensory. The process with which the ekphrast engages is one that moves from composition/transposition to exposition/demonstration (Gandleman 1991). In encountering an ekphrastic poem, Al-Joulan (2010) argues that the reader must mentally re-perform the ekphrast's composition by imagining "the sensual, emotional, intuitive, and intellectual aspects of the things communicated" (Al-Joulan 2010, 39).

The ekphrastic object

Several art theorists and contemporary philosophers have rethought the art object as an event or a performance to be encountered differently each time, rather than as a static object to be analyzed (e.g., Grosz 2008; Zepke 2014; Ziarek 2004). The common thread joining each of these perspectives is an assertion that the art object has an affective force. For Grosz (2008), that affective force comes from the earth: "It emerges when sensation can detach itself and gain autonomy from its creator and its perceiver, when something of the chaos from which it is drawn can breathe and have a life of its own" (Grosz 2008, 7). In a similarly Deleuzian-inspired conceptualization, Zepke (2014) refers to art as abstract machine. It is not a representation, because:

> nothing exists outside of its action, it is what it does and its immanence is always active. In the middle of things the abstract machine is never an end, it's a means, a vector of creation. But despite the abstract machine having no form, it is inseparable from what happens: it is the "non-outside" living vitality of matter.
>
> *(Zepke 2014, 2)*

Perpetually in movement, the ekphrastic object moves in and out of the poet's consciousness throughout the ekphrastic encounter (Kennedy 2012). Ekphrasis relies on this spatial-temporal play of forces, as it is these forces that *compel* the ekphrast's composition (Ziarek 2004). Marion (2002) refers to this phenomenon as "summoning." By this he means that artworks, as saturated phenomena, impose themselves on us, that is, they "show" themselves (Mackinlay 2010). They do this by revealing what is visible but also what is not visible. The invisible is the phenomenon that wells up inside us, affecting us and giving itself to us, such that the invisible appears to us in "the event of the visible happening" (Marion 2002, 73). This is what defines a saturated phenomenon—its capacity to give of itself again and again in excess of interpretation. As Marion states, "The painting continues to be seen in a fitting way as long as a look exposes itself to it as to an event that wells up anew—it continues to be seen as long as one tolerates that it happens as an event" (Marion, 2002, 72). When a painting is regarded with an analytic eye that attempts to classify or bracket it in some way, it does not itself appear, because "to see is to receive, since to appear is to give [oneself] to be seen" (Marion 2004, 80). In this sense, the ekphrastic object is not an object at all: It is not the artwork that the ekphrast responds to, rather what the artwork *does*.

Doing fieldwork differently

Dewsbury (2003) asserts that non-representational research is a call to witness the already-witnessing world. Doing so means being open to the "eventhood" of the world—the world we know without thinking; the world we can only know because we are already a part of, not apart from, it; the world we come to know in an embodied and affective sense if we allow ourselves to remain open to experiencing it as an event. As such, non-representational research is an attempt to encounter the world, just as the ekphrast encounters the ekphrastic object. Just as it is for the ekphrast, a researcher of the "non-representational" is compelled to witness the world in ways that allow it to reveal itself.

With this in mind, I undertook two years of fieldwork during which I engaged with sites and practices of therapeutic art making in the city of Melbourne, Australia. Using my own visual art practice as a starting point, I approached other artists for whom art making was a therapeutic practice. This included: Amanda Robins, a visual artist who engages in "slow art"; Swagata Bapat, a trained Indian dancer and practitioner of 5Rhythms™ dance therapy; Artist as Family, an artist collective who practices poetic permaculture; Lucy Sparrow, a fiber artist; and three practitioners of drain art (graffiti in storm water tunnels). Toward the end of this time period I was invited to join the General Assembly of Interested Parties, an artist collective that engages in sonic improvisation and liminal theater in public space, a practice which is more than therapeutic for most of its practitioners.

The intention of the research was to deliberately subvert what might otherwise be considered an ethnographic method. I did not, at any point, question my fellow artists about their practice or take fieldwork notes. As much as possible, I fully immersed myself in each practice, dwelling in the atmosphere of these fieldwork encounters (see Anderson and Ash 2015), practicing with or alongside other practitioners, and collecting both "sensory data" (video, audio, and stills) and embodied knowledge. In doing so, I selectively attuned to materials, movement, sensations, rhythms, vibrations, affects, transversal relations between actors and actants (human and non-human), and the spaces created in, around, between, and through them. I did so non-reflexively (see Lynch 2000), attending instead to the felt intensities of the fieldwork. These intensities are (still) held in the body as sensations and in memory as affects. As McCormack suggests, "Thinking in terms of affect thus holds on to the notion of a field of sensible experience without placing limits around that field by identifying it through discursive categories as a personally captured emotional state" (2003, 495). In practice, this meant eschewing "I" as a subject in order to foreground the bodily and affective intensities of therapeutic art making as a *practice*.

It was this "reservoir" of fieldwork experiences that later became the foundation for a series of arts-based explorations of the therapeutic-ness of art making, which in turn led to the production of a body of ekphrastic artworks (visual, sonic, and poetic), which in turn generated practice-based insights that were later published. This process of thinking through practice is described in full detail in a short monograph (Boyd 2017a); however, mostly due to practical constraints, the poetic accounts of fieldwork were omitted from the publication. Instead, I elected to publish the ekphrastic research poetry separately as a collection (Boyd 2015) and to write about it academically for the journal *ACME* (Boyd 2017b). In an attempt to elevate the processes involved in practice-based research, I did not reproduce the artworks I created in the monograph, and only 3 out of 28 poems were republished in the *ACME* paper. In a similar vein, I present just two conjoined poems in this of chapter (see Figure 10.1).[2]

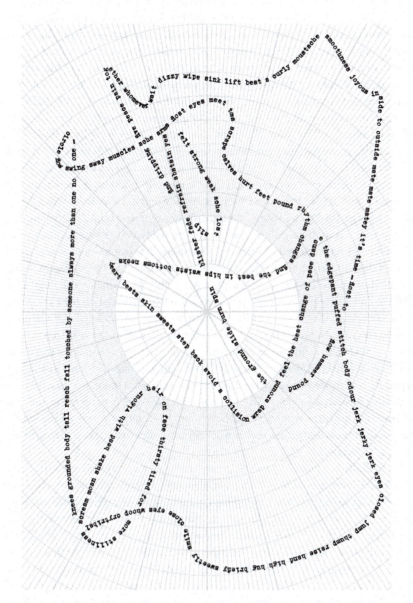

FIGURE 10.1   Two conjoined poems that re-present the "eventfulness" of a 5Rhythms™ dance class. Copyright Candice Boyd.

Ars poetica: forces, flows, and vital materialisms

Ultimately, I undertook ekphrastic poetry as an artistic experiment in affective knowledge production. As Lorimer (2015) argues, a non-representational style of writing is estranged from ethnography: "It is a heightened awareness, for seeing and feeling . . . an attunement to the shape-shifting quality of sense and the properties of *distributed* agencies" (emphasis added; Lorimer 2015, 186). I, therefore, strove in my research to create a poetic account of fieldwork centered on the materialities and immaterialities of therapeutic art practices as they had assembled and dispersed, spatially and temporally. People were present in these accounts, unnamed—their subjectivities co-mingled with sensations, performances, emotions, and the agency of things. As Dewsbury (2003) suggests, "the immaterial is something—only it cannot be imposed (known), it can only be exposed (felt)" (Dewsbury 2003, 1923).

I have very little poetic training, so in formulating an approach I was guided by several bodies of literature. The first comprised the philosophical underpinnings of the research project: process-oriented philosophy, Merleau Ponty and post-phenomenology, object-oriented ontology, and new materialisms. These philosophies demanded an approach that embraced notions of presentational immediacy and pre-reflexive experiences of subjectivity (see Whitehead 1978; Merleau-Ponty 1962). The second comprised non-representational theory in that my approach was motivated by a desire to enliven fieldwork experiences that were embodied and remembered (Thrift and Dewsbury 2000). The third was a technique of postmodern, avant-garde poetics known as *parataxis*. In this regard, I was particularly inspired by the work of poet Matt Hill. In a single anthology, Hill (2008) assembled 79 prose poems. As readers learn from material on the back cover of the collection, the work is Hill's "attempt to map out some of the territory between the mundanely surreal and the ordinarily strange" in what he refers to as a hard-edged life. In doing so, he embraced a dual definition of parataxis: Its formal definition as a style of writing without coordinating connectives and its implied definition as a "mode of experience." Parataxis is generated spontaneously; it is not planned or formulated prior to its emergence on the page. It occurred to me that this spontaneous form of generating poetry might facilitate a type of ekphrastic responding to fieldwork experiences that remained attentive to their forces, flows, and vital materialisms.

Aesthetically, I wanted to produce poetry with a pictorial as well as a verbal element. This is perhaps because I am fundamentally a visual artist even when I am working with sound, and it is also perhaps because of the Western world's (and the discipline's!) appetite for visual knowledge. Known as visual (or concrete) poetry, this form of poetry combines a visual composition with a textual one (Bohn 2010). Both need to have meaning. In combining parataxis with visual poetry, I first created a text line using graphic design software that would "house" the poem. Within the program, the text line could be stretched, curved, bent, or made to overlap with itself. It was also possible to create bounded shapes and write within them. I used both forms of visualization in the anthology and mixed these with disjointed, horizontal arrangements of words on lines. In each case, the choice of visualization reflected something of the practice I was writing about—the attributes of the site or the space or something about the path of the movements that were traced by the practices as they were performed. For instance, the two conjoined poems that described the experience of participating in a 5Rhythms™ dance class (presented in Figure 10.1) traced the outline of my movements over the course of a class, weaving in and out between approximately 70 other bodies that were also moving through the space. Another three conjoined poems reenacted three separate events of fiber art making on a text path that recreated the shape and the movements involved in stitching and knitting, both alone and in assembly (see Boyd 2017b).

Poetry is praxis that produces sensations. As Clay notes, "The poem is not composed, first and foremost with signifiers and significations, but (with due recognition that these are produced through language) with rhythms, sounds, images, feelings, and perceptions" (2010, 37). Writing on a text path is a curious way of doing this, as the writing starts in the middle and "grows" outwards along the text path until it reaches each end of it. In the case of the poems on 5Rhythms™ dancing, I created two text paths and joined the ends of each one together. This meant that as each poem came to an end, the two ends of the text line joined up with each other. With text paths prepared, I would intermittently close my eyes in order to conjure up fieldwork memories as images in my mind but also as sensations still present in my body. Creating each poem then became a dialogue between fieldwork memories as ekphrastic object and the composition as it unfolded in a kind of reciprocal process of attending and composing, feeling and remembering, thinking and sensing.

Having chosen the aesthetic (visual poetry) and form (parataxis), I generated 28 poems that told something of the materialities and immaterialities of therapeutic art making, and which, in turn, enabled me to theorize these accounts with respect to non-representational theory and process-oriented ontologies (Boyd 2017a). As with all practice-based research, practice knowledge is articulated in exegetical form but is also contained in the artwork to be exhibited, audienced, experienced, and engaged with by interested readers, viewers, and listeners (Barrett 2004). As with geopoetics more broadly, it demands to be read.

Closing remarks

Ekphrastic objects are not static; they are events or performances to be encountered. It is on this basis that I have attempted to illustrate how practices encountered through non-representational approaches to geographical fieldwork act as ekphrastic objects that can be responded to poetically by "researcher as ekphrast." As Magrane (2015) suggests, geopoetics as a process of thinking and creating new knowledge has validity whether or not the poetry itself is ever presented or judged to be "worthy." What is important in a research context is the fruitfulness of the method in generating insights into the phenomenon under investigation. In my case, the use of ekphrastic poetry as method helped me to reformulate "the therapeutic" in ways I would not have envisioned otherwise.

Notes

1 Non-representational theory originated at the University of Bristol in the 1990s with the work of Nigel Thrift. The work was later elaborated by several of his students. While a full explication of the theory is beyond the scope of this article, a key text is Thrift (2008) and an excellent introduction is available in Anderson and Harrison (2010).
2 Interested readers can obtain a copy from me directly or via The National Library of Australia. Interested listeners can access a poetry reading here: https://soundcloud.com/dr-candice-boyd/visual-poetry-reading.

References

Al-Joulan, Nayef Ali. 2010. "Ekphrasis Revisited: The Mental Underpinnings of Literary Pictorialism." *Studies in Literature and Language* 1: 39–54.

Anderson, Ben, and James Ash. 2015. "Atmospheric Methods." In *Non-Representational Methodologies: Re-Envisioning Research*, edited by Phillip Vannini, 34–51. New York: Routledge.

Anderson, Ben, and Paul Harrison. 2010. "The Promise of Non-Representational Theories." In *Taking Place: Non-Representational Theories and Geography*, edited by Ben Anderson and Paul Harrison, 1–34. Farnham: Ashgate.

Barrett, Estelle. 2004. "What Does It Meme? The Exegesis as Valorisation and Validation of Creative Arts Research." *Text* 3, Special Issue: 1–7.

Bohn, Willard. 2010. *Reading Visual Poetry*. Madison: Fairleigh Dickinson University Press.

Boyd, Candice P. 2015. *Forces, Flows, and Vital Materialisms in Therapeutic Art Making*. Melbourne: Author.

Boyd, Candice P. 2017a. *Non-Representational Geographies of Therapeutic Art Making: Thinking Through Practice*. London: Palgrave Macmillan.

Boyd, Candice P. 2017b. "Research Poetry and the Non-Representational." *ACME: An International Journal for Critical Geographies* 16: 210–223.

Bruhn, Siglind. 2000. *Musical Ekphrasis: Composers Responding to Poetry and Painting*. Hillsdale: Pendragon.

Clay, Jon. 2010. *Sensation, Contemporary Poetry and Deleuze: Transformative Intensities*. London: Continuum.

Clüver, Claus. 1997. "Ekphrasis Reconsidered: On Verbal Representations of Non-Verbal Texts." In *Interart Poetics: Essays on the Interrelations of the Arts and Media*, edited by Ulla Britta Lagerroth, Hans Lund, and Erik Hedlin, 19–34. Amsterdam: Rodopi.

Dewsbury, John-David. 2003. "Witnessing Space: 'Knowing Without Contemplation'." *Environment and Planning A* 35: 1907–1932.

Ehrenworth, Mary. 2003. *Looking to Write: Students Writing Through the Visual Arts*. Portsmouth: Heinemann.

Gandleman, Claude. 1991. *Reading Pictures, Viewing Texts*. Bloomington: Indiana University Press.

Grosz, Elizabeth. 2008. *Chaos, Territory, Art: Deleuze and the Framing of the Earth*. New York: Columbia University Press.

Hill, Matt. 2008. *Parataxis*. New York: BlazeVOX.

Kennedy, David. 2012. *The Ekphrastic Encounter in Contemporary British Poetry and Elsewhere*. Farnham: Ashgate.

Lorimer, Hayden. 2015. "Afterword: Non-Representational Theory and Me Too." In *Non-Representational Methodologies: Re-Envisioning Research*, edited by Phillip Vannini, 178–187. New York: Routledge.

Lund, Hans. 1982. *Text as Picture: Studies in the Literary Transformation of Pictures*. Lewiston: Edwin Mellen.

Lynch, Michael. 2000. "Against Reflexivity as An Academic Virtue and Source of Privileged Knowledge." *Theory, Culture, Society* 17: 26–56.

Mackinlay, Shane. 2010. *Interpreting Excess: Jean-Luc Marion, Saturated Phenomena, and Hermeneutics*. New York: Fordham University Press.

Magrane, Eric. 2015. "Situating Geopoetics." *GeoHumanities* 1: 86–102.

Marion, Jean-Luc. 2002. *Being Given: Toward a Phenomenology of Givenness*. Translated by Jeffrey Kosky. Stanford: Stanford University Press.

Marion, Jean-Luc. 2004. *The Crossing of the Visible*. Translated by James Smith. Stanford: Stanford University Press.

McCormack, Derek P. 2003. "An Event of Geographical Ethics in Spaces of Affect." *Transactions of the Institute of British Geographers* 28: 488–507.

Merleau-Ponty, Maurice. 1962. *Phenomenology of Perception*. Translated by Colin Smith. London: Routledge.

Moorman, Honor. 2006. "Backing into Ekphrasis: Reading and Writing Poetry about Visual Art." *English Journal* 96: 46–53.

OED (*Oxford English Dictionary*) n.d. s.v. "Ekphrasis." Accessed May 2, 2018. https:// en.oxforddictionaries.com/definition/ekphrasis.

Thrift, Nigel. 2008. *Non-Representational Theory: Space, Politics, Affect*. London: Routledge.

Thrift, Nigel, and Dewsbury, John-David. 2000. "Dead Geographies—and How to Make Them Live." *Environment and Planning D: Society and Space* 18: 411–432.

Whitehead, Alfred North. 1978. *Process and Reality (Corrected Edition)*. New York: The Free Press.

Zepke, Stephen. 2014. *Art as Abstract Machine: Ontology and Aesthetics in Deleuze and Guattari*. New York: Routledge.

Ziarek, Krzysztof. 2004. *The Force of Art*. Stanford: Stanford University Press.

11

THE TOPOPOETICS OF DWELLING AS PRESERVATION IN LORINE NIEDECKER'S *PAEAN TO PLACE*

Tim Cresswell

I think I was traveling. I think I was in a bookstore in New York City. It might have been St. Mark's bookshop in Manhattan, but that store may have closed by then. Nevertheless, I am fairly sure I was in New York when I came across the plain, cream Wave Books edition of a poem called *Lake Superior* by Lorine Niedecker. It was newly published and on a display table toward the rear of the shop. I picked it up and fell in love with the condensed spare poetic of the poem, with its references to geology, to travel, and to the settler colonial history of the upper Midwest of the United States. I was equally smitten with the inclusion of a hundred pages of notes on the poet's travels and research on the area: The hundred pages of notes had been used to produce a few hundred words of poetry.

Once I had read *Lake Superior*, I could not get enough of Niedecker's work. Inevitably, perhaps, I discovered her poem *Paean to Place* in her *Collected Works* (2004a). As with *Lake Superior*, I was dazzled by Niedecker's ability to use short lines, few words, and lots of white space to chart a form of dwelling—of being in place. As I learned more about the poet, I became interested in how a poet who lived almost her whole life in a particular place—a place on few people's mental maps—was able to capture something of what being in place meant to her. I became interested in how the form of her poems—her poems *as* places—reflected both the places to which her poems referred and the more general quality of dwelling in place.

I have never lived anywhere for more than seven years. One question I am frequently asked in relation to my work as a poet and as a geographer is if my life as the child of an air force family and subsequent travels as an adult have influenced the content of my writing. The answer must be "yes."

EMERGENCY

There were moments, sure, like the time we caught
the hedgehog in the hole we dug at the bottom of the garden
covered with twigs like the lion-traps in *Tarzan*.

But mostly it was lazy loops on bikes around the agglomerate
playground with its plastic swings and broken glass.
No 'peat bogs.' No 'fireweed' or 'lemongrass'—

trees were all council saplings, wire-sheathed
to keep vandals out. The soil mostly rocks.
The opposite of 'boreal.' I never saw a badger or a fox

but perched at the end of the runway at the air force base
I watched planes arrive and leave. Before texting or tweets
we reached for torches under our sheets,

pulled curtains back and flashed across the cul de sac—
Dot. Dot. Dot.   Dash. Dash. Dash.   Dot. Dot. Dot.[1]

In addition to directly reflecting on places I have lived, I think much of my poetic practice reflects a lack of connection, given I am only tangentially in place, or in a place where I do not, and never will, belong. The poems in my first collection, *Soil* (2013), center on themes of displacement and misplacement: Flowers make homes on the polluted soil of abandoned lead mines, mountain ash (rowan) are planted in towns, a fox climbs a skyscraper, and the narrator finds himself in a large bourgeois house of an Oxford academic, far removed from the air force homes of his childhood. The central sequence of the collection, called "Soil," attempts to rework the substance most often associated with reactionary and Romantic notions of dwelling into a metaphor for constant mobility.

SOIL 9

the material in the B horizon usually
derives from
the C horizon and is thus local

translocation
      of materials from the A horizon occurs
as worms and other organisms
            move deeper

B horizons
    have subscripts which indicate what materials
have moved and mingled
Bh   for instance   indicates
                    the presence of humus—rich organic matter[2]

My initial and continuing attraction to Niedecker's poetry is rooted, I think, in an approach to dwelling in place that mobilizes elements of the landscape both living and inert and does so through careful attention to shape and space. But while my fixation has been on troubling notions of dwelling, rootedness, and identity through reflection on my own displacement, Niedecker examines rootedness and dwelling through a constant return to what Iris Marion Young, in a feminist take on dwelling, has called the work of "preservation" (Young 1997). Preservation describes the constant work of maintenance that is necessary for dwelling to occur. It involves continual repetitive care. In more everyday terms, it is housework. Preservation is a gendered set of activities that has most often been associated with women in patriarchal societies. Drawing on Heidegger, Young contrasts preservation with the act of building—an act that has been given more heroic and masculine connotations. The constant and reiterative act of preservation seems particularly apt for Niedecker's poetry and is reflected in her biography.

Niedecker spent almost her whole life in and around Blackhawk Island in Wisconsin, with just a few years away in New York City and Madison, Wisconsin. The Wisconsin town was not a place on the literary map. There, she worked as a janitor for a spell and looked after a very modest house that was frequently flooded. She had admirers among influential poets associated with objectivism, such as Marianne Moore, Louis Zukofsky, and William Carlos Williams. They, like me, admired her sparse, condensed, style. Outside of this circle, she was mostly unknown. Her two original collections (in 1946 and 1968) were published in the UK. Her work only started to become widely known in 1985 when the poet Cid Corman edited a selection of Niedecker's poems (Niedecker 1985). Following several further editions of her selected poems, a biography, and collections of critical essays, she is now regarded as a major poet within literary circles; nevertheless, she is still little known to the public (Willis 2008; Penberthy 1993; Niedecker 2004a).

In this chapter, I approach *Paean to Place* through the lens of topopoetics. Topopoetics works with the double meaning of topos (as "place" and, in rhetoric, as "proper form") to explore how poems become places at the same time as they point to, or refer to, the varied experiences of place-making and being-in-place. The central concern of topopoetics is not the frequently remarked upon relationship between poetry and "sense of place." Rather, it is the relationship between the spatial construction of the poem on the page and a wider idea of both poems *as* places and poems as attempts to account for a general sense of being in (or out of) place (Cresswell 2017). A topopoetic approach involves connecting the form of the poem (as its own kind of place) with the acts of making places in the world. I focus here on the spatiality of *Paean to Place* and its relationship to the watery and liminal evocation of the work and effort of dwelling in place in the poem. The poem is only tangentially *about* Blackhawk Island, the setting for the poem. In a more direct sense, it *enacts* the fluid and repetitive *work* of living in place and the *effort* of preservation. This chapter, then, connects the spatial form of the poem (itself a kind of place that results from effort) with the theme of being in place that the poem enacts. Such a topopoetic exploration necessarily connects the shape of the poem to the spatiality of its contents—in this case, the multi-faceted portrayal of vertical and horizontal motion, the references to *in between* forms of existence, and the plethora of surfaces that the poem utilizes.

Paean to Place declares itself as a poem about "place." It is not a paean to "a" place—but place in general. As a "paean" it is designed to celebrate; it is a song of praise. While the title suggests that celebration is of place-in-general, the poem is simultaneously about a specific place in Wisconsin: Blackhawk Island on the Wisconsin River, where Niedecker was raised and where she lived for most of her life. Blackhawk Island is not actually an island. It is a strip of marshy land between Mud Lake and Rock River near the point where the river enters the larger Lake Koshkonong. It is a space where the boundaries between land and water are constantly changing. Floods were and are a repeated fact of life. "The Brontes had their moors. I have my marshes," Niedecker once wrote to the poet Louis Zukofsky (Penberthy 1993, 4). The poem (like others of Niedecker's) reflects on the narrator's marshy life "by water" both in content and through its sinuous form (Pinard 2008).

Paean to Place is a thin, long poem made up of 41 five-line stanzas. The stanzas are grouped in sets of two, three, or four, with wider spaces breaking the bunches. Left margins are staggered throughout, with the poet using, for the most part, three different margins for lines in the horizontal space of the poem. There is also irregular rhyming throughout. The stanzas do not stand by themselves as complete thoughts or sentences. Within each stanza there is a break between thoughts, and each thought flows on to the next stanza across the dividing space. The breaks in the flow of thought are often abrupt, creating a sense of montage in the poem. While the poet uses capital letters, she does not add punctuation. The form achieves vertical movement in a number of ways: the enjambment; the flow of thoughts between stanzas; and the overall long shape of the 41-stanza poem, with its relatively short lines throughout. The staggered margins enact a varied sense of horizontal movement as the words move away from, and toward, the left-hand side of the page. Pinard notes a similar spatial strategy in another poem by Niedecker—*My Life by Water*—arguing that it can "be read as a fluid map of the speaker's own creation 'by water' as if she has been conceived, indeed parented, by water" (Pinard 2008, 27).

While the reader is most likely to encounter *Paean to Place* running down one page and onto the next (as it appears in edited volumes), Niedecker also had a vision of the poem with each stanza on a separate page. She produced a handwritten version of the poem in an autograph book as a gift and expressed an interest in seeing the poem as a "little book all by itself" (quoted in Young 2003, para. 1).

The poem was eventually printed, in her original handwriting, in 2003. Reading the stanzas one page at a time, with a blank page between each stanza, shifts the reading of the poem into a more horizontal mode, with the action of turning the page embodying and amplifying the significance of the space (and time) between stanzas. The poem remains a watery account of intertwined rivers and lives. The combination of vertical and horizontal movement space in the poem produces, in spatial form, the action of repeated flooding. Water rises and falls and, as it does so, it moves into and off the land. Geographies are reconfigured on a regular basis. Place, here, is liquid and uncertain.

As Elizabeth Robinson writes of *Paean to Place*, "The poem enacts its own geography and geology, demonstrating downward movement through its lineation" (2008, 126). The lack of punctuation, the almost constant enjambment, leads the reader down the page. The poem can also be thought of as a kind of archipelago. Just as islands in an archipelago are separated by water yet linked by trade and ecosystems, each stanza in the poem is spatially separated from the next, yet still simultaneously linked through the logic of sentences. This pattern is consistent throughout the poem and is quickly established at the outset as the form is reflected in the content.

> And the place
>   was water

> Fish
>   fowl
>   flood
>   Water lily mud
> My life

> in the leaves and on water
> My mother and I
>   born
> in swale and swamp and sworn
> to water

The first two lines stand by themselves and immediately set up a paradoxical understanding of dwelling in a watery place: "And the place" suggests a degree of solidity and certainty while "was water" immediately undercuts it. We rarely think of place in terms of water. The watery parts of our world can be thought of as the very antithesis of sites of identity and belonging. Water is also distinct from "territory," with the latter's roots being in the apparent certainties of land—of "terra" (Peters, Steinberg, and Stratford 2018). "Place" and "water" set up an immediate play of both certainty and the absence of certainty, thereby evoking life on land that floods. The first stanza sets the pattern for all that follow. The first four lines are an alliterative list of liminal watery beings and matter—creatures linking land, water, and air. The "water lily mud" manages to suggest a profound mixing of matter. The image is one of mud as water mixed with earth, a lily as a plant that dwells in a muddy watery way: It presents a hybrid form of dwelling that appears to lack the certainty of a properly rooted (arboreal) existence. The stanza shifts in the last line to a new logic that flows over into the beginning of the next stanza—"My life/in the leaves and on water"—not unlike a water lily. In the second stanza, the sense shifts immediately: "in the leaves and on water/My mother and I/born/in swale and swamp and sworn/to water." In each of these stanzas another pattern is established. The indented lines create three left margins throughout the poem. The one exception is the single-word line "born" that is indented one step further than any other line in the poem. As the poet, August Kleinzahler, notes, "The spaces between words and lines, usually emphasized in the typography, lineation, and enjambments, functioned for Niedecker as a reminder of the silence from which the poems emerged, by which they were pervaded, and to which they returned" (2017, 170). In the case of the word "born," the space/silence around it heightens the narrator's own existence as one that emerges out of, and returns to, silence. The close connection between birth and land (as in the term "natal" and its link to "nation") is immediately disrupted by more alliterative neither/nor land/water nouns: "swale" and "swamp" connected in sound—but not in meaning—to "sworn" and again to "water."

Another pattern established in these first few lines is a sense of verticality. The swampy space clearly has a horizontal extent that varies with time: Water redrawing the land daily and seasonally. But there is, simultaneously, a mobile vertical space as water rises and falls. The first stanza's short lines perform a list that moves down sonically to "land" with the flood/mud rhyme. The contents of the list reflect this vertical space: fish (under water), fowl (on the water and in the air), flood (water rising and falling), and water lily mud (at the bottom of water and the foot of the list). Mud is also categorically low as a substance, as is "swamp" later on. Swale and swamp are heavy, weighty, low-lying, sodden elements of the landscape. They also exist in "in-between" states, that is, in states that are neither solid nor liquid. The work done by the nouns in the poem is supported by the constant use of various prepositions that establish spatial relation: "in" the leaves, "on" water, "in" the swamp, "to" water.

Niedecker sets up many of the strategies she uses through *Paean to Place* in these first few lines. The poem is full of creatures that inhabit spaces where one element meets another. Through the remaining stanzas we meet, in order: carp (a bottom-feeder), marsh marigold (a native of wet and marshy places), canvasbacks (a diving duck), sora rail (a waterbird), duck weed (a free-floating plant), and water-bugs (insects that skim the water surface). This collection of flora and fauna are marked by their existence on fluid margins—on surfaces where air meets water or water meets mud. This is further underlined by repeated references to substances that are neither/nor, such as marsh, mud, and fog—combinations of water and air or water and earth.

It is not just fauna and flora of vertical edgelands that gesture toward rising and falling but also various forms of human activity.

> My father
> thru marsh fog
>   sculled down
>   from high ground
> saw her face
>
> at the organ
> bore the weight of lake water
>   and the cold—
> he seined for carp to be sold
> that their daughter
>
> might go high
> on land
>   to learn
>   . . .

Here we are introduced to the narrator's father sculling (a shallow form of surface inhabitation not unlike the water-bugs and other surface creatures of the poem) out of marsh fog (a liminal water/air substance). He is sculling "down / from high ground." This is a strange image, as sculling does not happen on "ground" and, additionally, has to occur on a flat surface. He bears the "weight of lake water" and seins for bottom-feeding carp so that that his daughter (Lorine) "might go high / on land." As the words descend on the page, and as they move from left to right and back again, we are given both literal and metaphorical movement in vertical and horizontal space. There are other ways in which the narrator gestures toward this movement. Later in the poem:

> I mourn her not hearing canvasbacks
> their blast-off rise
>   from the water
>   Not hearing sora
> rails's sweet
>
> spoon-tapped waterglass-
> descending scale-
>   tear-drop-tittle
>   Did she giggle
> as a girl?

Here, the way the words create a space on the page mirrors the "blast-off rise" of the canvasbacks (ducks) from the water and the "descending scale" of the sora rail's call. The descending scale is further mirrored still later in the poem in a "Peewee-glissando," that is, in a sonic movement between high and low that maps onto the contradictory high/low notion of sublime/slime.

> Maples to swing from
> Pewee-glissando
>   sublime
>   slime-
> song

The poem, as Rachel Blau DuPlessis notes, "uses montage as method, suturing disparate materials together, sometimes with syntactic fragments of breaks unfilled by explanation. It makes a pulse of argument out of its gaps. Its organizational strategy is essentially nonnarrative, a discrete series, although it is generally chronological" (2005, 404). This technique creates a doubled spatiality of words linked in and across space. The blank space between stanzas, along with the jagged and irregular blank space created (negatively) by indentation and lineation, also does work by emphasizing the fluid and changing nature of life in a flood zone. Niedecker noted the active role played by blank space (figured as silence) in her poem, *Wintergreen Ridge*.

>   and silence
>   which if intense
>   makes sound

The irregular blank space is the other half of the irregular stanzas. The words and the silence flow into each other like water and land, but it is the space that is more present, that takes up more of the page. Indeed, one of the aspects of Niedecker's poetry that first attracted me that day in a bookstore in Manhattan was her economy. The lines are very short, making the blank space all the more present. In her poem, "Poet's Work" (see the following excerpt), Niedecker referred to the process of putting words on the page as the "condensery," that is, a process of condensing what needs to be said to often extreme brevity. The poem, *Lake Superior*, though long for Niedecker, is the product of hundreds of pages of journal notes on travel, histories of settler colonialism, and, most of all, geology (Niedecker 2013).

While *Paean to Place* begins with the paradoxical juxtaposition of "place" and "water," it ends on the edge:

> They fished in beauty
>   It was not always so
>   In Fishes
> red Mars

rising
rides the sloughs and sluices
    of my mind
        with the persons
on the edge

Between the certainties of place and the fluidity of water there are ever-changing edges. By the time we get to the end/edge of the poem, we have encountered all manner of edges and surfaces—liminal zones where one kind of matter meets, and often mixes with, another.

Niedecker presents us with a particularly fluid and contingent form of dwelling-in-poetry. The poem is clearly about immersion in place. As Rachel Blau DuPlessis notes, "[Niedecker] does not claim to 'sail' to any exotic place but to be within the place she is—the goal is saturation, not transcendence" (DuPlessis 2005, 408). "Saturation" is an apt term for this form of dwelling, with its finely balanced solid/liquid being. Liquid about to become solid or solid about to become liquid. Substances on the edge of being something else. "Dwelling," in the work of Heidegger and elsewhere, is rarely this contingent, this consistently negotiated. Dwelling suggests a certain solidity—the deep soil of Heidegger's famous cabin in the Black Forest. It suggests something deeply rooted and arboreal (Heidegger 1971). There is nothing obviously rooted about the kind of dwelling explored in *Paean to Place*. The flora, as we have seen, are more likely to be duckweed, a species that floats on the surface of the water, than oak trees. This is not thick dark forest soil but "water lily mud" and "slime." It is a kind of unrooted rootedness. The contingency of the images in the poem—and the kind of life the poem relates—is, like so much else in the poem, in a creative tension. The contingent rootlessness contrasts with evident rootedness of a life in one place told through the poem—a life by water, lived in mud, floods, and fog on the ever-changing topography of earth and water.

The contingency of dwelling by water is evident in another way too. Heidegger's account of authentic dwelling is centered on the process of building—of building a place in which *to be*. Niedecker lived in a cabin too. Her cabin, on Blackhawk Island, was a modest house that was, as the poem makes clear and as I mentioned earlier, frequently flooded. It was a cabin that was holding on to a tenuous existence, like a pond weed. Niedecker's relation to the cabin, the place she inhabited for decades, is not established through building but through the necessary work of maintenance and preservation. *Paean to Place* is scattered with references to the difficult labor of existence on Blackhawk Island, to the labor of the narrator, her father, and her mother.

The unending processes of work in an uncertain landscape are clearly gendered. The narrator's father is fishing, hauling. He "bowed his head / to grass as he mowed." The narrator's mother works with him, sometimes crossing the line that divides man's labor and woman's work:

> She helped him string out nets
> for tarring
> And she could shoot

The couple join in the work of maintaining a household and a tenuous livelihood:

> she
>
> who knew how to clean up
> after floods
> he who bailed boats, houses
> Water endows us
> with buckled floors
>
> O my floating life
> Do not save love
> for things
> Throw *things*
> to the flood

The "floating life" is not a life of "things." It is a life that involves repeated cleaning after floods, of bailed-out houses and buckled floors. It is, the poem suggests, a life where the process of labor trumps the achievement of "things."

The iterative labor of housekeeping is a kind of dwelling that reflects the unstable topography in which the labor takes place. The work of caring for buckled floorboards is a result of the mobile margin between solid and liquid, land and water. Water is, itself, doing ceaseless work as it moves in and out of land. Security is always temporary. This is not the monumental harmony of dwelling in the forest soil but something altogether more fragile that requires constant labor.

Iris Marion Young, in her essay "House and Home," provides a revisionary feminist take on Heidegger's concept of *dwelling*. Dwelling, in Heidegger, she argues, consists of building and preservation. Despite this twofold definition, it is building that gets all the attention, associated, as it is, with the work of men. Preservation, on the other hand, is the world of housework and the world of women. Young reflects on the rejection of both "home" and housework in feminist writing and particularly in the work of Simone de Beauvoir, for whom domesticity was a penalty of endless imminence in the face of man's transcendence. Women, Beauvoir argued, remain in imminence while servicing the transcendence of men.

"In the existential framework Beauvoir uses," Young writes, "transcendence is the expression of individual subjectivity. The subject expresses and realizes his individuality through taking on projects—building a house, organizing a strike, writing a book, winning a battle" (Young 1997, 148). Young questions Beauvoir's rejection of both home and the work of preservation suggesting that "while preservation, a typically feminine activity, is traditionally devalued at least in Western conceptions of history and identity, it has critical human value" (135).

The temporality of preservation is distinct from that of construction. As a founding construction, making is a rupture in the continuity of history. But recurrence is the temporality of preservation. Over and over the things must be dusted and cleaned. Over and over the special objects must be arranged after a move. Over and over the dirt from winter snows must be swept away from the temples and statues, the twigs and leaves removed, the winter cracks repaired. The stories must be told and retold to each new generation to keep a living, meaningful history.

(Young 1997, 153)

Young's revaluing of preservation—in this case, through housework—is based on the necessity of continually making and remaking a meaningful world. It is grounded in place-making not as a once and for all achievement of building but as an iterative process of maintenance (see also Pratt 2004).

One word to describe this form of dwelling-as-maintenance is "effort"—a word Niedecker uses as the poem takes a metaphysical turn:

> Effort lay in us
> before religions
>   at pond bottom
>   All things move toward
> the light
>
> except those
> that freely work down
>   to oceans' black depths

Niedecker maps the "effort" of labor onto the vertical movement through space that she has established throughout the poem. It is "effort" that comes before the abstractions of religion. We humans have to work in order to dwell. The fragile nature of the Blackhawk Island cabin with its buckled floors underlines the fluid time-space of Niedecker's marshy world.

There is one further kind of labor in *Paean to Place*: The labor of writing.

> I was the solitary plover
> a pencil
>   for a wing-bone
> From the secret notes
> I must tilt
>
> upon the pressure
> execute and adjust
>   In us sea-air rhythm
> "We live by the urgent wave
> of the verse"

The image of the narrator as a plover is telling. As with other living things in the poem, the plover inhabits and crosses over the elemental margins: earth, water, and air. The plover is a fairly unremarkable wading bird, not a storied species like a nightingale or a skylark. It characteristically makes a bobbing movement as suggested by the tilting, executing, and adjusting in the poem. These also suggest a writer at work "in the urgent wave / of the verse." These lines are a quote from an essay by Robert Duncan in which he both reflects on the ever-changing rhythm of the natural world and links the heart to the tides:

> There have been poets for whom this rise and fall, the mothering swell and ebb, was all. Amoebic intelligences, dwelling in the memorial of tidal voice, they arouse in our awake minds a spell, so that we let our awareness go in the urgent wave of the verse.
>
> *(Duncan 1995, 2)*

The image of the narrator/writer as plover connects the liminal wader to the liminal space of the littoral zone, to the space of "rise and fall," "swell and ebb." The work of writing maps onto the work of water on an energetic earth. Niedecker's writing takes its place beside other forms of labor, such as the tarring of nets and the cleaning of a flooded cabin. In this way, writing as work sits within a set of equivalences that refuses the kind of elevation often associated with the writer as disembodied thinker. In addition to writing, Niedecker spent her life supporting herself through various kinds of manual labor, including working as a janitor in the nearby Fort Atkinson Hospital.

Niedecker certainly figured her writing as work and made that quite clear in the short poem, "Poet's Work," in which she describes learning a "trade" when she sat at her desk in order to "condense" (Niedecker 2004b). For Niedecker, the work of poetry is the work of reducing, or condensing, the particularities of her world to a very spare set of carefully chosen words. Her lines are short, and the white space of the page framing the poems' jagged edges does as much work as the words. The work of producing the poem is simply work, from which there is no "layoff." It is a "trade" that sits beside the other kinds of manual labor that feature in *Paean to Place* and in other poems in her body of work.

The form of dwelling that *Paean to Place* relates and enacts is achieved through both the spatial form of the poem and the semantic content. It is an iterative form of dwelling that centers constant work from which there is "no layoff." It resembles what Iris Marion Young calls "preservation" as much as it does the "building" that Heidegger emphasizes. The poem connects a landscape in flux—the marshy landscape where water and earth exist in an uneasy process of changing vertical and horizontal space—to the constant press of labor. This is an unmappable place, a place that cannot be defined in terms of fixity and boundedness. Inhabiting this landscape is equally fluid in the sense that it is constantly evolving as a result of both human practice and natural processes. The landscape involves constant effort. The poem enacts this place through its archipelagic spatial form of short lines in short stanzas with connecting sentences, through its staggered margins, and the irregular interplay of text and blank space. This form is inhabited by indeterminate substances (mud, swamp, fog) and flora and fauna that exist in and between earth, water, and air (pond weeds, lilies, bottom-feeders, water-bugs). Humans dwell in this place, not through the construction of finished dwellings but through the labor of dwelling that such a landscape necessitates. One form of that labor is the unending "trade" of writing—itself configured as iterative and embodied in the form of a plover; itself an inhabitant of the ever-changing space where elements meet.

Niedecker's relationship to place is very different from mine. She lived almost her whole life in a single, off-the-beaten-track place in rural Wisconsin. I have lived mine around the world in intervals of three to seven years. My life has been lived as the son of an air force family, as a student, as a father and husband, and as a relatively itinerant academic. Hers was lived as a woman—as a daughter and a wife. She spent years looking after a modest wooden cabin that was frequently flooded. She worked as a janitor in a hospital. I have certainly done housework, but nothing that comes close to that of Niedecker's experience. What I hope to have shown is how a topo-poetic reading of *Paean to Place* reveals the artful connection between the creation of a poem as a spatial arrangement on a page, the semantic content of the poem as a praise song for being-in-place, and the theme of preservation—that is, the effort of place-making (and poem-as-place-making)—that pervades the poem.

Notes

1 Tim Cresswell, "Emergency." Previously published in *Front Porch Review,* January 2015.
2 Tim Cresswell "Soil 9" from *Soil*. London: Penned in the Margins, 2013, 48.

References

Cresswell, Tim. 2017. "Towards *Topopoetics*: Space, Place, and the Poem." In *Place, Space and Hermeneutics*, edited by Bruce Janz, 319–331. Cham, Switzerland: Springer.
Duncan, Robert. 1995. "Towards an Open Universe." In *Robert Duncan: A Selected Prose*, edited by Robert Bertholf, 1–12. New York: New Directions.
DuPlessis, Rachel Blau. 2005. "Lorine Niedecker's Paean to Place and Its Fusion Poetics." *Contemporary Literature* 46 (3): 393–421.
Heidegger, Martin. 1971. "Building Dwelling Thinking." In *Poetry, Language, Thought*, translated by Albert Hofstadter, 247–363. New York: Harper Colophon.
Kleinzahler, August. 2017. *Sallies, Romps, Portraits and Send-Offs: Selected Prose, 2000–2016*. New York: Farrer, Straus and Giroux.
Niedecker, Lorine. 1985. *The Granite Pail: The Selected Poems of Lorine Niedecker*. Edited by Cid Corman. San Francisco: North Point Press.
———. 2003. *Paean to Place*. Kenosha, WI: Light and Dust.
———. 2004a. *Collected Works*. Edited by Jenny Penberthy. Iowa City: University of Iowa Press.
———. 2004b. "Poet's Work." In *Collected Works*, edited by Jenny Penberthy. Berkeley, CA: University of California Press.
———. 2013. *Lake Superior*. Seattle WA: Wave Books.

Penberthy, Jenny. 1993. *Niedecker and the Correspondence with Zukofsky 1931–1970*. Cambridge: Cambridge University Press.

Peters, Kimberly, Philip Steinberg, and Elaine Stratford, eds. 2018. *Territory Beyond Terra*. London: Rowman & Littlefield.

Pinard, Mary. 2008. "Niedecker's Grammar of Flooding." In *Radical Vernacular: Lorine Niedecker and the Poetics of Place*, edited by Elizabeth Willis, 21–30. Iowa City: University of Iowa Press.

Pratt, Geraldine. 2004. *Working Feminism*. Edinburgh: Edinburgh University Press.

Robinson, Elizabeth. 2008. "Music Becomes Story: Lyric and Narrative Patterning in the Work of Lorine Niedecker." In *Radical Vernacular: Lorine Niedecker and the Poetics of Place*, edited by Elizabeth Willis, 113–130. Iowa City: University of Iowa Press.

Willis, Elizabeth, ed. 2008. *Radical Vernacular: Lorine Niedecker and the Poetics of Place*. Iowa City: University of Iowa Press.

Young, Iris Marion. 1997. "House and Home: Feminist Variations on a Theme." In *Intersecting Voices: Dilemmas of Gender, Political Philosophy, and Policy*, 134–164. Princeton, NJ: Princeton University Press.

Young, Karl. 2003. "Notes and an Appreciation on Lorine Niedecker's Paean to Place." Accessed January 3, 2019. www.thing.net/~grist/ld/ln/ky-ln.htm.

12

POKING HOLES IN THE COLONIAL CANOE

Creative writing as intervention in a 19th-century travel writing narrative

Sophie Anne Edwards

"We started off in swift and gallant style, looking grand and official, with the British flag floating at our stern."

(Anna Jameson 1838, 523)

"Yesterday afternoon there came in a numerous fleet of canoes, thirty or forty at least; and the wind blowing fresh from the west, each with its square blanket sail came scudding over the waters with astonishing velocity; it was a beautiful sight."

(Anna Jameson 1838, 192)

Stroke #1: tracing routes

In the 1840s, a new type of boat emerged in northwestern Lake Huron, at Mnidoo Mnising (Manitoulin Island) and the Straits of Mackinac. It was known as the Mackinaw or "Indian" boat, and integrated design elements from the traditional Algonquin canoe and European sailboats. The invention of this hybrid vessel is generally attributed to the Algonquin of Mnidoo Mnising and the French-Canadian boatbuilder Hyacinth Chenier. The vessel, and its local variations, became the most important craft on the Great Lakes during the fur trade and was used by local Indigenous fishers until the 1940s (Ratcliffe 2009). This hybridity, experimentation, uptake, and remaking of technologies for a specific local expression (alongside colonial uses of the vessel) presents an interesting metaphor for hybrid approaches to writing with, against, and through historical texts and their incumbent narratives.

Anna Brownell Jameson (1794–1860) appeared on the waters of Lake Huron around the same time as the Mackinaw. She traveled by carriage, steamer, bateau, and canoe from Toronto to Mnidoo Mnising in the summer of 1837. She was witness to important moments in the history and making of this particular geopolitical-cultural landscape. Jameson's *Winter Studies and Summer Rambles in Canada* (a later version issued as *Rambles Among the Red Men*) was to become a widely read and authoritative text on nineteenth-century Upper Canada (in her time and ours). Jameson also created 49 drawings and watercolors that documented her journey; these images were never published with the book. Key images—canoe, landscape, "Indian"—were created and circulated as etchings by Jameson, and are frequently published in contemporary histories, travel, and trade publications.

The images, collected into an album, were auctioned off after her death, and are now housed at the Toronto Reference Library. Copies of the etchings are housed at the Royal Ontario Museum and in personal collections. Together, the book and the images have contributed to the geographical imagination of the North Channel. Similar ways of seeing and knowing the land continue to resonate in local discourse and narratives, as well as in contemporary visual imaginaries of the region.

Imagining (her) place

drawing down the lines of
imaginative geographies
sliver streaks in a canoe's wake,
imageandtextimageandtext
ripple, circulate outward in spiraling tendrils with their own trajectories
over place, across time
repeatrepeat

Edwards 2017a. Imagining her place.[1]

Stroke #2: anna-logue—things seen through comparison

In this chapter, I position geographically attenuated visual/textual, poetic/academic experiments as a form of hybrid geopoetics that provides a vessel for me to intervene within and disrupt a particular historical text. I present a series of experimental "strokes" and examples of some of my experiments to intervene in Anna Jameson's narrative, and I muse on and raise some questions about hybrid forms and the possible shoals of this experimental geopoetics. I use "Anna-logue" as a way to shift Jameson's travelogue into an experiment that is continuously changing form, seeing things through comparison, and attending to textual–spatial relationships.[2]

Mei-Po Kwan argues that hybrid geography becomes a location or positionality between binaries (consider the disciplinary divide in geography), and that the "fluid identities it allows may also facilitate the creation of productive connections between these two geographical traditions" (2004, 759). Kwan claims that for a vital geography, we need "geographies and geographers of the third kind: those that cut across the divides between the social-cultural and the spatial-analytical, the qualitative and the quantitative, the critical and the technical, and the social-scientific and the arts-and-humanities" (760).

My geopoetic experiment puts me in a canoe with Anna Brownell Jameson. I create ruffles in the water with my paddle, use a draw stroke to pull us onto another tack, so that we might not gloss over the image, the text, so that we might see the image and text differently and together. We might consider how Chinese-Canadian writer Fred Wah "privileges the apprehensive potential of language on the move at the expense of meaning making . . . in order to unsettle, paradoxically, the intention of language to establish meaning" (2006, 186). He is not so interested "in clarity and closure, but [rather] in openness and unpredict-ability" (186).

Through my evolving experiment to intervene in Jameson's text, I seek a way to disrupt the meanings and authority of the original text. The original text maintains its authority through narrative cohesion, authorial voice, and the logic of temporal and geographical progression. Since I want to question the narrative, I attempt to break down this cohesion, attempt to make the text less established; I break down the text visually; I layer text, allowing words to bump against them-selves. The experimentation permits me to disrupt the authority of Jameson's text, to challenge the narrative progression of the book over time and through place, by placing her own words in different ways—by disordering them and see-ing them differently, both visually and textually. Anna's travelogue becomes an Anna-logue that pays attention to the symbols, the spatial position, the context of her text.

Variable spatial position/things seen through comparison (Sophie Edwards 2017b)

Canada is | not
wilderness   civilization
                                   home and country
                                       | she is

an uprooted tree ()
dying at the core ()
altogether comfortless
waiting, writing   writing
ink freezes
thoughts stagnate
as the ink in her pen

this must not   do
winter studies   /   summer rambles

Stroke #3: poking holes in the colonial canoe

My work speaks to and engages with the continually re/negotiated process of belonging and living as a settler within the territorialized land of Mnidoo Mnising, and the narratives that re/construct this belonging. Hybrid geopoetic experiments allow me to engage with my multiple co-existing and co-articulating positions as a community engaged curator, visual artist, geographer, scholar, woman, and resident/settler within this particular "contact zone," that is, within Mnidoo Mnising in Northeastern Ontario, Canada (Treaty 54 and Wikwemikong Unceded). My artistic and scholarly practices aim to disrupt the "constructed visibility" (Gregory 1998, 82) of representational narratives that inform and form the geographical imagination of Northern Ontario.

Poking holes in the colonial canoe

poke holes in the canoe
loosen the boards
topple the boat
pull myself up from the water
drag my wet jeans over the gunnels as I straddle the canoe
Masta and the other voyageurs leaning to counter my weight
I'd like to gaze into her eyes
slip my hand under her skirt
unsettle her en-cushioned bottom
scratch into her notebook
draw figures into her landscape
would my figures figure differently?
we are not wholly distant
we are the boat topplers
the eaters of history.

Edwards 2017–18. *Poking holes in the colonial canoe.*

Jameson traveled through northern Lake Huron and the Territories of the Three Fires Confederacy of the Odawa, Ojibwe, and Pottawotomi tribes prior to the formal European settlement and active displacement of the First Nations Peoples. This region, now comprising lands that are subject to a number of Treaties (including Treaties 11, 45, 54, 61, and the Williams Treaty territories), continues to be shaped by the same colonial processes that began in the early-nineteenth century. Jameson did not have a direct role in colonial administration; yet, her connection to key figures—not the least of which was her husband, Robert Jameson, the Attorney General of Upper Canada—and the wide circulation of her travelogue attest to the articulations between formal colonial administration and the informal narratives found in travelogues, books, and images. It is important to trace the colonial narratives that claim Territories and justify colonization, including those found in Anna Jameson's own writing. It is vital to poke holes in these persistent ways of seeing and constructing place.

Jameson dismissed *Winter Studies and Summer Rambles* as "idle" "fragments" written for her "own sex" (1990, 9–10). Despite her gendered self-deprecation, she nevertheless used her position and her writing as vehicles to weigh in on many colonial matters, ranging from the planning of the city of Toronto to the colonial management of Upper Canada—the latter being a region she assessed as an "inexhaustible timber-yard and granary of the mother country" (11). With its range of detailed descriptions and arguments, the book and Jameson's images became documents of "authentic" firsthand witness to the state of the emerging nation.

Winter Studies and Summer Rambles reproduced ideas about wilderness, "Nature," "empty lands," the "dying Indian," and other narratives central to British colonial rule. Excerpts from *Winter Studies and Summer Rambles* and Jameson's images are used in contemporary travel and trade books not only to document how things *were*, but also to document how things *are* today. Like Jameson, artists (visitors and residents of northeastern Ontario and Mnidoo Mnising) continue to create watercolor paintings and drawings of "Indians" in "costume," as well as paintings of "empty" landscapes that erase the violent history of colonization and the people of these traditional Anishinaabeg territories. The following experiment is an example of this rearticulation of Romantic, colonial sentiments: Jameson's text arranged on the left-hand side, and the uptake of her text by a contemporary writer, Margaret Derry, on the right.

1838/2007 (Sophie Edwards, 2018a)

It is for his interest, and for his
worldly advantage, that the rednature in her freshness and innocence
man should be removed out of his
[European's] way, and be thrust
back from the extending limits of
civilization—even like these forests
which fall before us, and vanishdefiled at once, and sanctified by contact
from the earth leaving for a while
some decaying stumps and rootsexpresses many people have felt—and still feel—
over which the plough goes in time,
and no vestige remains to say that
here they have been. Jameson 1839, Derry 2007

Stroke #3: visual/textual interventions

Over time, a discursive language shift occurred alongside a shift in Territorial sovereignty. Prior to 1837, "tributes" acknowledged the contributions of the Anishinaabeg (and other Nations) to Britain during the War of 1812, as allies in a nation-to-nation relationship.

When Anna Jameson arrived at Manitowaning on Mnidoo Mnising in the summer of 1837, she wrote about the annual "gift-giving." The language change from tribute to gift-giving and the location of the gift-giving were symbolic. A year earlier, the Bond Head Treaty designated Mnidoo Mnising as a Territory for the Anishinaabeg, with the proviso that the Anishinaabeg invite and encourage settlement of other tribes on the Island. This "voluntary" displacement would open more lands for British settlers in "abandoned" areas. Jameson concluded that the "intentions of the government are benevolent and *justifiable*" (497). She pronounced that "the Indians know neither sovereignty nor nobility" (463)—a narrative that justified displacement.

After 1854, the "gift-giving" was discontinued. In 1862 the signing of the Manitoulin Treaty (accompanied by cannon fire and threats) revoked the Bond Head Treaty. The government claimed that the "Indians" had neither occupied nor farmed the Island to the extent expected of the earlier Bond Head Treaty. Small reserves were carved out of the prior Territory. The language shift continued. Within contemporary discourse, "handouts" is a derogatory term used to describe the financial contributions to Indigenous peoples' health care, housing, and education.

Animating the Archive was an archival intervention project I curated in 2014 through funding from the Ontario Arts Council. The project aimed to trace links between displaced Scots that had settled on Mnidoo Mnising and the related displacement of local Anishinaabeg. The project artists included Michael Belmore (Anishinaabe), Amanda Thomson (Scottish artist-geographer), Elizabeth Reeder (American writer living in Scotland), Heather Thoma, and me (the final two based on Mnidoo Mnising).

My interest in language and geography led me to link Treaty maps and the legacy of language. Through personal communication with Alan Corbiere (Ojibwe historian M'Chigeeng First Nation on Mnidoo Mnising and PhD Candidate at York University), I learned of his tracing of the shift from tribute to gift-giving. I responded to this tracing and the changing Territories through the creation of a series of visual/textual images. These images presented a way to show the changing discourse and related Territorial losses of the Anishinaabeg. The series consists of three two-layered panels that were hung in a row in the museum exhibition space. Each two-layer panel consisted of a base map drawn in graphite on the bottom layer. Each base map showed the diminishing Territory of the Island Anishinaabeg with each subsequent Treaty. Each map was covered by a word written on a sheet of vellum: tribute, gift, handout. The images were exhibited in the final exhibition, which I curated in 2014. The viewer was invited to place a hand on the vellum and press the word onto the lower paper to reveal the map. The hand touched the word, and imposed the word on the changing Territorial landscape.

These images don't replace or "stand in" for the complexity and history behind the idea; rather, they enable a visual (and tactile, when experienced as art objects) way of communicating, or signaling, a complex spatial relationship between language and territory. The tactility of the pieces also helps visitors think about how our bodies claim space, as well as the hand-to-hand exchange of the tribute/gift/handout.

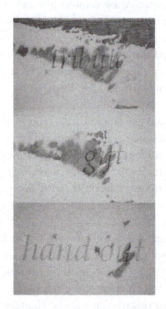

FIGURE 12.1 Edwards, S. 2014, *550 × 200 chain lengths*. Text and drawing on paper and vellum. 55 × 20" each.

Source: Photographs by Sophie Edwards.[3]

Stroke #4: the page as visual space

The *chain lengths* works may not be considered poems. They do, however, speak to language and the relationship of words to geography and territory, and they allow me to experiment with an intervention in received language. In the context of where I live, exhibiting the images in a gallery space in a settler archive allowed me to visually demonstrate the history of a particular way of understanding and claiming place. As an installation, the work became an intervention in a settler archive within the claimed Territory drawn on the maps. The works do not function in the same way as printed images, a point which raises interesting questions about the relationship between the body, poetics, and the land, and the page as a visual space, and how geopoetics function differently depending on the context of its publication and display, and what each medium and context allows.

> Re-structure
> horizontal (disallowed)
> format of requirements
> reduces, forms
>
> the personal
> renders the word
> acceptable (a woman's book)
>
> the personal
> renders the work
> unacceptable (a woman's book)
>
> double-sided (disallowed)
> assemblage disassembles meaning
> of linear/pre forms
>
> materialist analysis
> dematerializes in layers
> of my settler longings
>
> Edwards 2018b. *The page.*

Visual experiments, such as the *550 x 200 chain lengths* series, invite us to understand the page as a visual space, a potentially productive consideration. Consideration of the page as a visual space leads to interesting questions about publication constraints and their influence on the written word. If I present these pages in a horizontal orientation to reference the album of Jameson's images, does it affect meaning? Would a different orientation or size invite the reader to consider the materiality of the page? What can be done or shown differently across a spread? Exhibiting spreads, multiple pages, and reading side-by-side permit one narrative to be read against the other in a nonlinear, non-hierarchical way, and allow two (or more) ideas to be visually read with/against each other. The hypertext may hold much more possibility for experiments with the page, but it simultaneously restricts the referential and tactile reading that the physical page invites. As I work on my geopoetic examination of *Winter Studies and Summer Rambles*, I consider what arrangement of the page, what images and layerings, might speak to the geographies of the book, to the territories of the places about which I am writing, and how the visuality of the words might speak to, and make uncomfortable, Jameson's and my geopoetics.

Stroke #5: creative writing as research

I have found that archival research is as much a logical, methodical process as an imaginative one. My mind wanders and wonders. The questions that lead me to a particular archive or along a particular tack often come from thinking creatively about the subject. Sometimes freeform, creative writing allows me to consider aspects of Jameson's writing that are not drawn out in her book. For instance, the voyageur paddlers. Who were they? What might they have thought about their own experience? What was their encounter with Jameson?

> Jameson has taken a skull from the cave and has tucked it
> into her travel bag. Each night we sleep, the four of us in
> the tent, with her and the skull. I am tired of travelling
> with these messages. With these government men.
> Assisting their wives over the gunnels. She is heavy-set,
> this one. Marie's warm white arm—I would nip into her
> now, if I could be by her side in the light of the kitchen
> stove. I stoke up the smouldering fire in the creeping
> shadow before retiring. There are shadows in the
> darkness. I've never liked this island. La Cloche Island is
> better, larger, open, close to the mainland. Some have
> stayed back. Maybe a Mackinaw boat next spring.
> Fishing—there is work to be had in that. Rumour is, the
> main island will be opened up. Land to be had too. Fold
> myself in the land, my own rotting skull. Who will dig
> me up, or set me out to dry on a cold cave floor. Ah, just
> sing a song over my body. Pierrot keeps good time.
> Keeping them warm, these visiting travellers, paddling
> faster to push past fogs of mosquitoes, slower to
> command a view. Jameson never mentions her husband,
> except when she has a command, but her coins work just

as well. McMurray's said nothing, but they all like a
trinket, don't they? Trinket, blanket, pipe, skull. I've not
seen the skull this last night. What is that sound? Marie
sings well too. The child must be much grown in these
few months. Maybe it won't join the other small bodies in
the cold hard ground. There is crackling. Would she leave
the dead and rotting babies, and come north with me? My
legs are burning, I must not scratch, this plant is an
agony. Since the first time, I seem to get this rash each
voyage. Jameson turns on the rock. Sighs. There is
flashing outside the tent. The rock is warming. The
crackling is louder. Outside the moss is burning; the fire
has jumped. I stamp it out. Sparks have caught in the tree
above. The shadows are shifting and travelling. I am tired
of travelling.

(Edwards 2018c, *Untitled*.)

The stories within the previous passage are not "true"; instead, they riff on the voice of Masta, one of the voyageurs on Jameson's passage from Sault Ste. Marie to Manitowaning. This method allows me to play with reports that Jameson removed a skull from a burial site. As an imaginative act, I can consider possibilities and wonder and respond to questions about Jameson and the people she traveled with and encountered. I can think about their choices, directions, and possible encounters. I can include voices that are missing or rendered invisible in Jameson's text. By writing creatively, I can integrate multiple stories—including those of the writer as researcher and the writer as human person, whose daily activities I can chronicle alongside the stories of those people and ideas who are the focus of my study. In the case of the work I present here, my creative writing became a form of imaginative research that led me back to the archives to seek out letters from the voyageurs with whom Jameson traveled.

Stroke #6: the hybrid canoe—productive possibilities and challenges

As the previous sections have indicated, I work with and between at least two genres. To expand on my canoe metaphor, this mode of practice resonates with the experience of an inexpert paddler whose newness to the activity causes her to waste energy: She does not know how to paddle straight and true; she constantly shifts her paddle back and forth across the canoe. The related French expression sums up inexpert modes of practice quite nicely: "Qui s'assoit entre deux chaises se retrouve sur le cul," which means if you sit between two chairs, you will eventually fall on your backside. Or, as many more experienced paddlers know, the chance of toppling the canoe is at its greatest if one stands with one foot on shore and the other in the boat. One must have one's weight firmly in one place, and one's center of gravity preferably low in the canoe. Masterful paddlers, like masterful writers, can shift around in a boat, move from one canoe to another, shift packs from shore to boat while standing in it, and even stand gondola style to look out for shoals. They are—like strong writers—seasoned feelers of their craft. Their bodies adjust and shift in subtle ways. They know their craft, their paddles, their bodies, how to read the water, and how to navigate to a destination.

I am aware that the unresolved, experimental nature of my work poses multiple challenges: to my process, to legibility, to the making of a clear and concise argument. I am both an emerging scholar and an emerging experimenter. I am shifting back and forth; I just may topple my boat.

In my geopoetic experiment, I do not fully know my destination, the direction I am headed, or the techniques I will need to arrive. I am learning as I go. It could make for a long trip. I might not be "successful": And therein lies the problem. In literary contexts, the writing must grip the reader somehow. The reader must feel the truth of the text (sometimes a sentence feels true when it cannot actually be parsed in a way that can logically demonstrate this truth). By contrast, academics must find a formal method to both weigh and defend their arguments.

It is difficult at the best of times to write a good academic argument. When writers choose to add creative experiments into the form, structure, language, and style of the text itself, they are faced with quite a challenge. Experiments, hybrid forms, and creative writing in academic work "might appear less attractive to many, or even simply not an option at all for those more precarious scholars" (Hawkins 2018, 8). If I am not an expert paddler, I might crash upon the rocks, topple the boat, and rend a hole in the hull as I attempt to navigate some tricky waters. It might serve me well to stay in the shallows or to paddle along the north shore out of the wind. These strategies would be easier and more straightforward.

Yet, I persist in this voyage, in the building and navigation of my craft. I feel that as an academic and creative writer, I must take risks and experiment. As an artist/creative writer, I have struggled with the linear, structured writing/reading of my academic work. And, as an academic, I seek the analysis that undercuts problematic representational and Romantic artistic practices. So, despite the risks, I push away from shore in order to move between genres. The process might be just as valuable as any particular destination. This hybrid approach raises questions about the process, structures, and demands of academic writing and its relations to colonial epistemologies: *How* I write may be just as important as *what* I analyze. Traditional academic writing positions the argument, presented in a linear structure, as the dominant, accepted form of knowledge production. However, there are multiple ways of knowing and perceiving, including those that are nonlinear, open-ended, story-based, or poetic. Not knowing is sometimes just as important as making claims. A broken structure can loosen a dominant narrative (and a claim to authority).

Critical geopoetics explores the relationships between language and place, between narrative structures and the structure of history and territory. The

Mackinaw boat thus serves as a productive metaphor for experimentation in form. The vessel integrated technologies and methods from two traditions (Algonquin and European) to create something specifically local and highly navigable. In a similar way, hybrid texts could potentially "transgress and displace boundaries between binary divisions and in so doing produce something ontologically new" (Rose 2000, 364).

I take seriously Fred Wah's notion that critical thinking is an act of exploration and discovery, and thus I posit that a hybrid text might open up historical texts and loosen the cohesive narrative of colonization. It might increase the drag—the resistance—on the current, on forward motion. These hybrid forms create space for my multiple positions as scholar, poet, and resident on Mnidoo Mnising, all of which are framed by my position as a settler within this Territory, and all of which are influenced by a love for the geological, morphological, ecological, and aquatic characteristics of this place that is at the boundary of Southern and Northern Ontario, where the Cambrian Shield slides below the sedimentary rock of the Bruce Peninsula. Through this creative–analytical mode of practice, I aim to complicate—and make uncomfortable—my multiple positions, all of which are connected to complex colonial processes. Hybrid texts have the potential to offer spaces for anti-colonial imaginings. My geopoetic experiment aims to poke holes in the colonial canoe, to disrupt the visual-textual legacy of Anna Jameson's narrative of the North Channel and a colonial way of seeing, knowing, and constructing place.

Notes

1 All poems included in this chapter are unpublished.
2 My thanks to Chris Turnbull for drawing my attention to the creative potential in the use of Anna-logue/analogue.
3 Originally published in the Exhibition Catalogue: Edwards 2014. *Traces*. Kagawong: 4elements Living Arts. Images used with permission by the artist.

References

Derry, Margaret. 2007. *Killarney: Georgian Bay Jewel*. Caledon: Poplar Lane Press.
Edwards, S. 2014. *Chain Lengths*. Manuscript in preparation.
———. 2017a. *Imagining her Place*. Manuscript in preparation.
———. 2017b. *Variable Spatial Position | Things Seen Through Comparison*. Manuscript in preparation.
———. 2017–18. *Poking Holes in the Colonial Canoe*. Manuscript in preparation.
———. 2018a. *1838/2007*. Manuscript in preparation.
———. 2018b. *The Page*. Manuscript in preparation.
———.2018c. *Untitled*. Manuscript in preparation.
Gregory, Derek. 1998. "Power, Knowledge and Geography: The Hettner Lecture in Human Geography." *Geographische Zeitschrift* 86 (2): 70–93.
Hawkins, Harriet. 2018. "Geography's Creative (Re)turn: Towards a Critical Framework." *Progress in Human Geography* 43 (6): 963–984. https://doi.org/10.1177/0309132518804341.

Jameson, Anna Brownell. (1838) 1990. *Winter Studies and Summer Rambles in Canada.* Toronto: The New Canadian Library.

Kwan, Mei-Po. 2004. "Beyond Difference: From Canonical Geography to Hybrid Geographies." *Annals of the Association of American Geographers* 94 (4): 756–763. https://online library.wiley.com/doi/full/10.1111/j.1467-8306.2004.00432.x.

Ratcliffe, John E. 2009. "The Mowat Boat and the Development of Small Watercraft on the Great Lakes." *The Northern Mariner* 19: 193–223. Canadian Nautical Research Society. www.cnrs-scrn.org/northern_mariner/vol19/tnm_19_193-223.pdf.

Rose, Gillian. 2000. "Hybridity." In *The Dictionary of Human Geography*, edited by Ron J. Johnston, Derek Gregory, Geraldine Pratt, and Michael Watts, 364–365. Oxford: Blackwell.

Wah, Fred. 2006. *The Diamond Grill*. Edmonton: NeWest Press.

13

THUKELA POSWAYO'S POETRY OF DWELLING

Emily McGiffin

Geography as a discipline is concerned with relationships between people and places and with the creative means by which people develop and represent those relationships. In keeping with its origins in the industrialized West, geography's modes of representation are predominantly visual; even within the shifting context of a digitizing world, maps, imagery, and texts tend to be two-dimensional media based on the written word. By contrast, for many people around the world, speech remains the primary vehicle for transmitting knowledge and sharing culture and memory (Turin 2012). Predominantly oral cultures recognize the performativity of speech, aware that "language is a mode of action and not simply a countersign of thought"—it is experienced physically, as a corporeal function, and is often recognized as a way of conveying or exercising power over people and things merely by uttering their names (Ong 2002). Oral geopoetics are not only a powerful mode of communicating the nature and history of places and strengthening human connections to them; these art forms are also particularly vulnerable to disruption through changes to landscapes and cultures (Turin 2012).

At the same time, oral poetic forms are both adaptable and enduring, and they continue to hold an important place in countries and communities that have experienced profound and violent disruptions. In South Africa's Eastern Cape, several centuries of upheaval have brought changes to oral poetry, yet the tradition of izibongo, an ancient poetic genre practiced by amaXhosa people, has continued uninterrupted (cf. Kaschula 1993). Known in English as "praise poetry," izibongo is part of a cluster of similar panegyric forms that are practiced throughout much of the region (Finnegan 2012). The English translation derives from its origins as a form of court poetry addressed to chiefs and kings, and to outsiders the poetry appears mainly to extol the virtues and accomplishments of its subjects. While this interpretation may be true in some cases, the most gifted and accomplished poets serve as community historians, political critics, entertainers, and healers. Their nuanced performances, which these days may occur at a variety of events ranging from festivals and meetings to weddings and funerals, combine praise, genealogies, and even sharp criticisms of their subject (Opland 1983, 1998; Kaschula 2002).

The amaXhosa are a diverse group of isiXhosa-speaking peoples—including people of the amaRharhabe, amaGcaleka, amaMpondo, abaThembu, amaBomvana, amaXesibe, and amaMpondomise kingdoms—whose traditional homelands cover much of what is now South Africa's Eastern Cape province. Iimbongi (singular imbongi) is the isiXhosa word for the poets who perform izibongo. Since the arrival of the written word in the amaXhosa homeland, distinctions have arisen between three groupings of these poets: iimbongi nosiba, who create written poetry during periods of quiet or solitary reflection; iimbongi namanhlange, who blend contemporary elements into their performances, such as rap, hip hop, or slam poetry rhythms; and iimbongi zonthonyama, who compose their oral poems spontaneously during performance delivered in a more or less traditional style (McGiffin 2017).

Izibongo performed in the traditional style are often deeply evocative both of place and of the ancestral presences that continue to inhabit those places, tying people into a lineage and a form of dwelling that extend into past and future. Through the poet's voice, language becomes a healing medium that strengthens connections to land and lineage, often producing or enabling a flow of emotions in listeners. The poetry is active; it performs both spiritual and material work as umoya—the spirit that inspires the poet to speech—moves from the landscape through the body of the poet and into the air, traveling in linguistic form to listening ears. There is an earthly physicality to the poetry—spoken most often in the open air at public gatherings, it is by nature an inclusive art form that welcomes its listeners into the communal experience of cultured and attentive dwelling.

In this chapter, I discuss the work of one of the poets I met during the year I spent living and traveling throughout South Africa's Eastern Cape province, studying the environmental politics of the izibongo genre. Thukela Poswayo is an imbongi originally from the rural areas near the town of Engcobo who now lives in the Eastern Cape city of Mthatha. Like other poets and healers, he accepted his poetic calling with reluctance, following a period of psychic disturbances that made his vocation apparent. Over many years of practice, Poswayo has developed his reputation as a gifted poet and is invited to perform at locations throughout the province.

I traveled to South Africa as a geographer and ecocritical literary scholar interested in questions of colonialism, uneven development, and literature as a form of agency and identity that can help bind people to their environments and communities. When I arrived in the country, I found that the rich vernacular language traditions that have been practiced for centuries are largely absent from literary and geographical scholarship alike. I found this oversight to be of particular concern in environmental and ecocritical scholarship because this literature is intimately tied to histories of dispossession, disrupted environmental relationships, the imposition of foreign cultures and economies, and an ongoing struggle for land and dignity. As I am neither South African nor Black and have only a limited familiarity with the language and culture, I clearly run the risk of erring or failing to do justice to the poetry and its authors. Yet izibongo is a vital component of contemporary South African literature and culture that has not received widespread acknowledgment in large part because of entrenched racial biases. It is a public medium, designed to provoke thought and discussion among its audiences, as well as a geopoetic medium particularly well-suited to consideration here.

Izibongo, iimbongi, and audiences

I met Thukela Poswayo in the summer of 2015, during a year in which I lived in rural, urban, and peri-urban township areas in South Africa's Eastern Cape Province while I carried out my research. At the time, I was staying in the village of Ngxo-tyana,[1] a rural community set in the rolling green countryside of the Eastern Cape coast. My research assistant and I had traveled nine miles in the back of a pickup truck that bumped slowly along the rutted coastal road to the town of Willowvale. There we climbed onto a minibus taxi and carried onward to the amaXhosa King's Great Place at Nqadu, some ten miles closer to the inland town of Idutywa that straddles the N2 highway. We had journeyed to Nqadu at the invitation of King Zwelonke himself, who suggested the event as an opportunity to see an imbongi perform. Poswayo's performance that day was part of a festive and informal series of music and dance performances interspersed with official proceedings that marked the launch of a holiday season anti-drinking campaign. He was the final performer, delivering a poetic address to King Zwelonke before the latter addressed the crowd of about 200 people.

The second set of performances took place on 11 March 2016 in the village of Gwedana, also in the Mbashe municipality, during a formal, traditional ceremony in honor of Chief Mthetho and his accomplishments. Poswayo performed two poems during this event, the first before King Zwelonke delivered his address to the assembled gathering of about 500 people and the second before Chief Mthetho delivered his. In early December 2016, I attended a third performance that was part of the festivities of a large wedding of several hundred people held in a rural village near the town of Engcobo. Poswayo once again performed two poems at the event, addressed to the groom who sat at the head table on a raised platform in the center of an enormous tent.

At each event, I made audio recordings of the poems, which ranged from three to eight minutes long. I brought the recordings to Makhanda (formerly Grahamstown) where they were transcribed and translated by Dumisa Mpupha, a Makhanda imbongi and mother-tongue isiXhosa speaker. I worked with these translations further, adjusting them for rhythm, clarity, and musicality. Across the performances I had recorded, I was able to observe similarities and differences in

Poswayo's performance style and the subjects, themes, and word choices of the various poems.

The context of the three performances varied considerably, ranging in formality and spirit from festive to official, contemporary to traditional; yet, they shared certain notable features. First, all were large gatherings of several hundred people who lived in the surrounding rural areas. These were not niche poetry events for a small, highbrow audience but performances that fit seamlessly into official proceedings at public events. Second, Poswayo did not fit the image of the imbongi described in much of the literature and shown in recordings that I had encountered. He did not wear any kind of head gear or special costume, and he did not carry a staff or spears—fairly common accoutrements for historical and contemporary iimbongi alike. Instead, at each of the performances, he wore a collarless, button-up shirt, ordinary slacks, and a pair of polished dress shoes. Each of his izibongo were addressed to the honored figures at the event and were performed in anticipation of that speaker's speech to the crowd. When I reviewed the translations later, I found repeated exhortations to King, Chief, and groom to be leaders with strength of character, that is, to be the eloquent and upright "men of backbone" that South Africa so badly needs.

Poetry of dwelling

My position as a white Canadian scholar traveling to rural amaXhosa communities to extract poetry for analysis raises problematic questions of ethics, power, and representation. I am an outsider not only to the communities and their cultures, race, and language, but also to the country and its history. Thus, one of my concerns in carrying out this study of Poswayo's poetry and of izibongo more generally was to begin to develop a decolonized approach to multicultural and multilingual literary scholarship that integrated the knowledge of people with a lived experience of the poetry. To this end, I interviewed 14 practicing iimbongi and 50 community members. I spoke with poets living in the rural Mbashe municipality as well as in the city of Mthatha in the former Transkei[2] region, in Zwelitsha township outside King Williamstown in the former Ciskei[3] region, in townships adjacent to East London, and in Joza township outside Makhanda. Together, these research locations spanned the three pre-1994 jurisdictions that comprise what is now the Eastern Cape Province. In both Joza and the Mbashe municipality, I spoke with 25 community members whom I selected purposively to reflect the diversity of ages and occupations in the community. Interviews were conducted either in English or in isiXhosa with the assistance of an interpreter, according to the participant's preference. During each interview, I asked poets and audience members to speak about their experience of izibongo, its subject matter, its role in the community, the frequency of performances, and where one is likely to see them. Overall, the research radically reconfigured my received ideas about poetry and literature, including my notions of what makes "good" poetry and what poetry and poetic language is and does.

The results of the interviews were revelatory. A wealth of unexpected spiritual and linguistic complexities emerged, along with a near-consensus from rural and peri-urban people alike that these poets and their poetry are deeply important. As well as serving as community historians, entertainers, and prophets, their words: "give people hope"; "revive a spirit of ubuntu," or care for humanity; touch people deeply; and remind listeners of the value of their shared language, culture, and identity as amaXhosa people. In particular, poets and audiences alike spoke about iimbongi as gifted healers who are able to move people with their words. According to one young man from Joza Township, "They heal us. Like when they talk isiXhosa, when they're rhyming their words, they bring us a knowledge that comes from them to us." Through my conversations with poets and their audiences, I found that iimbongi zomthonyama are widely considered to be gifted and even prophetic public orators capable of receiving and transmitting messages from ancestral or holy spirits. The receptivity to these messages enables iimbongi to burst forth with complex verses composed on the spot.

Along with the translated transcriptions of performances that I recorded, these interviews affirmed that the affective power of izibongo derives from details of the craft: an imbongi's choice of words and metaphor, the rhythm of the language, gesture and voice, and the layers of cultural significance that inhere in the poetic tradition and its performance. These elements combine to create a particular expression of society, culture, spirituality, and politics that is stronger and more powerful than the words alone would suggest. In this way, iimbongi help tighten connections to landscape and heritage not only by conveying information but also by enabling people to access deeper emotions about these things.

As with other literary genres, the meaning of izibongo is inseparable from its form. The acts of transcription and translation result in a fundamentally different text than the original performance. Not only the original language, but also sounds, rhythm, cadence, and gesture are absent from the printed English text. Also lost are the nuanced figures of speech that are particular to place and culture and do not translate well into English. Nevertheless, the translated lines remain rich in details that illustrate amaXhosa relationships to place and homeland, locations of layered, multigenerational dwelling. In Poswayo's poetry, although each poem was original and composed spontaneously for the event at hand, various themes and repeated motifs emerged over the three performances. The following excerpts illustrate the themes of rivers, cattle, and genealogies that are central aspects of Poswayo's poetry, of amaXhosa traditional culture, and of contemporary rural life.

The following selection contains the opening stanzas of Poswayo's first performance for King Zwelonke at the December 2015 event. With his opening lines, he draws his audience into active participation in the poem. He catches their attention again a few lines down with the phrase, "Ndithi ndimbi ngapha ndimhle ngapha," a familiar reference to the uthekwane, or hamerkop, an auspicious waterbird bearing layered cultural significance who hunts for its food by turning its head from side to side to peer into the water, first with one eye then with the other.

Imbongi: A Zwelonke!
Abantu: A Zwelonke!
Imbongi: A! Zweloooonke!
Abantu: A! Zweloooonke!

Iyakhumbulana mntane nkosi.
Iyakhumbulana thole leduna.
Ndelula amehlo ndayibona imilambo,
Ndanga ndinga qhayisa ndixele
    uthekwane
Ndithi ndimbi ngapha ndimhle
    ngapha.

Ewe kaloku thole leduna
Siyinqamle imilambo.
Sawubona uMbhashe.
Sayinqaml' imilambo
Sawusel'uMgwali.
Sawubona uMthatha, sawubona
    uMthamvuna.
Salibona iThukela, salibona iCongo,
Salibona iZambezi, salibona
    iLimpopo,
Siyibonile imilambo
Ngoba neLubhelu siyibonile.

Ewe kaloku ndibiza ngabom.
Ngoba kaloku ukuze kulunge
Ndithi Zwelonke
Funeka ndiyibiz'imilambo ye Afrika.

Ewe kaloku kwakudala,
Le mini yayi saziwa iyakuz'ifike
Ungeka qashulwa, engekakuzali unyoko
Wawusele uzelwe.
Ke kambe okwethu kukungqina
Sithi nal'ithole lika Xolilizwe.
Nants' inkonyane yohlanga
Eyabizwa ingekaveli.

Ewe kaloku! Le ndawo ukuyo
Kwakukhe kwahlala omnye umntu kuyo
Imbongi: Hail Zwelonke!

Audience: Hail Zwelonke!
Imbongi: Hail! Zweloooonke!
Audience: Hail! Zweloooonke!

We miss each other, child of a chief.
We miss each other, son of a bull.
I stretched my eyes and saw the rivers.
I wish I can boast like the thekwane

Saying I'm ugly on this side, beautiful
on that.

Yes then, son of a bull,[4]
We crossed rivers.
We saw the Mbhashe.
We crossed rivers
And drank the Mgwali.
We saw the Mthatha, we saw the
    Mthamvuna.
We saw the Thukela, we saw the Congo.
We saw the Zambezi, we saw the
    Limpopo.
We've seen the rivers,
Why, we've even seen the Lubhelu.

Yes, now I name them deliberately.
So that all may go well
I say Zwelonke;
I must name the rivers of Africa.

Yes, you see, even long ago
It was known this day would come.
Before you were conceived, before
You were born to your mother, you existed.
Now we need only witness,
Saying that this is Xolilizwe's[5] calf.
This is the calf of the nation,
Named before he was born.

Yes of course! The position you hold
Was held by someone before you,
As there will be someone after.
Kusezaw'hlal'omnyumntu

Yiyo lo nto funek'uchul'ukunyathel' uchule	That is why you must act with care,
Uwabal'amanyathelo	Counting your steps,
Ngob'umlambo owela kuwo kwedini	For the river you're crossing, young man, Is full of slippery stones,
Uzele amatye agcwel'ucolothi.	Deep pools to the north, deep pools to
Ngentla ziziziba, ngezantsi ziziziba!	the south.[6]
Chula ke ukhangele ngaphesheya	Be steady then and look across[7]
Ngoba kaloku ukwenza kwakho namhla	Because what you do in this time
Kuyakulandela ngemihla sewungasekho.	Will follow you when you depart.

The second stanza of the previous selection is deceptively simple. Far more than simply a list of rivers, the lines carefully name the lifeblood of amaXhosa society that forms the essential point of connection between people and their landscapes and histories. Rivers are inhabited by ancestral spirits; thus, by naming them, Poswayo makes an allegorical reference to the ongoing dwelling of these spirits on lands and waterways. Rivers also gave shape to historical patterns of amaXhosa migration, settlement, and seasonal transhumance. By beginning with the names of nearby rivers before moving to names of rivers progressively farther away, Poswayo maps both spatial and temporal distance, stepping backward through time and linking the present company to their forbears, tracing the amaXhosa migration from northern regions of the Eastern Cape and beyond. The sequence concludes with the mythical land of Lubhelu from which the amaXhosa are purported to have traveled over 700 years ago (Gérard 1971).[8]

The following excerpt is taken from Poswayo's first poem performed during the ceremony for Chief Mthetho. As with the first excerpt, Poswayo begins with an opening that draws his audience into a call and repeat. And, once again, he opens with the image of rivers. In this poem, however, the imagery shifts immediately to that of the multi-colored Nguni cattle, which hold deep cultural and ceremonial significance for amaXhosa. Cattle and rivers remain the central elements throughout the selection to which the poem returns as it flows on to references to ancestral names and other creatures who inhabit the landscape.

Imbongi: A! Zwelonke	Imbongi: Hail! Zwelonke
Abantu: A! Zwelonke	Audience: Hail! Zwelonke
Imgongi: A! Zweloooonke!	Imbongi: Hail! Zweloooonke!
Abantu: A! Zwelonke!	Audience: Hail! Zweloonke!
Mntan'omhle	Honorable one,[9]
Amehl'am ath'akukhangela ndabon'imilambo	Opening my eyes, I saw rivers.

Ndathi ndakujonga ngaphesheya
Ndazibon'iinkomo zako kwenu
Ndazibon'iinkomo zako kwethu

Ndiyazaz'ezako kwenu
Ndiyazaz'ezako kwethu
Ndiyazehlula ngemibala
Kuba zingaphesheya kwemilambo
    zicacile.
Zicacile zizakuhle
Zibonakala ngok'tyhobo

Ewe kaloku!
Yithi khe ndicaphule ndenjenje
Ngoba kaloku ukuze kulunge
    maLawundini
Vumani kuba sendiliphethibhozo
Ndabel'izizwe
Vumani kaloku
Ndabel'iQamata
Ndabel'umhlaba kalok'omagqagala
Ndabela kaloku umhlaba kaloku wakulo
    Daliwonga
Kuba kaloku kulapho zaphuma
    khon'iinkomo
Zaqweqwema zadl'amathafa
Zafika kwaChotho zamila
Zabuya nentombi
Yafika yazal'amadodana

Namhla ke kuthi mandithi
Ukuze kulunge kum
Zikho kalok'iinkomo zakulo Mvuzo
Zikho kalok'iinkomo zakulo
    Thambekile
Nawe nkonyana yakulo Thambekile
Yithi ndithi rhuthu
Nang'umwangalala ndikwabele

Ewe kaloku!
Ukuze kulunge maLawundini
Vumani kaloku ndiwele kalok'imilambo,
Ndiwel'uMbashe, ndiwel'iXuka
Ndinyuke kaloku ngoMkhonkotho

When I looked across
I saw your family's cattle,
I saw my family's cattle.

I know those of your family
I know those of my family.
I know them by their colors[10]
For even across the river they are
    distinct.
They are beautifully distinct
They appear now, charging.

Yes then!
Let me say something
For in order for things to improve,
    maLawundini,[11]
Allow me, for I already have the knife.
I distributed among nations.
Allow me then,
I gave to Qamata.[12]
I gave the land with its dry boulders.
I gave the land from Daliwonga's[13] family

From which the cattle came,

Running to the fields.
They reached Chotho and stopped,
Returning with a girl.[14]
She arrived and gave birth to young
    boys.

Today I want to say only
That things are well with me.
There are the cattle from Mvuzo's family,
There are the cattle from Thambekile's
    family.
And you, calf of Thambekile's family,
Let me take out
These scattered coins to share with
    you.[15]

Yes then!
So that things will be well, maLawundini,
Allow me to cross these rivers.
I crossed the Mbashe, I crossed the Xuka

Ndakufika phezu kwentaba	And ascended Mkhonkotho.[16]
Ndivul'amaphiko	I arrived at the top of the mountain, Opened my wings
Ndime kaloku ndixel'intsikiz'im'emaweni	And stood on the cliffs commanding as a hornbill
Iqhayise'amahahane	Boasting to the hadedas,
Isithi mna ndihluthi	Saying I am full,
Ndihluth'amaqonya	I'm full of maqonya.[17]
Kazi wena hahane uyakurhayisa ngantoni na	I wonder, what would you boast about, hadeda,[18]
Kuhluth'intsundwan'enje	When the other is so full too?

The first of the two poems performed at the third performance I attended—the wedding in December 2016—is addressed to the groom. This poem opens with the image of the poet as an old man calling for his stick to aid him in an arduous journey across rivers and valleys and the steep, rough ground of the Engcobo countryside—the groom's ancestral homeland as well as Poswayo's.

Amathun'anabile	It is growing late.
Thina nto zaziy'ixesha lokulala lixesha lokulala	We who know the time to sleep, it is time to sleep.
Ndiniken'iintonga zam ndiwel'imilambo	Give me my sticks to cross rivers,
Ndiniken'iintonga zam ndiwel'imifula	Give me my sticks to cross valleys.
Ndithambek'amathambeka	I walk down the hills,
Ndixawuk'imixewuka	Stepping over the rough ground,
Ndigob'izagobe	Bending when I need to—
Ndikhomb'imibombo kokwethu	I face the land that is ours.
[...]	[...]
Ulumkile kwedini	You are wise young man,
Usibonisil'izint'ezilumkileyo	You've shown us wise things
Kuba phakathi kwamadod'ubungakhetha kuwo	For among men you could choose from
Ukheth'ukukhetha kwaNxasana	You chose from the Nxasana family
Thina nto zaziyo asothukanga nto	We who know them are not surprised
Kuba siyaz'olu sapho	For we know this family
Lwakhe lwenz'izimanga ngeny'imini	Who have done wonders in their day:
Lalunguz'imilambo	They looked for rivers
Lawela nezizwe	And crossed them with nations:
Awel'amaMpondomis'ekhokelwa ngamaNxasana	The amaMpondo crossed, led by the Nxasana clan,
Awel'amaMfeng'ekhokelwe ngamaNxasana	The amaMfengu crossed, led by Nxasana clan,
Zatshw'izinto zalunga	And things were well.

The poem's concluding lines, in the following excerpt, invoke the conjugal love that has brought together husband and wife, and indeed the entire company gathered to celebrate their union. By speaking and sharing such love, the groom can display the same leadership with which his ancestors led the amaMpondo and the amaMfengu and in doing so will bring the metaphorical benediction of rain to a suffering land.

Kwathi ngexesha lokuphila kwalo mihla	What happened during our lives in those days?
Zenzek'izint'ezintle kokwethu	Beautiful things took place in our country,
Kwamnandi kwaba chosi kwabahele	People were happy and well.
Thetha k'ukuthand'umfazi wakho	Speak about your love for your wife
Ukuz'iphind'ibuy'in'imvula	So that the rain may return,
Liphind'eli lizwe lintshul'intw'ezintle	So that this country can grow beautiful things
Siphinde sibuye kokwethu	So that we may be at home again.

What does it mean "to dwell"? The *Oxford English Dictionary (OED)* lists various definitions, including its archaic usages, "to tarry, delay" and "to abide or continue for a time in a place, state, or condition." In current usage, it can mean both "to occupy as a place of residence; to inhabit" and "to remain (in a house, country, etc.) as in a permanent residence; to have one's abode; to reside, 'live'." The *OED* notes that the most frequent current use of the word is figurative, as in "to spend time upon or linger over (a thing) in action or thought; to remain with the attention fixed on; now, esp. to treat at length or with insistence, in speech or writing." Yet perhaps most significantly, it also means "To continue in existence, to last, persist; to remain after others are taken or removed." All of the various subtleties of meaning resonate in Poswayo's poetry, yet the last is the most poignant. Dwelling, for the amaXhosa of South Africa's Eastern Cape, is a verb fraught with struggle and tension. Over the course of a century and a half—from the nineteenth century wars of dispossession through the imposition of colonial and apartheid laws and mass relocations—the colonial and apartheid South African states disrupted and infringed upon amaXhosa dwelling to the extent of their considerable capacities. Their primary aim was to coerce rural, self-sufficient people off their lands so as to make both their land and, especially, their labor available for the pursuit of profit. As the Chamber of Mines noted at the turn of the last century:

An abundant supply of cheap labour drawn from the coloured races is of supreme importance, and without this aid there do not appear to be any great potentialities for the shareholder, the white mine *employé*, or the country at large. The burning question is how this vital factor in the general prosperity can be provided as the mining industry demands. The only remedies seem to be: (1) More legal and moral pressure to compel a great number of natives in

British possessions to work, and for longer periods. (2) To extend the present recruiting area with the utmost rigour.

(Bourne 1903)

Scholar Jeff Opland, who has written extensively about iimbongi and their shifting role through the colonial and apartheid periods, explains how izibongo articulate the tragedies of dispossession: "The names of their ancestors held particular significance within the system of ancestor veneration, for not only is an individual descended from individuals, ritually he *is* his ancestors, *and his ancestors are identified with their dwelling places*" (Opland 2005, 50, emphasis added). That is, to invoke ancestral lineages in the poems is to invoke their homelands and vice versa. In his discussion of a poem by the late David Livingstone Phakamile Yali-Manisi about the ancestral chief Mfanta and his heir Manzezulu, Opland goes further:

> The sacral chief *is* the people: his strength is theirs, and his well-being is ensured by the sympathetic attention of his ancestors to his affairs and to the affairs of the chiefdom. But this ritual relationship is ruptured because Mfanta lies buried on Robben Island, where he was dumped by the whites for resisting white encroachment and fighting for the rights of his people. *And so* Manzezulu's people are destitute, and they will remain destitute and troubled as long as Mfanta's bones lie restless in foreign soil, unappeased by sacrifice.
>
> *(Opland 1987, emphasis original)*

Through this lens, the violence and destitution wrought by apartheid and persisting in its aftermath can be seen as the direct result of ruptured spiritual connections between people and their ancestral homes.

Poswayo's invocations of homeland and lineage are reminders of this past and of the sacred responsibilities binding families to their landscapes and the generations yet to come. His poetry contains features unfamiliar to English speakers—particular introductory and concluding phrases, genealogies and praise names, and local idioms, to name just a few. With these poetic devices and the multilayered significances they contain, izibongo illustrates the fluidity of poetry and the ways in which it can act on people in material and tangible ways. Thukela Poswayo's poetry celebrates dwelling not as a fixed or permanent condition but as a fluid state of being that encompasses the seasonal migrations and long, incremental migration that are part of amaXhosa history. Rooted in traditions that see landscape, cattle, and people as a close triumvirate surrounded by benevolent ancestral spirits, Poswayo's poetry speaks to the simultaneous dwelling of entities beyond humans and the importance of each to the integrity of the whole.

Notes

1 The Roman letter "q" denotes a palatal click, "x" a lateral click, and "c" a dental click. Combinations of consonants (e.g., nq, ngc, xh) indicate variously voiced or aspirated clicks.
2 This former "Bantustan," or Black land reserve, has a complicated political history that includes being offered nominal independence by the apartheid government. As a result

of its history, it remains an almost entirely Black region whose culture, language, and settlement patterns differ markedly from other parts of the Eastern Cape.

3 The Ciskei is another former Bantustan area, located across the Kei River from the Transkei. Again, the history of this region can be seen in its contemporary geographies.

4 E.g., the son of a warrior, a brave person.

5 Zwelonke's father.

6 A reference to the people who dwell at the bottom of the river pits (a site of initiation of spiritual people).

7 I.e., to the ancestral realm.

8 For a more detailed discussion of the poem and its geopoetics, please see McGiffin 2017, 2019.

9 Mntan'omhle: Literally, "Beautiful child."

10 Cattle colors are often particular to clan.

11 "MaLawundini" is someone without a tradition or custom who wants to fit in somewhere. The term is sometimes offensive, but it is not meant to be so in this case. It could mean that the King may feel like amalawundini because the imbongi is going to say things he may not understand.

12 Refers to the town of Qamata (iQamata) rather than the god (uQamata).

13 The family includes Kaiser Daliwonga Mathanzima, whom the apartheid government installed as chief of the Transkei.

14 I.e., The cattle were the lobola (bride price) for the girl from Chotho.

15 "Umwangalala" refers to dispersed objects generally; it is interesting to note the correspondence between money, grain, and cattle in Kropf's (1915) definition: "Grain thrashed out and lying spread on the floor; small money scattered about, cattle dispersed." The word also captures the fact that the coins are abundant, befitting to the recipient's status.

16 Near the town of Centane.

17 Amaqonya are large green and silver caterpillars of the emperor moth that feed on the mimosa thorn-bushes (Kropf 1915). Traditionally, the caterpillars, which can grow to ten centimeters, are an important food source for people and birds alike. As Kropf describes, "The boys kill it by inverting the head and thus pressing out the intestines; they then roast and eat the remainder" (359). Although a different species, amaqonya are analogous to the more familiar mopane worms, a major protein source throughout southeastern Africa.

18 The mispronunciation in the isiXhosa (the correct word is uyakuqhaisa) is for alliterative effect, echoing the smoother aspirated h's of "hahane" and "kuhluthi" rather than interrupting these mellifluous sounds with a hard palatal click.

References

Bourne, H., and R. Fox. 1903. *Forced Labour in British South Africa: Notes on the Condition and Prospects of South African Natives Under British Control*. London: PS King & Son.

Finnegan, Ruth. 2012. *Oral Literature in Africa*. Cambridge, UK: Open Book Publishers.

Gérard, Albert S. 1971. "Xhosa literature." In *Four African Literatures: Xhosa, Sotho, Zulu, Amharic*, 21–49. Berkeley, Los Angeles and London: University of California Press.

Kaschula, Russell H. 2002. *The Bones of the Ancestors are Shaking: Xhosa Oral Poetry in Context*. Lansdowne, South Africa: Juta.

———. 1993. "Imbongi in Profile." *English in Africa* 20 (1): 65–76.

Kropf, Albert. 1915. *A Kafir-English Dictionary*. South Africa: Lovedale Mission Society.

McGiffin, Emily. 2019. *Of Land, Bones, and Money: Towards a South African Ecopoetics*. Charlottesville: University of Virginia Press.

———. 2017. "Oral Poetry and Development Ideology in South Africa's Eastern Cape." *Third World Thematics: A TWQ Journal* 2 (2–3): 279–295. doi:10.1080/23802014.2017.1402666.

OED (Oxford English Dictionary Online), s.v. "Dwell." Accessed October 15, 2018. www.oed. com.ezproxy.library.yorku.ca/view/Entry/58765?rskey=SAif0a&result=2&isAdvanced =false#eid.

Ong, Walter J. 2002. *Orality and Literacy*. London: Routledge.

Opland, Jeff. 1983. *Xhosa Oral Poetry: Aspects of a Black South African Tradition*. Cambridge: Cambridge University Press.

———. 1987. "The bones of Mfanta: A Xhosa Oral Poet's Response to Context in South Africa." *Research in African Literatures* 18 (1): 36–50. (emphasis in original).

———. 1998. *Xhosa Poets and Poetry*. Cape Town: David Philip Publishers.

———. 2005. *The Dassie and the Hunter: A South African Meeting*. Scottsville: University of KwaZulu-Natal Press.

Turin, Mark. 2012. "Foreword." In *Oral Literature in Africa*, edited by Ruth Finnegan, xvii–xx. Cambridge, UK: Open Book Publishers.

14

ISLOTE POETICS

Notes from minor outlying islands

Urayoán Noel

In his classic *Insularismo* (1973), Antonio S. Pedreira reflects on the "rico," or rich, in Puerto Rico. Against tourist gazes that see only an "Island of Enchantment" framed through the "rhetorical exuberance" of that "rich" port, Pedreira juxtaposes "nuestra existencia agria" ("our [Puerto Ricans'] bitter existence") (1973, 98) with the botanical metaphor of the "vegetative languor" (98) of the Puerto Rican temperament.[1]

Key to Pedreira's argument is his invocation of the "pobres islotes" (poor islets) surrounding the main island of Puerto Rico, with names such as "Caja de Muertos" (Coffin Island) and Desecheo (a corruption of "Cicheo" or "Sikeo," Its Indigenous Taíno name enmeshed with the Spanish word "desecho," meaning waste or garbage). Recovering these "poor islets" as spaces of death and waste is a way for Pedreira to provide "una expresión honrada de nuestra realidad" ("an honest expression of our reality") beyond the "optimismo metafórico" ("metaphorical optimism") that metonymically renamed the island of San Juan Bautista for its rich port (98), with San Juan becoming its capital. This renaming was an act of ideological misreading that echoes in the "enchantment" that Puerto Rico markets to tourists to this day.

While remembered for its meditations on Puerto Rican islandness and/as isolation ("aislamiento" evoking "isla"), *Insularismo*'s mapping of existential geography dismisses the outlying islands, or "islotes," within the Puerto Rican archipelago in the name of a centripetal movement from the coast inward that excludes the open sea but begins to manifest itself vigorously away from the coast ("tiene fuerza centrípeta: excluye mar afuera, pero empieza a manifestarse vigorosamente costa adentro," 98). By contrast, Antonio Benítez-Rojo (1996) imagines the repeating island, whose (post?)colonial spatial logics echo across the Caribbean, through the metaphor of the tropism (1996, 4), as in plants turning toward the sun. The turning-toward hinted at in the biological metaphor of the tropism grounds, for Benítez-Rojo, the possibility of a non-mimetic archipelagic poetics of repetition, akin to Édouard Glissant's in its insistence on a rhyzomatic difference, of islands no longer atomized but turned toward each other, far from the vegetative languor of Pedreira's cursed islandness.

Here, I focus on two short, innovative small-press poetry books whose thematic emphasis is on outlying Puerto Rican islands: Joanne Kyger's *Desecheo Notebook* (1971) and Nicole Cecilia Delgado's *Amoná* (2013). Both books, lacking page numbers and invested in experimental and gendered projects of remapping, complicate Pedreira's centripetal island and the archipelagic as variously theorized by Benítez-Rojo, Glissant, and Martínez-San Miguel. Although I follow Martínez-San Miguel in insisting on an archipelagic approach to Puerto Rican and Caribbean Studies, I want to emphasize how Pedreira's *islote*, for all its problems, can help us nuance a geopoetic approach beyond island/archipelago binaries. An *islote* is a small island, and for Pedreira a poor one, yet its marginality is a necessary corrective to discourses of cosmopolitan optimism. At the same time, Pedreira hints at the second meaning of *islote* as a crag—a broken or projecting rock that is inaccessible or inhospitable. In this sense, an *islote* is not unlike the Taíno word *Sikeo*, that "high, mountainous land" that was translated as "Cicheo" (Fernández de Oviedo y Valdés 1855, 465) by Spanish chronicler Gonzalo Fernández de Oviedo y Valdés (1478–1557) and eventually into the violently symbolic "Desecheo" of death and waste.

I want to recover the cragginess and smallness of the *islote* as necessary to an archipelagic poetics, in keeping with a vision of the Caribbean as defined by what José Lezama Lima (1969, 113) called the "pedregosidad" ("rockiness") of the Americas and by the *mornes* ("craggy hills") of Glissant's poetics of relation, those found "rising abruptly behind the Caribbean beaches in Martinique" (1997, 207). Beyond biotic logics, *islotes* are marked by death and waste under colonialism, paradoxically walkable spaces whose craggy heights resist easy occupation and translation. Against austerity politics and ecocolonialism as they have shaped Puerto Rico and the Caribbean over the past two decades, these *islotes* seem less like outliers and exceptions than a ground for new geopoetic imaginaries. In emphasizing the "minor and outlying" quality of these *islotes*, I echo the term "US Minor and Outlying Islands" as it has long been used to organize US empire in and beyond the Caribbean. I do so critically, and I follow Martínez-San Miguel's invaluable *Coloniality of Diasporas* (2014b) in marking at least two observations: how postcolonial approaches tend to center a narrow nationalism, and how progressive politics in Puerto Rico and across the Caribbean tend to privilege a hermetic islandness. I also suggest how poetry can recover these marginal *islotes* to underscore the limitations of conventional poetics of place in and beyond the Caribbean.

Joanne Kyger's dream notebook

In an interview with Paul Watsky, Joanne Kyger (1934–2017) recalls traveling to Desecheo in 1971, invited

> by Peter Warshall, a Harvard Primate Research student, along with three or four other students, to study the rhesus monkeys that had been left there a

few decades earlier in order to build up a troupe of monkeys for use as laboratory animals to test the Salk vaccine.

(2013, 95)

While Warshall and his team were seeking to understand how the monkeys "had
adapted and survived on essentially a desert island, an island with no water and very
little rain," Kyger was pursuing her own "course of study," armed with a copy of Jung's
Memories, Dreams, Reflections in an effort to "[investigate] the self" (Watsky 2013, n.p.).

Kyger adds that there was only one other woman on the trip, so that *Desecheo
Notebook*'s Jungian "course of study" occurs not just "near the end of the so-called
psychedelic revolution" but also against the backdrop of a writing life that, as Linda
Russo (2013) notes, constantly negotiated survival and group membership in the
masculine/homosocial space of postwar New American Poetry countercultures.
Kyger's invocation of Desecheo's role in US laboratory experiments and vaccine
testing (part of a long tradition of scientific/medical colonialism in Puerto Rico)
links her course of study and the group's study of the rhesus monkeys as competing
yet overlapping explorations of *islote* survival.

Later in the Watsky interview, Kyger describes how a chance meeting with poet
Robert Creeley in Bolinas, California, informed the dream reportage of *Desecheo
Notebook*. Kyger complained to Creeley about a bout of writer's block, with Creeley replying "You can't try" and Kyger concluding: "You can't push it," later adding, "It turns out reporting dreams made up much of the content of the writing
I did on the island" (96). The idea of this notebook not as a description of a trip to
Desecheo, but as a reporting of dreams summoned forth by this inhospitable *islote*,
allows for a valuation of Desecheo as "nonsite." With its depiction of deserted/
desert-like island landscapes, its Jungian dreamscapes, and its problematizing of the
page and the gendered self as defined by a plenitude in emptiness, *Desecheo Notebook*
invites a reading informed by Robert Smithson's concept of nonsite, one that values
the "empty" space of the *islote*, the outlying island, as an imaginary for revisionist
forms of community—even as Kyger's book in other ways reproduces the colonizing Pedreiran tourist gaze mentioned earlier.

Lytle Shaw has analyzed how Bolinas—the legendary hippie-era poets' colony
in Northern California where Kyger lived in the early 1970s and that is referenced
throughout *Desecheo Notebook*—functions as a nonsite in the work of Kyger, Creeley, and other poets who relocated there from the bustling poetry communities of
New York and San Francisco. As Shaw puts it, these poets saw in Bolinas an escape
from the city and into nature but also the possibility, partly as a negation of urban
life, "to resist the lure of the local and practice instead an immersion in a kind of
expanded present tense" (2013, 121). Here, I want to claim the materiality and
irreproducibility of Kyger's notebook (its personal allusions, coded languages, and
references to her Bolinas peers, especially) as a search for an outlying island that
neither turns inward (toward Puerto Rico's main island) nor outward (toward the
larger Caribbean). Here, Shaw's distinction between place and site matters, especially the way in which place "falsely grounds and organizes the fluid and dispersed"

(6) while site-specific poetics such as Kyger's "always coincide with other claims about discursive and historiographic sites" (259).

This distinction is evident in the poem on the back cover of *Desecheo Notebook*. While it begins by describing Desecheo as a place "off West Coast Puerto Rico / 1 1/2 miles long 1 mile wide" (1971, n.p.) it goes on to evoke its standing as "the end point of / an upheaval that happened / in the Caribbean."[2] Kyger here is, in their purest senses, neither a nature/ecopoet seeking to document and preserve a place whose value is always a priori nor a hippie/travel poet seeking out an open plural politics through a sited writing. In her writing, Desecheo reveals its dialectical status as nonsite that leads back to the "closed limits" of her "inner coordinates" (Smithson 1996, 364). In this context, the book's prefatory map is significant: It is a "Plan of the Aguada Nueva de Puerto Rico" from *A General Topography of North America and the West Indies* (1768) by Thomas Jefferys, cartographer to King George III. Published just before the American Revolutionary War (1775–1783), this map embodies a logic of empire, especially as it lumps together North America and the West Indies as colonial spaces. Desecheo's location at the northwest corner of the map echoes not only Smithson's liminal framing but also his dialectical understanding of site against the nonsite center, which in this map is the middle of the ocean, off the Mona Passage—an area historically associated with pirate activity, seismic events, and dangerous crossings.

While there is little acknowledgment of that history in Kyger's book, we can read the "upheaval" against that history. Desecheo is a mappable island (Columbus

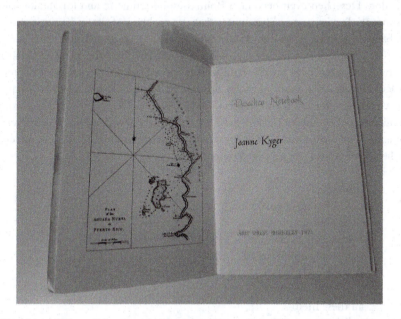

FIGURE 14.1 "Plan of the Aguada Nueva de Puerto Rico," from *A General Topography of North America and the West Indies* (1768) by Thomas Jefferys, as it appears in Kyger's *Desecheo Notebook*.

explored it in 1493), but it also lacks surface water; it appears, like nearby Mona and Monito islands, geologically distinct from the main island of Puerto Rico. Since the 1980s, Desecheo has been a US National Wildlife refuge, its nature preserved only by virtue of Puerto Rico's colonial relationship and at the expense of the access to the island itself. Of course, this nonsite approach to Desecheo does not preclude a tourist gaze, as the waters surrounding the island have gained cult status among divers and snorkelers; still, although local fishers informally take tourists to Desecheo, restaging *Desecheo Notebook* today would be an act of trespassing. The ironic force of Desecheo's nonsite is magnified when considered alongside the recent history of the island of Vieques, which over the past 20 years has become a complex metonymy of Puerto Rican decolonial and environmental politics in the neoliberal age, devastated by Hurricane María yet subject to attempted "boutiqueification" as an upscale resort island largely under US multinational control. (This is a recent history that provides the backdrop to Delgado's 2016 artist book, *subtropical dry*.)

Kyger's embrace of Desecheo as nonsite seems partly an attempt to complicate a utopian politics of site. Thus, she writes "I want to go back / I want to get back / out of this / fairy tale land." She ends this page with the oblique line "This write often," an instance of the sort Shaw theorized, in which Bolinas poets complicate place not only by meditating on their writing of place but also on the materiality of the writing, that is, on the microlevel "this"-ness (2013, 5) of writing in space and time. Kyger invokes Bolinas in ways that interrupt the closed space and time of Desecheo ("news of Bolinas"), but the speaker never makes it back to Bolinas, nor does Desecheo even become a Bolinas-esque refuge in and for an alternative (homosocial) community. Shaw notes that the problem of site in Kyger is also the problem of writing in the shadow of her superstar friends and neighbors such as Ginsberg, Creeley, and her ex-husband Gary Snyder (the latter invoked by Kyger as someone who "lives on an island in front of a black lake," perhaps a reference to Kitkitdizze, his homestead in the foothills of the California Sierra Nevada and a key reference for US nature and ecopoetry). Kyger's Desecheo contains, through Jungian memory, Snyder's island; her Desecheo embodies how the speaker's search for a new "quality of mind" (Kyger's words) in this communal escapade is undercut by her awareness of her gendered difference:

Thursday
        the quality of mind
like I am keeping track of
3 or 4 people at one time
            bathing naked
    A letter to Bobbie Creeley
        no place to sit
    all these men
        I just want a place
    for myself
                (Kyger 1971)

The overcoming of the self so prized by the era's homosocial alternative poetry communities hinges on a privilege unavailable to the speaker. Buddhism, in the poetry of Ginsberg and other traveling male poets of the era, often revealed itself in the understanding of the body as decaying matter ("desecho?"), but the speaker's body cannot be simply undone ("deshecho"), as references to sex and Tampax litter the idyllic tropics of Desecheo—a litter echoed in the craggy topographies of Kyger's lines. The book ends not in Bolinas or a reanimated Desecheo, but in an imagining of New York, juxtaposed against the italicized *wild* of Desecheo:

> Friday's nature
> > can a *wild* animal
> exist anymore
>
> .
> wilderness
>
> .
> words. spirits guides.
>
> .
> further away
> > into crammed people land
>
> .
> > New York
> > > (Kyger 1971)

Against Pedreira, *Desecheo Notebook* posits a centrifugal poetics moving outward, away from the iconicity and singularity of site toward the referential field of nonsite, as Desecheo's dreamscapes are enmeshed with Bolinas and New York, privileged locales of 1960s/1970s poetic counterculture, in a semiotic/psychic/geopoetic wilderness. If the New York-Bolinas trajectory of Kyger and many of her poet peers echoes the imperial logic of westward expansion famously articulated by Frederick J. Turner (1893), Kyger's North-South iteration of the geopoetic *islote* of Desecheo can be read both as an attempt to interrupt that frontier logic and as an echo of the Beat-era imperial travel fantasy of going south to Mexico or South America, often understood as places of chemical and sexual liberation and lawlessness. Writing about *Desecheo Notebook* in the context of Kyger's travel poetry, Jonathan Skinner (2000) characterizes Kyger as a poet of "captivity" but adds that it becomes impossible to tell what in the book is "dream or reportage, or reported dream." My own interest is in opening Kyger's poetics of captivity to an *islote* poetics that might critically revise Pedreira's privileged if pessimistic gaze, even while noting the problematic aspects of Kyger's engagement with Desecheo. Like the rhesus monkey study that brought her to Desecheo, Kyger's book is an investigation of creative survival and living together, even as it hinges on the colonial violence that settled and unsettled the *islote*.

Nicole Cecilia Delgado's anomalous *islote*

Amoná (2013) by Nicole Cecilia Delgado is a poem set on the nature reserve, Mona, an outlying Puerto Rican island that shares the Mona Passage with its smaller neighbors, Monito and Desecheo. When it appeared in 2013, the book appeared to be an anomaly: first, it was a homemade edition, personally distributed to friends and colleagues; second, it was more akin to the innovative, socially engaged poetry in Puerto Rico that is still largely rooted in and around San Juan and that typically emphasizes an urban setting and/or sensibility. Delgado's work emerged in the heyday of *poesía urbana* or urban poetry in Puerto Rico—the early-to-mid-2000s, when poets such as José Raúl "Gallego" González, Guillermo Rebollo Gil, and Hermes Ayala were crossing over with daring (if largely homosocial) translocal poetics informed by the urban vernacular traditions of salsa, hip hop, and Nuyorican poetry (Noel 2011), and by contemporary breakthroughs in slam/def poetry and reggaeton. Working across the "broken Souths" that Michael Dowdy (2013) maps in the shadow of neoliberalism and globalization, these poets affirm a class- and race-conscious vision aligned with the "from below" perspective the late Juan Flores (2009) finds in so many Nuyorican/Diasporican cultural productions. While Delgado's work is largely in solidarity with these urban/vernacular poetics, her trajectory and evolution are very different.

Delgado began publishing in the early 2000s when she was a student at the University of Puerto Rico's Mayagüez campus, on the west coast of Puerto Rico, near Mona and Desecheo—and far from the urban poetics then booming in San Juan. She collaborated with reading series and publications in the area, co-founding and editing the journal *Zurde*. She moved upon graduating to New York, soon becoming associated with the poetry scene in and around the Nuyorican Poets Cafe and specifically with the young Colombian poets galvanized around the late poet and activist Ricardo León Peña Villa. Editorial Palabra Viva (Medellín/New York) published Delgado's 2004 debut, *Inventario secreto de recetas para enrolar las greñas con cilindros de colores*. It was at this time that she learned how to make small artist books under the East Harlem-based Puerto Rican poet and book artist Tanya Torres.

Although a youthful book, *Inventario* anticipates Delgado's *islote* poetics in poems such as "islas." Here, the speaker proclaims "Yo soy *de las islas*" ("I'm *from the islands*," 2004, 65, emphasis original). The italics mark an archipelagic difference, since Puerto Ricans tend to use the singular "la isla" to refer to the entire archipelago, echoing Pedreira's violent insularist logic. Since Delgado is writing *from* New York, we can also read the plural here as encompassing the islands of New York City and their diasporic histories. Two lines down, the speaker attempts to distinguish her plural islands from "el insularismo pedregoso y fatalista" ("the rocky and fatalistic insularism") of her ancestors, embodying the tensions of *islote* poetics even as they risk uncritically reinscribing Pedreira.

Delgado would later enroll in and quickly abandon graduate school in upstate New York, ironically echoing Pedreira, who briefly lived in New York while studying medicine at Columbia University. She eventually settled in Mexico, where she was involved with feminist, Indigenous, and community-based poetry collectives and where she discovered the alternative *cartonera* publishing movement. This comprised no-frills and largely hand-to-mouth decentralized collectives publishing chapbooks with recycled cardboard covers and often featuring colorful stenciling or drawing. The *cartonera* movement began in Argentina in the early 2000s as a creative response to a (US-backed) debt and austerity crisis. Upon her return to a now debt- and austerity-ravaged Puerto Rico, Delgado and poet Xavier Valcárcel established the first *cartonera* in the Caribbean, Atarraya Cartonera (2009–2015). More recently, Delgado founded the Risograph publisher La Impresora, which she currently runs with Amanda Hernández. Conceived as an "imprenta-escuela" (part press, part school), La Impresora now also sponsors the Feria de Libros Independientes y Alternativos (FLIA), Puerto Rico's independent and alternative book fair, which Delgado founded in 2012. Through these projects, Delgado provides free youth writing and alternative publishing workshops across Puerto Rico, while publishing innovative writing from across Latin American and the Caribbean, including diasporic/Latinx communities in the United States. La Impresora has become a crucial initiative in the months since Hurricane María, working toward off-the-grid publishing and modeling independent publishing as a decolonial hemispheric practice. Delgado's work has inspired and informed recent initiatives such as Anomalous Press's 2018 "Puerto Rico en Mi Corazón," a bilingual broadside series for hurricane relief in Puerto Rico bringing together poets from the archipelago and the diaspora.

Delgado's *Amoná* (2013) is a miniature (3 x 2.3 inches) foldout book about a camping trip to Mona, whose Taíno name gives the book its title. Like *Desecheo Notebook*, *Amoná* revises masculinist poetic/political genealogies, in this case rewriting the monumental trees of the poem "Arboles" (1955) by the iconic Puerto Rican poet and nationalist, Clemente Soto Vélez, (1905–1993) that serves as a midpoint epigraph. The mid-point is key given both the foldout book format and the location of Mona: The island is at the intersection of the Atlantic Ocean and the Caribbean Sea; it is also halfway between Puerto Rico's main island and Hispaniola ("Amoná" is purported to mean "what is in the middle" in Taíno). In "Arboles," as it is cited in *Amoná*, Soto Vélez celebrates "la canción" ("the song") of the trees as the transformation of "la persona universal" ("the universal person"), and Delgado reclaims Soto Vélez as part of a poetics of resistance to ecocolonialism informed by activism related to the US military occupation of the outlying island of Vieques. One of Delgado's first La Impresora publications was her 2016 artist book, *subtropical dry*. The publication documents a camping trip to Vieques in similarly eco/geopoetic terms. At the same time, it explores the limits of the page. In this case the book is blue-green and approximately 4 x 4 inches. It is largely empty, has unnumbered pages, and is covered in a case featuring a topographic map of Vieques. These spare, blue-green pages reclaim "el agua entre las islas" ("the water between islands") as their archipelagic "territorio" (2016, n.p.). When I hosted Delgado at New York University a few weeks after Hurricane María, she observed, only partly in jest, that she has become something of a "camping poet," and she connected this and her artist book practice to her desire to think about space—including the space of writing—in new, different ways, and especially in the context of neoliberal austerity and environmental colonialism.

Much of *Amoná* is familiarly ecopoetic: meditations on plastic bottles in the ocean, drawings of trees on the book cover, the Taíno-folklorized title. Yet, Delgado understands Mona in terms of its colonial geography. She reflects on its old lighthouse (purportedly designed by Gustave Eiffel of Parisian fame) and on the tensions between the privileged exteriority of place and her own inner search. As with Kyger, the speaker's quest for silence and inner peace is confounded at every step, in this case by her fellow campers, storms, saltpeter, inner voices, and the overload of the information age. All these invasions are scored in unnumbered pages of spare poetic prose: "Imagino el papel de los periódicos, los píxels de una computadora, el diálogo de algún oráculo" ("I imagine the newspapers' paper, a computer's pixels, some oracle's dialogue," 2013, n.p.) Even as Delgado concludes by claiming Mona, recovering "la playa y la cueva que también fueron nuestra casa" ("the beach and the cave that were also our house") in an obverse of Pedreira's insularist move, her book shares in Kyger's recovery of the "wild" inner coordinates of nonsite. This is particularly clear when Delgado imagines utopia through the insightful pun "delirar en la Cueva de Lirio" ("to rave in Iris Cave," perhaps Mona's most famous cave, although Delgado's antipoetic wordplay plucks its flower).

FIGURE 14.2 Foldout detail of Mona, from Delgado's *Amoná*.

Source: Photograph by Abey Charrón.

Initially, *Amoná* is framed by Soto Vélez's trees and by the phallic photographic triptych of the old lighthouse, framed by tall, thin trees against a cloudy sky (see Figure 14.2). Yet Delgado leads us away from the phallic/masculinist politics of the national/nationalist revolutionary poet into the gendered recesses of the cave/rave, and to an anti-epilogue that imagines the speaker snorkeling naked and encountering the flora and fauna of Mona (e.g., soursop, the endangered Mona ground iguana). For Soto Vélez, the semiotic power of trees has to do with their privileged spatiality, with a music that can traverse "sobre su meridiano" ("across its meridian"), akin to the Romantic loftiness of the Atalaya de los Dioses ("Watchtower of the Gods"), the avant-garde poetry group he cofounded in 1923. By contrast, Delgado's book ends modestly, with a photograph of a couple hiking through waist-high overgrowth, with no trees surrounding them. The phallic Romantic/nationalist poetics of privileged vision of the island/nation gives way to an *islote* poetics grounded in a difficult embodiment that is shaped by the accidents of Mona's landscape (dry, subterranean); this poetics paradoxically centers the irreproducible *islote* as the midpoint ("Amoná") of an archipelagic Caribbean. Like Kyger's Desecheo, Delgado's Mona is gendered but spare; it is not lush and vegetative in ways that we associate with reproduction. In asking us to imagine a more egalitarian gendered community without procreation, both poets point to a feminist ecopoetics shaped by colonial histories—one that cannot be reduced to Earth as Mother.

Geocriticism has largely focused on fiction, with Westphal's (2007) glossing of real and fictional spaces providing a framework for scholars such as Tally (2011). Even the burgeoning field of geopoetics has stressed the narrative dimensions of poetry, as in Eshun and Madge's embodied "storytelling" (2016). Inasmuch as geopoetics reflects the still largely Anglo-European, male, and heteronormative orientation of geography, it risks an instrumental recolonizing of space, foreclosing intersectional and locally rooted approaches. We see this clearly in the culturally specific "islote." Kyger and Delgado help us engender new modes of spatial critique and belonging; they experiment with notebooks or poetic prose while complicating narrative/oral geopoetics through an insistence on the materiality and visuality of writing and the book.

Both *Desecheo Notebook* and *Amoná* complicate circulation. The former remains out of print, though it is included in Kyger's 2002 selected works, *As Ever. Amoná* only ever circulated minimally, and it is not represented in Delgado's selected works, *Apenas un cántaro: poemas 2007–2017*. Both reimagine political space beyond urban, homosocial poetry movements. Both associate feminist poetics with a nonsite politics of the inner journey, with a processual and intertextual writing that challenges mappable space at the limits of the page, often experimenting with empty space and minimal text, so that the page embodies the desert/deserted yet sedimented space of the *islote*. Both move us away from a narrative geopoetics to one that makes meaning in and from crags—to one that creates sense in the detritus, silences, and violences of colonial logics as these fall out of fact and become undone (*des-hecho*).

Islote poems

(porous transcription of poems improvised while walking along Randalls Island on a windy day)

1. www.youtube.com/watch?v=1wjSoRzJ3Aw
I'm back on Randalls Island across from the Bronx Kill creek fatal waterway that separated
island from mainland or should I say mainland from archipelago behind me is Manhattan an
island in front of me Queens part of another island a long one airplanes overhead invisible
perhaps through the trees but audible a roar through rustling leaves that signals littorals
Atlantic fractures fissures broken voices bloody echoes of our black Atlantic our slave
Atlantic our indigenous but this is also an *islote* more than a topography a choreography
bodies of empire in movement and repose *islote* which translates as islet no pilot for
this eyelet no visionary poetics we're done with that we're doing our own thing our own song
like ducks in the kill animales de otro estuario el más precario *islote* is lot but not a lot really
especially its second meaning that craggy place uprock clamation accidents of the Americas
airfare welfare who pays the fare nothing fair about it this hemisphere I say we deal with it
this *islote* is not allotted it just is lottery of meaning José Emilio Pacheco in Vancouver scratch
that even if they were aware that's my second Caribbean Atlantic accidents of voice we are
smartphones connect us micromegaphones shouting down the voice inside trains behind me
planes before me voice behind factories slide into ecocide seeing how well viewfind
anthropo scene just need to think back to the *islote* scrappy craggy place where meanings end
and fish die some eaten by birds pretty soon the bards will die it's about damn time que se
pudran que se hundan barges and tugboats float on to the Atlantic it's this way and that way
it's no way always back to the *islote* no eco location accepting echo locution la colocación
del cuerpo junto al lindero la ribera lo escarpado lo liminal como el limo el desfondamiento
del pensamiento emptied out dig deep into the *islote* more crag sag of voice a geology not
related to any other yes archipelagos matter geopolitically think of New York Manhattan
Staten Long North and South Brother former Welfare and still that way Rikers shut it
down Rikers shut it down put down the phone shut down the institutions of the voice work
through the crag take note how close they are Rikers Island and LaGuardia airport the
privileged mobility of certain diasporas what we do in academia including academic poetry is
fetishize spatialities turn them into specialties specialization is its own nation the specious kind
the only one I'm trying to be less specious more inauspicious like the crag the *islote* putting
the spit back into my speeches splotches of oil from refineries like that one no time for fineries
time in

2. www.youtube.com/watch?v=lv4DeR9lZ74&t=207s

we are a dorsal fin in waters full of rare earth runoff endorphins made from melted smartphone
parts memories of quartz in a quarry that never was dorsal fins or just el fin del tiempo ponerle
fin al poema sin endosarlo this won't be one of those poems that endorses sunsets for poetic
effect let images follow courses post-colonial remorses on the networking sites between our
eyes our islets neuronal nations we are the only kind that ever was no fatherlands including
poetry paters some of whom I admired this isn't Whitman friends I'm homosocial but not
like that I'm not from one of these states no states at all never were but I'm not Pedreira's
pedregosity is that a word? either pedregosidad like latinidad how ironic when words that
are supposed to mark a condition we want to diagnose in order to decolonize and tear down
patriarchy all end in -dad latinidad same dad as austeridad austerity I'm recording this to
video to save it for posterity no selfie stick use my forepaw instead get it? *forepawsterity*
that was really bad like the *islote* air in any case this poem ain't for pa rest in peace nor
ma how's Florida? how's the cat? no pa no ma no ma no pa no mapa unmappable
because it has no parents it's not apparent sin parentesco sin característica aparente disappearing
and reappearing neural flash of island that's the *islote* my only sea is in my epilepsy propulsion
of my convulsion toward meaning yes that's the Manhattan skyline behind me flutter of the
trees low resolution but enough poems about skylines enough urban poems I wrote enough of
those my sky is in the islet my skylet I got no skills that pay the bills no skillet to fry
rhymes in islote eyelet I let I let I let dejo que me pase deja que todo pase let it run
its course Boricua style we'll smile it would be nice to throw a party maybe it will bring
the tourists back post María my mother's name and grandmother's but neither goes by that
other names for austerity what if austerity is our futurity it can't be that's one reason I improvise
against austerity against authority let flow become meaning happenstance rants if I'm cut off
from my family who live on other islotes the little family I have let this song this improvisation
be equal to the words families mutter to themselves wondering how they'll make it out or make
it through what it all means islote brings me back to land something our Caribbean crews
have not quite figured out what is territoriality in a small place Kincaid what does sovereignty
Yarimar Bonilla mean when your diaspora is way larger than your home islands a whole bunch
of islotes each one tethered to each other that's why I'm open to archipelagos Martínez San
Miguel Glissant the island can't contain us even though I clarify when asked que soy Boricua
de la isla Puerto Rican from the island no such thing then many islands many conquests many
tongues many Caribes one is deceived if one thinks there is just one Caraïbe caravans of
migrants here too trains indigenous histories here too hegemonic whiteness antiblackness
aquí también executive mestizaje euphemisms ditto but something's changing I see it as clear
as I see the sunset the hurricane changed us or no fuck us I can't speak for anyone else the
problem with we intentions Sellars it became my song the sound that runs through my lungs
becoming the sovereignty of the islote I embody new age as that might sound the nation won't save
us the manicured islands won't either and I say us as in the French on won't save one we're
our strength or our struggle and the way out is the shores coasts litorals littorals we
walk our lit oral flow on and off the page denied a language enslaved reembodied
recirculated a shouted secret every craggy place even those like Desecheo that I've never been
too really off limits but I imagine through my fellow poets poetas ellas leading the way as
they always have not just every crag but also statues and monuments and laws and
jurisdictions that define el islote not just physical or geographic or psychic or biotic space

affective space but also conceptual space the limits of this imprecise untranslatable islet
craggy place small island but for that very reason we must enter it on its own terms risk
getting lost understand your own relation to the space navigate distractions imperial navigation
metaphor intentional including sirens and airplanes overhead back to this Randalls Island
city island island city urban archipelago that was once separate from Wards Island got filled in
last century now one island two names Ward Randall what kind of name for a Boricua poem
is Ward Randall but then what kind of a Boricua poet is Urayoán Noel walking along a baseball
diamond kind of cuts a path toward home there is no home you're out always out you
know who knew that Julia de Burgos island poet feminist decolonial poet slash nationalist
that's a tricky one lived and organized in New York many parts of the Bronx often considered
Latina wrote of a farewell island a welfare island across languages self-translation which
as we know is the only nation que como sabemos es la única nación no se traduce se tropieza
the same way the indigenous Taíno Sikeo becomes Desecheo wasteland empty space yes
translation works both ways conquistadores translated Taíno to Spanish and left us with the bill
killed something these words can't fill translation from above and from below Julia translated
by a government spying on her and her subversive nationalist politics see Harris Feinsod not
that my islote is that kind of subversion just a version but also an attempt at geopoetic inversion
of the logic of islandness and its opening up to archipelagos and hemispheres all the familiar
moves of Caribbean Studies and hemispheric studies what if I'm just here this islote not
something to fetishize essentialize mourn or celebrate slot that becomes me slut of memory
to be

Notes

1 All translations are mine.
2 There are no page numbers in either *Desecheo Notebook* or *Amoná*, so none have been
   provided here. The text of *Desecheo Notebook* is reproduced in its entirety (albeit in a radically compressed poem, without the page breaks and map) in Kyger's 2002 collection *As
   Ever: Selected Poems*.

References

Benítez-Rojo, Antonio. 1996. *The Repeating Island: The Caribbean and the Postmodern Perspec-
    tive*. Translated by James E. Maraniss. Durham: Duke University Press.
Delgado, Nicole Cecilia. 2004. *Inventario secreto de recetas para enrolar las greñas con cilindros de
    colores*. Medellín: Editorial Palabra Viva.
———. 2013. *Amoná*. N.p.: n.p.
———. 2016. *Subtropical Dry*. N.p: n.p.
———. 2017. *Apenas un cántaro: Poemas 2007–2017*. Edited by Mara Pastor. Bayamón: Edi-
    ciones Aguadulce.
Dowdy, Michael. 2013. *Broken Souths: Latina/o Poetic Responses to Neoliberalism and Globaliza-
    tion*. Tucson: University of Arizona Press.
Eshun, Gabriel, and Clare Madge. 2016. "Poetic World-Writing in a Pluriversal World:
    A Provocation to the Creative (Re)Turn in Geography." *Social and Cultural Geography* 17
    (6): 778–785.

Fernández de Oviedo y Valdés, Gonzalo. (1535) 1855. *Historia general y natural de las Indias, islas y tierra-firme del mar oceano.* Madrid: Real Academia de la Historia. Accessed December 26, 2018. https://archive.org/details/historiageneraly01fern.

Flores, Juan. 2009. *The Diaspora Strikes Back: Caribeño Tales of Learning and Turning.* London: Routledge.

Glissant, Édouard. 1997. *Poetics of Relation.* Translated by Betsy Wing. Ann Arbor: University of Michigan Press.

Kyger, Joanne. 1971. *Desecheo Notebook.* Berkeley: Arif.

———. 2002. *As Ever: Selected Poems.* Edited by Michael Rothenberg. New York: Penguin.

Lezama Lima, José. 1969. *La expresión americana y otros ensayos.* Montevideo: Arca.

———. 2014b. *Coloniality of Diasporas: Rethinking Intra-Colonial Migrations in a Pan-Caribbean Context.* New York: Palgrave Macmillan.

Noel, Urayoán. 2011. "The Body's Territories: Performance Poetry in Contemporary Puerto Rico." In *Performing Poetry: Rhythm, Place and Body in the Poetry Performance,* edited by Arturo Casas and Cornelia Graebner, 91–110. Amsterdam: Rodopi.

Pedreira, Antonio S. (1934) 1973. *Insularismo: ensayos de interpetación puertorriqueña.* Rio Piedras: Edil.

Russo, Linda. 2013. "How You Want to Be Styled: Philip Whalen in Correspondence with Joanne Kyger, 1959–1964." In *Among Friends: Engendering the Social Site of Poetry,* edited by Anne Dewey and Libbie Rifkin, 21–42. Iowa City: University of Iowa Press.

Shaw, Lytle. 2013. *Fieldworks: From Place to Site in Postwar Poetics.* Tuscaloosa: University of Alabama Press.

Skinner, Jonathan. 2000. "Generosity and Discipline: The Travel Poems." *Jacket 11,* April. http://jacketmagazine.com/11/kyger-skinner.html.

Smithson, Robert. 1996. "A Provisional Theory of Non-Sites." In *Robert Smithson: The Collected Writings,* edited by Jack Flam, 364. Berkeley: University of California Press.

Tally, Robert T., ed. 2011. *Geocritical Explorations: Space, Place, and Mapping in Literary and Cultural Studies.* New York: Palgrave Macmillan.

Turner, Frederick J. 1893. "The Significance of the Frontier in American History." *Annual Report of the American Historical Association,* 197–227. www.historians.org/about-aha-and-membership/aha-history-and-archives/historical-archives/the-significance-of-the-frontier-in-american-history.

Watsky, Paul. 2013. "A Conversation with Joanne Kyger." *Jung Journal* 7 (3): 94–116. www.tandfonline.com/doi/abs/10.1080/19342039.2013.813830.

Westphal, Bertrand. (2007) 2011. *Geocriticism: Real and Fictional Spaces.* Translated by Robert T. Tally. New York: Palgrave Macmillan. Accessed December 26, 2018.

15

THE UNBENDING OF THE FACULTIES

Learning from Frederick Law Olmsted

Jonathan Skinner

Earth scene

Rather than stake out a specialized niche for creative writing or literary studies, the premises of ecopoetics, like geopoetics, inherently challenge disciplinary, generic, and professional boundaries, while at the same time resisting the administration of vague interdisciplines. The premise that creative work be aware of and informed by geological scale is primary and shared, with the "Anthropocene" corollary of human impact at geological (planetary) scale. Some suggest we consider a "Capitalocene," or "Plantationocene" era, to more precisely name the social, political, and economic formations at work in this scaling up of the human "footprint" (Haraway 2015).

While some argue that the "self-conscious Anthropocene" is a recent development, possibly more recent than the "Gaia" theory with which it is sometimes associated (Keller 2017), awareness of geological scale goes back to the formation of modern geology with Charles Lyell and James Hutton, amongst others, to Thomas Burnet's *Sacred Theory of the Earth*, to Renaissance cosmographers, and to poetic treatments of cosmogony in origin myths at the dawn of literatures around the world (including Hesiod's *Theogony* for Western literature). Ecology is thinking at "Earth magnitude" (Kahn 2013), and, according to poet Ed Roberson, "the nature poem occurs when an individual's sense of the larger Earth enters into the world of human knowledge" (Dungy 2009, 4). Rather than ask how ecology affects the writing and reading of poetry, we might instead consider how Earth magnitude unmoors our inherited sense of poetry, as a "literary" kind of making, to rethink the materials, sites, and scope of poetic practice.

Instead of adding more poets to the canon helmed by Walt Whitman and Emily Dickinson, for instance, ecopoetics might make a case for considering Frederick Law Olmsted, the "park maker" widely credited as founder of landscape architecture, to be one of the major American poets of the nineteenth century, a poet whose "distant effects" we live with today, as they continue to evolve in city-scale works across North America (Rybczynski 1999, 15). This would entail a poetry returned to its root sense of "making" (etymologically, from the Greek, ποιεῖν), exceeding the sense in which, for instance, Olmsted's proposed design for Montreal's "mountaintop park" affords, in his words, "successive incidents of a sustained landscape poem" (1997a, 214), or the sense in which "composing" with trees extends the agency and scale of such making (Olmsted 1997a, 204, 261; 2015, 987), since such a composition, depending on the species planted and the scope and administration of the design, will not be "finished" within the maker's lifetime (Olmsted 1997a, 13–14, 92). Such a case is made ecopoetically, that is to say, through a kind of writing and reading (or to borrow a term from Jed Rasula, "wreading") between text, its social dimensions, and site that seeks to manifest the meaning of park as geophysical poem in language both poetic and discursive, rather than through the received terms of academic criticism or expressive poetry (Rasula 2002, 11). For in such terms Olmsted's parks clearly are not poems, until they can be made poems, experienced, read, and written as poetry at the scale of the earth.

Ecopoetics shares with geopoetics a field of action, and of inquiry (making of such action a poetics), that by its nature exceeds the containment of any single discipline. In this both are modes of posthumanist inquiry, that is to say, modes of inquiry into the limits of the humanities as constitutive of humanist possibilities— limits that can only be registered through contact with other disciplines (Wolfe 2010, 115–118). In an age when those limits are brought into sharp relief by "scaler dissonances" characterizing what is now called the Anthropocene (Keller 2017, 38), such inquiry gains new urgency. As Cary Wolfe argues, after Niklas Luhmann, posthumanist inquiry is transdisciplinary when disciplines evolve in the contact zones afforded by networks of distributed reflexivity, rather than by thematizing conditions of their own possibility through weak forms of interdisciplinarity or pluralism (Wolfe 2010). An ecopoetics such as Olmsted's acts through the project of the "park" as an exclosure for democratic aesthetics within capitalist prospects. While its debt to the humanism of eighteenth-century English landscape design and Victorian "civilizing" mores is overt, it also assembles disciplines, under the emerging rubric of "landscape architecture," for unfinished compositions of an imagined future, contending with the "magnitude of geological change" (Smithson 1973). The planning, construction, and administration of Olmsted's parks, massively distributed over space and time, multiply contact zones for disciplines. Lacking in the posthuman dimensions of ecology, his parks nevertheless create posthuman spaces for the coexistence of bodies variously shaded by the human. Olmsted's "landscape architecture" arguably orients the human sensorium, psychology, and society to dimensions of earth magnitude or geopoetic scale. Asking us to reenvisage the role and scope of poetry at the scale of the earth, geopoetics comes into focus when we read Olmsted's works as we might read poems. Ecopoetic practices, such as walking and assembling textual materials and genre hybrids in pursuit of edge effects (Skinner 2001, 2018), offer means to this reading, also a form of writing that we might consider more properly as a practice of "wreading" through the social and environmental effects of Olmsted's earthworks.

A poetic ramble through Olmsted's parks, on paths shaded or lit by his many writings on parks, leads us to Enlightenment "neohumoral" somatic ideas, to the Victorian "miasma" theory of disease, to Romantic understandings of psychology and mental health based on Lockean philosophy of sensation, to the "moral treatment" of English asylum landscape design, and to a communicative theory of democracy, all inscribed in the designed shapes and materials of Olmsted's urban landscapes (Tremenheere 1841; Eddy 2008b, 192, 197; Szczygiel and Hewitt 2000; Halliday 2001;Tuveson 1960; Hawkins 1991; Gopnik 1997):"An influence . . . that, acting through the eye, shall be more than mitigative, shall be antithetical, reversive, and antidotal" (Olmsted 1997a, 244). Such influences make these parks more than ornamental arrangements and more than the descriptive imitations of picturesque tradition they are sometimes taken to be. Rather, we approach them as translations of Earth magnitude into effective therapeutic action, sites for social healing, with a vision for poetry equal to the emerging forces of the Capitalocene, mobilizing all the charms of the aesthetic (passive or "receptive recreation," Olmsted 1997a, 73–80): As Raymond Williams puts it, locating a division I will return to later, "the moment came when a different kind of observer felt he must divide these observations into 'practical' and 'aesthetic' . . . this need and position are parts of a social history, in the separation of production and consumption" (Williams 1973, 120–121). It is a return to poetry as the "all of making" that conjoins music, art, rhetoric, architecture, design, gardening—with, instead of drama, "commonplace civilization" on display (Kocik 2013, 54; Gopnik 1997). Already in the eighteenth century, to cite Williams again, the proscenium frame and movable flats were being developed simultaneously with country manor landscapes (Williams 1973, 124).

Prospect

Grasslands bring you imperceptibly higher until the earth curves; you can see all around, while much remains hidden. What seems flat dips and rises precipitously as you set your course. A grassland unfolds over time much as a forest does, in detail. Yet the horizon, only appearing at the edge of the forest, spurs and inspires you. Another kind of horizon opens along the strata of the earth's crust. These are "geo" horizons, of time and space. Earth magnitude inspires a distant prospect, a view that brings earth scale to mind. What of sightless, feeling, undulating, many-armed soil citizens? Cthonic consciousness (in the "Chthulucene") might do without horizons (Haraway 2015). But while we have much to learn from what lies beneath, poking our heads into the airy envelope gives us our kinds of mind. What's oceanic about the prairie involves a meeting of sea, soil, and air. Plus fire, sun burning in the cells of our foreheads.

Olmsted cited temporary blindness caused by sumac poisoning as his excuse for dropping out of college (Martin 2011, 17–18). He did not seem drawn to exhausting powers of sight on sublime magnitudes, yet distance moves his landscapes:

"A small catch of smooth blue horizon, at an infinite distance in the plain coun-try . . . Here the eye ultimately composes itself; and that way too the road hap-pens actually to lead," noted Thomas Jefferson in his famous passage on the view from Harper's Ferry (1787). Capability Brown's crowdsourcing of features, sheets of water, grass expanses and clumps of trees responds to symphonic and anticipates cinematic art, the serpentine lake with its leading promontories, the grand sweep-ing drive composing a sequence of passages, the massing eye of the masses. To this eye Olmsted brings nostalgia for the Connecticut hills he wandered as a child, an agrarian landscape of not exactly Jeffersonian liberty, rather, solitary escapes from boarding school discipline.

Olmsted was drawn to the imagination of "an unlimited range of rural condi-tions," to ease suggested in the shade of the overhanging browse lines of cedars of Lebanon and London planes (Scobey 2002, 235). As an American farmer in England entering the vale of Chester, Olmsted "looked down upon a beautiful rich valley, bounded on the side opposite us by blue billowy hills. In the midst of it was the smoke and chimneys and steeples of a town" (Olmsted 2002, 121). More George Inness than Thomas Cole then. Therapeutic principles would, however, lead Olmsted to privilege the beautiful over the picturesque, as he described the nearby Eaton Park:

A gracefully, irregular, gently undulating surface of close-cropped pasture land, reaching way off illimitably; dark green . . . very old, but not very large trees scattered singly and in groups—so far apart as to throw long unbroken shadows across broad openings of light, and leave the view in several direc-tions unobstructed for a long distance. Herds . . . quietly feeding near us, and moving slowly in masses at a distance; a warm atmosphere, descending sun, and sublime shadows from fleecy clouds transiently darkening in succession, sunny surface, cool woodside, flocks and herds, and foliage.

(147)

Here there are no abrupt precipices, rather gentle undulation, unbounded calm, depth-giving alternation of effects, including a kind of pastoral concentration (browsing flocks and herds) that sets the landscape in motion. It is primeval savan-nah and livestock pasture in one, silent synecdoche for the human masses.

Olmsted read, and reread, Johann Georg Zimmerman's *Solitude*, a Rousseau-esque treatise on the cure of mountain scenery and the Swiss middle landscape (Roper 1973, 35). To be sure, his aesthetics cultivate the "ha ha" moment, unen-closed power's need for disclosing views, inseparable from vanity strips of seclusion and concealment (Williams 1973, 120–126). (Pope: "half the skill is decently to hide" [in Williams 123].) But Olmsted was averse to epiphany, shy of the pressures of conversion that possibly precipitated the crisis leading, when he was a child, to what may have been his mother's suicide after attending a religious revival (rather than, as the official story goes, an accidental overdose of laudanum) (Martin 2011, 8–9). Horace Bushnell, a neighbor in Hartford and minister to Olmsted's friend

Loring Brace's family, preached an unorthodox, immanentist Christianity, reject-
ing the necessity of conversion while contending that Christian faith would be
adequately shaped by environment, which had a way of exerting "unconscious
influence" (170):

> The chief end of a large park is an effect on the human organism by an action
> of what it presents to view, which action, like that of music, is of a kind that
> goes back of thought, and cannot be fully given the form of words.
>
> *(Olmsted 1997a, 260–261)*

Like the country manor landscapes he admired, Olmsted's designs sought a
perpetually renewed sense of possibility rather than obvious greatness. But what
relation did they bear to "the mathematical grids of the enclosure awards"? As
Raymond Williams argues,

> they are related parts of the same process—superficially opposed in taste but
> only because in the one case the land is being organised for production,
> where tenants and labourers will work, while in the other case it is being
> organised for consumption—the view, the ordered proprietary repose, the
> prospect.
>
> *(1973, 124)*

In England Olmsted also discovered, more generally, a nature made for and with
humans, and in particular, a "park for the people" that brought rural prospects
into the heart of a shipbuilding suburb of Liverpool (Olmsted 1997b). Inspired by
Birkenhead Park, and without resolving evident contradictions, Olmsted realized
country manor views for the urban masses.

Olmsted was closer to Andrew Jackson Downing and his vine-clad Gothic cottage
than to Thomas Jefferson's Athenian temples. But the twentieth-century effects of his
parks are closer to the Hudson River School sublime than to Downing, framing his
landscapes with city towers rather than a Cotopaxi or Catskill ridge. Olmsted spoke
openly of the therapeutic value of parks—a project sharpened in, if not impelled by,
the contexts of the Civil War. Malcolm X noted that reading Olmsted, in the volume
of his antebellum reports on the "Cotton Kingdom," helped him to better understand
the horrors of slavery (Martin 2011, 404). In the redemptive park vision set forth in
perhaps the most comprehensive statement of his park poetics, "Public Parks and
the Enlargement of Towns," his 1870 contribution to discussions about plans for the
Boston park system, Olmsted cites Lincoln's Second Inaugural Address ("With malice
toward none, with charity for all"), itself an echo of Galatians 3:28, in his description
of the communicative democracy he envisaged as the highest purpose of his parks:

> all classes largely represented, with a common purpose, not at all intellec-
> tual, competitive with none, disposing to jealousy and spiritual or intellec-
> tual pride toward none, each individual adding by his mere presence to the

pleasure of all others, all helping to the greater happiness of each. You may thus often see vast numbers of persons brought closely together, poor and rich, young and old, Jew and Gentile.

(Olmsted 1997a, 75)

"White and black" are not explicitly conjoined in Olmsted's language, but his history of unvarnished reporting on slavery, through his firsthand accounts of travels across the antebellum South on commission for the *New-York Daily Times* (1852–1854), and the echo in the previously quoted passage of Lincoln's speech at the end of the Civil War, are good indications he had racial healing in mind. Olmsted co-planned and supervised the building of Central Park with war on the horizon. Landscape architect Austin Allen, student of John Brinkerhoff Jackson, suggests that Olmsted (famously given to insomnia) suffered recurring nightmares about the Middle Passage (personal communication, 20 August 2008), from a traumatic memory of the brutal shipboard flogging of a young boy, for cursing, that he witnessed during his brief stint as an apprentice sailor on the *Ronaldson*, fused with no doubt equally or more traumatic memories of the lashing of slaves (amongst other brutal treatment and social degeneracy) that he witnessed while touring the Southern States (Martin 2011, 33, 88). Compounding this would be Olmsted's guilt about the removal of Seneca Village (through the first use in US history of eminent domain to create a large park), a 264-strong village of impoverished immigrants, Indigenous residents, hardscrabble bone boilers and hog farmers, and many enfranchised (property owning) former slaves, occupying the western edge of the future park. The property owners were paid fairly (according to Olmsted's biographer Justin Martin, one of Seneca Village's founders received $2,335 for his house and property on land he had purchased for $125 in 1825) but by other accounts it was not a peaceful removal (129–30). Allen reads the destruction of Seneca Village as a violent origin for Olmsted's assertion of the democratic park, pointing out that "it is the Seneca Villages that have now surrounded the Olmsted parks. Fifty-seven percent of the use of Central Park is by minorities" (Allen 2008).

Olmsted also led the Sanitary Commission during the Civil War. His theories about the "sanitary" influence of trees (e.g., absorbing excess moisture from the soil, providing barriers to and shading the ground to prevent the heating and release of malarial "gases") developed through his contact with health professionals (Olmsted 1968). They are entangled with "miasma" theory and period fears about "putrefaction" in the soil (Szczygiel and Hewitt 2000), also sharing in the more general nightmarish sense of "foul liquid and meat" in the ground, from overcrowded cemeteries, but also from visions of the hundreds of thousands of war dead left to rot on the battlefields of the South, as expressed most strongly by Walt Whitman in "This Compost":

O how can it be that the ground itself does not sicken?
How can you be alive you growths of spring?
How can you furnish health you blood of herbs, roots, orchards, grain?
Are they not continually putting distemper'd corpses within you?
Is not every continent work'd over and over with sour dead?

Where have you disposed of their carcasses?
Those drunkards and gluttons of so many generations?
Where have you drawn off all the foul liquid and meat?
I do not see any of it upon you to-day, or perhaps I am deceiv'd,
I will run a furrow with my plough, I will press my spade through the sod
    and turn it up underneath,
I am sure I shall expose some of the foul meat.

(1856)

We now have a more positive attitude to soil (one that Whitman's poem actually works out), but his necropoetics speaks to our sense of a morbid planet, saturated with human "leavings," so that more than ever we seek green shelter from the stacks of "sour dead" rising at the feet of the Angel of History (Benjamin 1940, 249).

Olmsted's sunny greensward, affording opportunities for "passive recreation," gregariousness, and the "commonplace civilization" of what one might call a communicative adjacency of "horizontal" leisure groups (in contrast to the fixed, segregated, "vertical" society he found in the South), needs to be read into and beneath the grass, for these darker notes (Gopnik 1997). In their results Olmsted's projects play into alienation and the separation of production and consumption: As landscapes of leisure they are destined to obscure true relations of production, and they play a role as engines of real estate speculation. But in their refusal to divide aesthetic from practical aims they carry more subterranean values.

Olmsted's comment that the therapeutic value of parks is the "foundation of all wealth" because it sustains the wage earning power of the workforce—"there is not one in the apothecaries' shop as important to the health and strength or to the earning and tax-paying capacities of a large city"—is more a rhetorical hook for the industrialists than a cynical admission of his parks' productive utility, but it's there (Olmsted 1881, 23). Of this therapeutic value, Olmsted remarks:

> to this process of recuperation a condition is necessary, known . . . as the unbending of the faculties . . . and this unbending of the faculties we find is impossible, except by the occupation of the imagination with objects and reflections of quite different character from those which are associated with their bent condition.
>
> *(Olmsted and Vaux [1866] 1968, 100–101)*

Such action on the imagination via the "faculties" can be traced to the "neohumoralism" of the Scottish Enlightenment:

> The causal powers of such principles were linked to a medical theory held that body tissue was made up of fibres that, depending on their composition, could be made to contract or relax. Contraction forced fluids out of the viscera and flesh, thereby making the body hard. Relaxation allowed fluids to seep in, thereby making it soft. If one wanted to maintain a healthy nervous

system, one had to make sure that the tissues remained properly balanced
between hardness and softness.

(Eddy 2008a, 5)

In François Fénelon's immensely popular *Instructions for the Education of a Daughter*
(1688, but still being printed and widely circulated, in a translation by George
Hickes, as late as 1805), "due bending and unbending of the Faculties both of Body
and Mind" are prescribed (1713, 281). Following these popular notions, Olmsted
designed his parks with a mind to directly affecting the human physiology (in an
essay on the "unconscious" effect of trees in urban environments he invokes, without naming it as such, proprioception, to describe how the body negotiates complex environments largely unconsciously [2013, 64]). In this Olmsted's approach is
modern, anticipating the Reform Park (Cranz 1982). But the science is medieval,
or perhaps yogic, in the oldest sense. And the method is poetic, if not musical; the
art of serial form in time that Olmsted invokes in his plan for Mount Royal, Montreal replaces "distinct spectacles" with "successive incidents of a sustained landscape poem, to each of which the mind is gradually and sweetly led up, and from
which it is gradually and sweetly led away, so that they become a part of a consistent experience" (1997a, 214). Though Olmsted was no ecologist—his choice of
plantings, responsible for introducing several invasive species, was made primarily
with regard to aesthetic effects with a less determined concern for ecology, habitat, or conservation—our veneration of his "greenswards" obscures Olmsted's basic
understanding that park designs be adapted to the site. The city of San Francisco
did not accept his plan for a linear "xeriscaped" park—one that would work with
the area's microclimates and with California species—because city planners wanted
a Central Park for the West Coast, hence Golden Gate Park's imitation greenswards
(Martin 2011, 273–274).

We have barely tapped the extent to which Olmsted's creative career, conceived
as a kind of thoroughgoing poetry and a poetics, models how we might rethink
human ("Anthropocene") social and ecological relations to our "post-natural"
planet, in all their unevenness and in constructed rather than given terms, and how
we might reenvisage the role and scope of poetry and poetics itself, as a geopoetics.
His parks' capacities to hold together all manner of contradictions, to articulate an
art of and for the body and its sensations at the largest meaningful scale, to respond
to traumatic historical and environmental circumstance, and to integrate political,
social, medical, psychological, aesthetic, and ecological theories into design elements model—however primitively—the extent to which a poet's range of interests
and competencies might interact, for better and for worse, with the work of designing a more sustainable human place on Earth. They are heterotopic realizations,
always in process, along the lines of Robert Smithson's, "Somehow to have something physical that generates ideas is more interesting to me than just an idea that
might generate something physical" (1972, 298–299). Yet, they are also better than
utopian imaginings at helping to keep the live contradictions and tensions legible,
on the surface of our version of an impossible future (the end of capitalism, less easy

to imagine than the end of the world) within the ever more limiting conditions of the possible.

To learn from Olmsted, I suggest that as (geo)poets we might write in and with his parks. The territory is not the map, and the experience of the territory, itself in process, is never fixed. Walking, reading, writing are the necessary recursive gestures, which a poem may be particularly well suited to embody. The following, written in and through a visit to Buffalo's Delaware Park, takes a preliminary stab at what such a poem might look like.

The Unbending of the Faculties

leaving the Cheerios-scented
nature preserve we pass
General Mills's Elevator

a billboard reading
"Instant Everything"

up the Thruway and Niagara
to Santasiero's for
Pasta Fasoola, Pasta
Santanesca
dry red table wine

past Cornershop:

"An 'outside' was always what I wanted
to get to, the proverbial opening
in the clearing, plain church with massed,
seated persons, the bright water
dense with white caps and happy children . . . "

to Bidwell Parkway

no neighborhood
more than "many minutes' walk"
from a PARKWAY where

some substantial recreative advantage
is incidentally gained
in passing through

or that was the idea: to put the city
in the park, not the other
way around

"to be done in fresco, considering
exterior objects,
some of them at a distance

and even existing only
in the imagination"

PARK
"a simple, broad, open space of clean

greensward,

with play of surface and a sufficient
number of trees about
to supply a variety of light and shade."

Buffalo's estimated canopy cover twelve
percent, the national average twenty seven

crossing a bay—the distances
                              hard to judge, what seemed empty
ripple with singularities:

a sort of tumulus ("par"
a Natural Regeneration Area,
a "Planetary Garden"
with insect soundscapes

"an agitation of flows around the planet
—winds, marine currents, animal and human migrations—through which
the species transported find themselves mixed
and redistributed."

what is the park to me
spots of time
on the edges of the fairway
I know a place

some mower with a sense
of humor left

a weedy island
city of clover and iris flag
datura thistle & daisies

hollyhocks
queen anne's lace
sumac, brown-eyed susans

big oaks on two sides
not Parkside's

green islands . . . commons . . . ornamental
fountains and flower plantations, traffic
circle plantings

not trees as trees
                  turf, water, rocks, bridges, as things
        in themselves

but Riverside's open concept islands,
with their greenswards, tree-shaded
nooks and obliquely placed benches,

spaces inviting the walker in
to tarry, landscapes screening
the houses on either side

whereas Parkside houses face off, in naked,
open competition.

The Park

A Promenade (Or treadmill
may be carried along the outer part
of the surrounding groves.

     No Parking On Grass

Other parks could hold stages, parade grounds, formal gardens.
The Park would be devoted
        to a natural
           landscape

                of wide rolling meadows, wooded thickets
penetrated by winding paths, and a large lake

     of islands and inlets.

The fairway feels striated, as one walks through. Crossing territories. Lines.

choosing an undeveloped (1868), sloping landscape traversed by the
peacefully meandering Scajaquada Creek: grassy areas with water and stands
of mature trees, the perfect place, for a natural oasis in the city

in the Bertie Waterlime, on glacier-churned bones of Neosiluric coral, not
far from the pool of eurypterids.

Stone bridges, in the northeast corner of the park, once spanned the
Quarry Garden—since filled with debris from the construction of Route
198—now sit on grass.

"Four Trees—upon a solitary
Acre—
Without Design
Or Order, or Apparent
Action—
Maintain —"

Survivors of the Columbus Day (2006) storm
where the snow did

THUNDER

   in the Episcopal Church
   as Blaser reading
   his poem for Creeley
   said the word "God"

into the Silurian, into Onondaga limestone, "the Devonian"
right where the tracks from Canada cross Main Street . . . "a precise
Route 20 limit of Western New York."

the sky towers over the towering trees, in Burchfield's
                                     The Three Trees

Flaneur of the lakeshore and foothills of Lake Erie, Alleghenies, Salem to
Buffalo. Sunflower mystic. Painter of nightmoths.

Steps into the ordinary kingdom of open country. Like Olmsted, Burchfield
walked all hours and watches . . .

with you, the boxcars shifting, and the thousand-eyed creatures.

The Scajaquada has since been replaced by a "meandering"
                            expressway (1960s NY State Route 198),

dividing the park in two

and imposing its 50 mph rush to join I-190 on the lake.

Then one bird, a song. Flattery? Or, *fuck off.*

"Scajaquada," a shibboleth
it meanders so
when you get pulled over
in Buffalo

all the art of a park . . .

"*of which the predominating influence is to stimulate exertion;
OR which causes us to receive pleasure without conscious exertion . . .*

*is to influence the mind
through the imagination*"

with a pastoral, hazy outline . . .
        obscurity of detail further away

*a wild flower on a grassy bank . . . while we have passed it by without stopping . . .
may possibly . . . have had a more soothing and refreshing sanitary influence . . .
than an imported flower blooming under glass*

and with picturesque, entangled . . . clumps
complexity of light and shade near the eye

The old "change of scenery" cure

presenting the mind

with designs that "*withdraw the mind to an infinite distance*"

from all "*objects associated with streets
and walls
of the city*"

to an infinite distance

"*the constant suggestion to the imagination of an unlimited range
of rural conditions*"

On the track, one experiences the park in relation to a lane, to the relative velocity of other walkers.

Catching up, passing, getting passed.
Bicycles. And beside me, Jack Foran has a theory
about the psychogeography of Buffalo—
trace of the Devil's Hole Massacre, the Seneca cession,
and a displacement of Elmwood Avenue mind and matter,
some ten meters west and east of center.

. . . scraped bare, rubbed against the sky, exposed. A place for watching clouds.

Humboldt Parkway, a green artery from The Park in the west to The Parade
to the east, made way for "33,"
four lanes pounding the suburbs to work, dividing them,
a moat, a border vacuum . . .

the color line the Conservancy's master plan crosses, with Delaware Park:
"Reconsider the golf course—maintain, downgrade to a
9-hole course, or remove."

Who plays here
who can't play in their own Olmsted park,
east of the Main Street line?
Push-back from the mayor's office, with the unions.
Maintenance with
or without art.

"This is Buffalo"—crushing an invisible cigarette butt, on the back of his
hand.
Drawing a line, with her finger and eyes—"that long northern light."

There are vast quantities of it in the air immediately above The Park,
reflected off the lake.

Locke:

"to establish sound minds in sound bodies—the foundation
of all wealth.

a condition is necessary, known . . . as the unbending of the faculties
which have been tasked,

with objects and reflections of quite different character
from those . . . associated with their bent condition"

the foundation of all wealth

the Albright Knox Art Gallery and the New York State Building
"deductions" from the park proper,

> entailed cutting down
> several hundred mature trees.

"men who have been breaking down frequently recover tone
and are able to retain an active and controlling influence
in an important business,

> *from which they would have otherwise been forced*
> *to retire."*

golf, football, and baseball secured a foothold in the Meadow
by 1899. Formal baseball diamonds appeared in 1914.

"school-girls . . . to be taken wholly, or in part, from their studies,
and sent to spend several hours a day rambling

> *on foot in the Park"*

therapeutic atmosphere made available
to the masses: a prophylactic, open
asylum

I could stand still for hours, on the edge of insomnia, nostalgia—where else to go?
in Olmsted's parks the open space system, the progressive era, and the reform
park are layered

> the foundation . . .

the greensward in the country manor plan

> saved
>> as a golf course

favoring the *"unconscious or indirect recreation . . .*
which goes back of thought, and cannot be fully given the form

> of all wealth

A note on the sources

This poem quotes language from Gilles Clément ("An agitation of flows," *Le jardin Planétaire*, 2009), from Robert Creeley ("An 'outside' was always what I wanted," "For Anya," *If I Were Writing This*, 2003), from Emily Dickinson ("Four Trees," *The Poems of Emily Dickinson*, 1999), from the writings of Frederick Law Olmsted (most of the language in italics is drawn from *Civilizing American Cities: Writings on City Landscapes*, edited by SB Sutton, 1997), from Charles Olson ("A precise/Route 20 limit," *Muthologos*), and from *Buffalo Olmsted Park System: Plan for the 21st Century* (Buffalo Olmsted Parks Conservancy, 2008). A previous version of this poem appeared in *EDNA: A journal of art in residence*, Issue 5: 2015. I am grateful to the The Millay Colony for the Arts for a one-month residency (2013) during which I was able to conduct the research behind the poem and this essay.

References

Allen, Austin. 2008. "Dreaming Spaces Anew." *Project for Public Spaces*. Accessed December 31, 2008. www.pps.org/article/austinallen.

Benjamin, Walter. (1940) 1992. "Theses on the Philosophy of History." In *Illuminations*, edited by Hannah Arendt and translated by Harry Zohn, 245–255. Reprint. London: Fontana Press.

Clément, Gilles. 2009. *Le jardin Planétaire*. Translated by Jonathan Skinner. *Gilles Clément*. Accessed October 20, 2019. www.gillesclement.com/cat-jardinplanetaire-tit-Le-Jardin-Planetaire.

Cranz, Galen. 1982. *The Politics of Park Design: A History of Urban Parks in America*. Cambridge, MA: MIT Press.

Creeley, Robert. 2003. *If I Were Writing This*. New York: New Directions.

Dickinson, Emily. (1863) 1999. *The Poems of Emily Dickinson: Reading Edition*. Edited by R.W. Franklin. Reprint. Cambridge, MA: Belknap Press of Harvard University Press.

Dungy, Camille, ed. 2009. *Black Nature: Four Centuries of African American Nature Poetry*. Athens, GA: University of Georgia Press.

Eddy, Matthew D. 2008a. "'An Adept in Medicine': The Reverend Dr William Laing, Nervous Complaints and the Commodification of Spa Water." *Studies in History and Philosophy of Biological and Biomedical Sciences* 39: 1–13.

———. 2008b. *The Language of Mineralogy: John Walker, Chemistry, and the Edinburgh Medical School, 1750–1800*. Oxford: Routledge.

Fénelon, François. 1713. *Instructions for the Education of a Daughter, Done into English, and Revised by Dr. George Hickes*. London: Jonah Bowyer.

Gopnik, Adam. 1997. "Olmsted's Trip: How Did a News Reporter Come to Create Central Park?" *The New Yorker*, March 31.

Halliday, Stephen. 2001. "Death and Miasma in Victorian London: An Obstinate Belief." *BMJ (Clinical Research Ed.)* 323 (7327): 1469–1471.

Haraway, Donna. 2015. "Anthropocene, Capitalocene, Plantationocene, Chthulucene: Making Kin." *Environmental Humanities* 6 (1): 159–165. https://doi.org/10.1215/22011919-3615934.

Hawkins, Kenneth Blair. 1991. "The Therapeutic Landscape: Nature, Architecture, and Mind in Nineteenth-Century America." PhD diss., University of Rochester.

Jefferson, Thomas. (1787) 2014. "Notes on the State of Virginia." In *Writings*, edited by Merrill D. Peterson, 123–326. Reprint. New York: Library of America.

Kahn, Douglas. 2013. *Earth Sound Earth Signal: Energies and Earth Magnitude in the Arts.* Berkeley: University of California Press.

Keller, Lynn. 2017. *Recomposing Ecopoetics: North American Poetry of the Self-Conscious Anthropocene.* Charlottesville: University of Virginia Press.

Kocik, Robert. 2013. *Supple Science: A Robert Kocik Primer.* Oakland: On Contemporary Practice.

Martin, Justin. 2011. *Genius of Place: The Life of Frederick Law Olmsted.* Cambridge, MA: Da Capo Press.

Olmsted, Frederick Law. 1881. *Mount Royal, Montreal.* New York: G.P. Putnam's Sons.

———. 1997a. *Civilizing American Cities: Writings on City Landscapes.* Edited by S. B. Sutton. Cambridge, MA: Da Capo Press.

———. (1851) 1997b. "The People's Park at Birkenhead, Near Liverpool." In *The Papers of Frederick Law Olmsted: Supplementary Series, Vol. 1: Writings on Public Parks, Parkways and Park Systems*, edited by Charles E. Beveridge and Carolyn F. Hoffman, 69–78. Reprint. Baltimore: Johns Hopkins University Press.

———. (1852) 2002. *Walks and Talks of an American Farmer in England.* Reprint. Amherst, MA: Library of American Landscape History.

———. (1882) 2013. "Trees in Streets and Parks." In *The Papers of Frederick Law Olmsted, Vol. VIII: The Early Boston Years, 1882–1890*, edited by Charles E. Beveridge, Ethan Carr, Amanda Gagel, and Michael Shapiro, 60–67. Reprint. Baltimore: Johns Hopkins University Press.

———. 2015. "Project of the Biltmore Arboretum [N.D.]." In *The Papers of Frederick Law Olmsted, Vol. IX: The Last Great Projects, 1890–1895*, edited by David Schuyler, Gregory Kaliss, and Jeffrey Schlossberg, 985–988. Baltimore: Johns Hopkins University Press.

Olmsted, Frederick Law, and Calvert Vaux. (1866) 1968. "Preliminary Report to the Commissioners for Laying Out a Park in Brooklyn, New York: Being a Consideration of Circumstances of Site and Other Conditions Affecting the Design of Public Pleasure Grounds." In *Landscape into Cityscape: Frederick Law Olmsted's Plans for a Greater New York City*, edited by Albert Fein, 95–127. Reprint. Ithaca: Cornell University Press.

Olmsted, Frederick Law, et al. (1871) 1968. "Report to the Staten Island Improvement Commission of a Preliminary Scheme of Improvements." In *Landscape into Cityscape: Frederick Law Olmsted's Plans for a Greater New York City*, edited by Albert Fein, 173–300. Reprint. Ithaca: Cornell University Press.

Olson, Charles. (1978) 2010. *Muthologos: Lectures and Interviews.* Edited by Ralph Maud. Vancouver: Talonbooks.

Rasula, Jed. 2002. *This Compost: Ecological Imperatives in American Poetry.* Athens, GA: University of Georgia Press.

Roper, Laura Wood. 1973. *FLO: A Biography of Frederick Law Olmsted.* Baltimore, MD: Johns Hopkins University Press.

Rybczynski, Witold. 1999. *A Clearing in the Distance: Frederick Law Olmsted and America in the Nineteenth Century.* New York: Scribner.

Scobey, David M. 2002. *Empire City: The Making and Meaning of the New York City Landscape.* Philadelphia: Temple University Press.

Skinner, Jonathan. 2001. "Editor's Statement." *Ecopoetics* 1: 5–8.

————. 2018. "Walking." In *Counter-Desecration: A Glossary for Writing within the Anthropocene*, edited by Linda Russo and Marthe Reed, 73. Middletown, CT: Wesleyan University Press.

Smithson, Robert. (1972) 1996. "Conversation in Salt Lake City: Interview with Gianni Pettena." In *Robert Smithson: The Collected Writings*, edited by Jack Flam, 297–300. Reprint. Berkeley: University of California Press.

————. (1973) 1996. "Frederick Law Olmsted and the Dialectical Landscape." In *Robert Smithson: The Collected Writings*, edited by Jack Flam, 157–171. Reprint. Berkeley: University of California Press.

Szczygiel, Bonj, and Robert Hewitt. 2000. "Nineteenth-Century Medical Landscapes: John H. Rauch, Frederick Law Olmsted, and the Search for Salubrity." *Bulletin of the History of Medicine* 74 (4): 708–734.

Tremenheere, Seymour. 1841. "Report on the State of Elementary Education in Norfolk." In *Minutes of the Committee of Council on Education: With Appendices, 1840–41*, 424–466. London: Her Majesty's Stationery Office.

Tuveson, Ernest Lee. 1960. *The Imagination as a Means of Grace: Locke and the Aesthetics of Romanticism*. Berkeley: University of California Press.

Whitman, Walt. (1856) 1996. "This Compost." In *Walt Whitman: Poetry and Prose*, edited by Justin Kaplan, 495–497. Reprint. New York: Library of America.

Williams, Raymond. 1973. *The Country and the City*. Oxford: Oxford University Press.

Wolfe, Cary. 2010. *What Is Posthumanism?* Minneapolis: University of Minneapolis Press.

16

BORNE-AWAY

Tracing a gendered dispossession by accumulation

Diane Ward

Introduction

A friend, a 40-year resident of Oakland, California, commented recently that the city has become a dumping site. I pointed out that objects piled alongside sidewalks (mattresses and box springs, IKEA chairs, gaping chests of drawers) comprise households and are the visual evidence of evictions. These piles of extracted lives have become part of the urban landscape along with tent encampments beneath overpasses and in unpaved patches along freeways and recreational vehicles parked end-to-end along blocks-long stretches of city streets. These household objects and housing-optional solutions reveal the presence of the bodies that occupied or occupy these spaces.

Another sight that has become familiar in large cities up and down the West Coast from California to Vancouver is the empty lots where middle-class houses and surrounding vegetation have been "scraped," a real estate term for the purchase and demolition of older homes that are quickly replaced by privately built, boundary-to-boundary single-residence structures to maximize the value of the lot. Where is the evidence of bodies in this latter case of displacement and shifts in neighborhood composition? Houses-homes are transformed into commodities through global financial processes, affecting individual everyday contexts at the center of which is the body—what Adrienne Rich calls the "geography closest in" (1987). The houses' ultimate occupants, often women, are devalued and jettisoned along with the life of the home. This chapter explores how a geo-centered and collaborative visual art and poetic exploration can conjure and engage with the spectral human presence held in material objects.

In what follows, I document my collaboration as a poet and geographer with West Coast-based artist, Ursula Brookbank. I studied visual art and began writing and publishing poetry in my teens. In my poetry, I have interrogated power, the body, and gender relations through a focus on the materiality of poetic form in a dialectic relation with the materiality of the world. The materiality of the world has always been a part of my practice, but that is not unique among poets. My poetic contribution continues to focus on bringing language into contact with the material world in a project of cognitive mapping laid over a spatial experience of the world.[1] In my poetry, foregrounding the surface of language has not lent itself to description of the landscape but instead to an excavation of the physical experience of landscape as it intersects with and is experienced through language.

Brookbank works in the mediums of film, video, photography, installation, and performance. She uses historic photographic methods—cyanotype, instant film, Super 8 and 16mm film, and overhead projection in her films, installations, and performances. These have been presented widely in the United States and elsewhere. In her project, *She World*, Brookbank constructs an archive of women's material effects as a way to explore the gendered materiality of the loss of a family home. She focuses on the tactility and texture of film, historic photographic methods, and overhead projection to explore different ways to present the material in the archive (www.ursulabrookbank.com/she-world).

At the time of writing, Brookbank's archive was in a converted, detached garage behind her house in the Mar Vista neighborhood of Los Angeles. In the backyard, which functioned as a studio workspace, a table set up beneath a tarp held several models of housing structures, all painted white (Figure 16.1, on the left).

FIGURE 16.1   Brookbank's studio workspace, 2018 (photograph by Diane Ward); Home-Life, 2018, forty 10.75 x 13.75" archival prints (detail).

Source: Photograph by Ursula Brookbank

Brookbank extended the archive project by photographing women's homes as they were being staged for sale by real estate brokers, sold to individuals or developers, and demolished to make way for new structures. The sales were often conducted by surviving family members. The demolition phase of the houses provided the subject matter for models of—memorials to—the demolished houses. Brookbank constructed these models out of discarded materials such as cardboard toilet paper rolls or tissue boxes and then photographed them (Figure 16.1, on the right).

In an email exchange on 21 April 2018, Brookbank explained that erasure is a big part of the project, along with witnessing and watching as a generation disappears. In our collaboration, we aimed to address this disappearing generation—the women and the houses, linked through the home-life—using poetry to promote a greater understanding of the shared affect of gendered dispossession, felt across socioeconomic difference.

Poetry extending the field of affect

I produced two untitled poems for the collaboration. Both include grayed-out text indicating alternate lines that I considered during the composition of the poems. I decided to retain all of these lines in the final text because they perform the role of traces of dispossession. The alternate text disrupts the flow and introduces a stalled space for affective experience. The text was exhibited at The Edouard de Merlier Photography Gallery at Cypress College in Cypress, California, 8–18 March 2018 alongside some objects from the archive and a selection of the photographs of the models. The poems were written by hand onto heavy-gauge archival paper, using pencils with different hardness to indicate the text and the traces (or alternate) lines. Brookbank and I produced a textual score (see "Untitled #2" later) and performed this at the exhibit opening party. This score for two voices consists of the text of my poems alongside a list of objects from the archive, the woman's name associated with each group of objects, and some descriptive text about the houses. Brookbank read this catalog of objects while I simultaneously read the poem. This performance text was later workshopped by the Los Angeles text performance collective, SoundWordLab, and performed by Pauline Gloss and me at the Poetics Research Bureau (www.poeticresearch. com/2018/05/thursday-may-10-celina-su-diane-ward-sa.html). In the performance, Gloss and I exchanged roles each time the word "SIGH" occurred, pivoting between a recitation of the archived items and the poem's capturing of the affect of the photographs of the house models. This was intended to perform the co-constitution of the traces of home-life captured in the photographs, the archive, and the performers' voices.

Untitled #1

I haven't figured [my figure doesn't right] [my figure isn't right]
to right this devastation.
Product of devotion.
Or, when I see another cry,
I cry.

This time, the solution determines the crisis.
Handheld surface of grim celebrations.
In death, the overcast, is a bright meta

Action is human or nonhuman
Constructive, rotting
But no stasis, silence
Never defined finitely
Embodied

Expectations of loss
Focus [core] [hub]
Shadow [darken] [flicker]
Edge [frame] [boundary] [bitterness tone]
Known Light source [grasped] [comprehended]

How could I walk through
Knowing what is nothing-there
My demand
Is not
Paying-demand
But looking
Long

As if the organs hesitate to embody
Light rounds the corners
Light Sources compete for standing
Concealment, no-concealment
The long fingers, named and faced

Who straddles these dissolving structures
Unfigured

Untitled #2

Child fell asleep under the SIGH
Whose skin is a neglected surface
Whose SIGH folds into itself
The tree bows
Dimensions refuse rationalization
But linger
Long

We dangled our very existence on a nonSELF homeLIFE, but in the
process of living off homeLIFE we also raze it

Days of death are counted in viva-city
Pending demolition permits
Places held by drying bodies
Matter in this form
Lifts itself above matter SIGH
To hover in never-touching boundary
To never reach settle SIGH

If only we could SIGH
Incredibly, its form
When I breathe, do you collapse
And you breathe with me
When you collapse SIGH
Deflate dissolve to enter
The tree gestures
Without demand

I cry when I SIGH
Harmonic ocular surface
Ocular door
Ocular face
Function void

SIGH moves through
The curtain
Wind on loosened shingles
Once-walls commence shifting
To stay SIGH to not be destroyed
To not be created
To Begin With

I am a mass shifting through
Out the forms
That split with my breaking
Hold up my hand–over devoting
SIGH

Where the hill may rest beneath table
Terrain smoothed in cloth–like logic
The surrounded re–laid over in remembering
The surface–ground of fibers forced along cardinal directions

(Ward 2018)

Framing the houses project

An extended sense of dispossession

For Harvey (2003), accumulation by dispossession involves the dispossession of livelihood, the commons, or collective wealth. This dispossession provides investment opportunities for overaccumulated capital, and is, in this sense, a crucial means by which capital realizes accumulation under neoliberalism (Perreault 2013, 1064). Flipping the terms, Perreault describes a process of dispossession by accumulation in the context of Bolivian mining. He defines three forms of accumulation: of territory by mining operations, of water and water rights by mining firms, and of toxic sediments on agricultural fields (1050).

Dispossession is commonly linked to practices of land encroachment as public or common assets are converted into private property. However, the concept has been extended to encompass an appropriation of labor and the "wearing out of laboring bodies" (Butler and Athanasiou 2013, 252) through insecure and low-paying jobs that have surfaced simultaneously with cuts to education, health, and life-supporting welfare programs (252). New forms of subjectivity are then produced so that human life is turned into capital, and debt becomes a tool of biopolitical governmentality, that is, "a political and moral economy of life itself" (267). The original meaning of "economy" refers, of course, to the management of the *oikos*, the household, including the family, the family's property, and the house. Following this linkage of home with an economy of life, I explore the loss of home-life, specifically the vibrant and gendered materiality of women's lives as they are bound up in the home. I use materiality in Bennett's sense of a "vibrant materiality that runs alongside and inside humans" (Bennett 2010, viii).

We can see the material evidence of the loss of public and community spaces through the process of accumulation by dispossession (e.g., in sold-off Works Project-era post office buildings). We also experience dispossession through an anticipatory loss of place as the transformed built environment introduces art galleries, upscale restaurants and shops, and updated residential architecture. For example, there is proliferation of microbreweries, many of which celebrate place: Their names and labels emphasize the local, but the burgeoning industry goes hand-in-hand with rising inequality as housing costs go up and availability goes down. There is a sameness to this specificity as each locale asserts and markets its unique sense of local.

The collaborative project described in this chapter addressed the transformation of housing in the heightened real estate market in Los Angeles. As social media and technology firms proliferate in Los Angeles (Google, Yahoo, BuzzFeed, Facebook, Snapchat, and so on), there is a visible transformation of the surrounding neighborhoods. Displacement of renters has been accomplished through the application of the Ellis Act, a California state law that permits landlords to take rent-controlled units off the market by changing their use to condominiums.[2] As houses have become repositories of personal financial accumulation, the wealth stored within them is made available only through the dispossession of home, and often therefore of the women whose lives are intimately bound to the home. Displacement and gentrification are most-often addressed in the context of disadvantaged and poor communities that feel the negative effects disproportionately. In using poetry to explore the affect of dispossession conveyed through individuals and the materiality of their homes and surrounding neighborhoods, I hope to promote a greater understanding of the shared spaces of dispossession, as a gendered process across socioeconomic lines.

Constructing the archive and its supplementary structures

Brookbank's ongoing project, the She World archive (www.ursulabrookbank.com/she-world), comprises items cataloged according to the women they are associated with. It includes sheet music, photographs, clothing, wigs, household *tchotchkes*, and myriad personal items such as hairbrushes and cosmetics. These spectral objects define a former life, exposing "the affective dimensions of the collected items . . . [and] their capacity for embodying the residual vitality of their original owners" (Brookbank and Finlinson 2018, 407). The materials in the archive embody the women who were teachers, stenographers, scientists, writers, housewives, Scout leaders, physiologists, club women, botanists, musicians. The houses are the sites where the value of these women's lives, symbolized by objects, was accumulated. For these middle-class women, this accumulated value eventually meant their own dispossession.

Moving within methodological spaces

Hawkins distinguishes between geographers who interpret and analyze art ("dialogues") and geographers who collaborate with artists or curators to carry out research and practice creative techniques ("doings") (Hawkins 2011, 465). Regarding the "creative geographies" of landscape, Hawkins (2011, 467) notes that geographers work as, or with, artists to develop critical-creative writing styles that move beyond description. The goal is to evoke "the experiences of being-in and moving-through landscape" while maintaining a critical stance toward the politics and ethics of an "embodied act of landscaping" (467). My contribution to this collaborative project—what I was "doing" as a poet-geographer—was to interpret, using language, the affect produced through human/non-human relations relations: the women, the archive of their objects, the houses, the discarded materials used to construct the models, the models themselves, the photographs of the models, and the artist and her camera.

As we extended the project with poetry, we introduced an experience of time to the ontology of human/non-human relations. Witnessing the affect of the spectral materiality of loss was not intended to "make present previously absent objects" (DeSilvey 2007, 420) or stories but to explore the "mutual enfolding of self and world . . . beyond the singular personal experience" (Hawkins 2011, 467). Regarding the linking of geographical research with a poetic approach, I am cognizant, like artist-geographer Trevor Paglen, that public discourse, political economies of ideas, form, content, and production of space are all interrelated. In an interview with Michael Dear, Paglen cites Allan Pred when he suggests the most effective approach may be to forget methodology and listen to the materials. They will suggest their own ways of being researched (quoted in Dear 2011).

Brookbank invited me to engage with the archive as a poet—what I initially thought of as "writing through" the objects she had collected. She has been accumulating objects from estate and yard sales since 2007 when she began building the She World archive. She has used the objects in her art performances and invited writers, musicians, and artists to engage with the archive for their own work. The archive struck me as something that called out for interpretation and narrative; its collection of objects comprises an archive that can be categorized (and narrated) in any number of ways. Brookbank's unusual arrangement classifies objects—photographs, personal hygiene items, books, clothing—according to which woman they belonged. Her approach is phenomenological in that she believes objects can convey a tactile and intensely personal connection to each woman and offer audiences a direct experience of each woman's life-world (Finlinson 2018).

As I was puzzling over how to engage as a poet with the archive, I noticed a table set up under a tarp just outside the door. On the table were several small, white structures, reminiscent of fourth-grade Social Studies projects modeling historically significant buildings; but these looked as though they had been left outside, weathered and decomposing. I asked after these, and Brookbank explained they were models she had constructed and then photographed of some of the homes in which the archived women's lives had been lived. I began to formulate a way to address Hawkins' "doings" through a geographical/poetic approach that involved the models.

My aim in this chapter is to explore the resistance to disappearance of the house and dispossession of home-life as experienced through the feminine body and its effects—through an intersection of the women, the artist, and the poet. In the collaboration, visual and poetic approaches sought to convey the affect of loss and dispossession. Here, I offer a methodological approach that invites a critique of the "site of the image" (Rose 2007, 19)—in particular, the conditions of its technological production—to show how artist and poet-geographer strive to break down boundaries between researcher and subject, while expanding the resulting artwork-scholarship into "something in excess of research" (Holly cited in Rose 2007, 21).

In Brookbank's models and photographs, there is no language- or image-based ethnography of individuals that have occupied these houses. There is no narrative to be constructed to understand what has happened, how this has happened. We can experience Brookbank's photographs as the sites of perception, a form of art as visual discourse that resists the imposition of a Foucauldian discipline creating particular subjects (Foucault 1972; Rose 2007). Discipline is exercised prior to these images at the time of the house's "staging."[3] The house is not a home, but a commodity for sale.

A poetic housing

Writing the intersection of home-life, lifeworld, and resistance to dispossession

My poems were written in response to Brookbank's photographs of the models of the housing structures mentioned earlier. Brookbank modeled the houses in various states of their demolition, using cast-off materials. She then photographed these models. The models do not represent the physical presence of the houses. The photographs, in turn, do not represent the physical presence of the models. While the mundane material objects in the archive are familiar to us (we know when and what they were used for) as domestic objects that embody the women associated with them, the houses no longer exist materially in the world.

I began my poetic intervention at this point: Using language as a poet-interpreter of the voice of things (Ponge 1972) and their traces, I created a poetic dialogue among three parties: Brookbank, who invites witness to loss; the affective presence of the material traces of home-life (in the objects, the models, the photographs); and my own perceptual experiences. My intention was to translate visual percep-tions into language to ensnare an embodied experience of erasure, loss, and dispos-session. It was my hope that, along with the visual affect, my poetic language could enable the experiences to linger and, ultimately, grow to frame a more humanistic experience of the dispossession of bodies, lives, and home-life.

Brookbank's photographs of the models are ethereal images of structures that seem to respire (Figure 16.1, on the right). Breathing in, the inflated organic forms undulate, they are not staying in place; breathing out, the skeletal structures of the buildings are revealed beneath the deflated illusory surfaces. The images call into question our assumptions about seeing and knowing: They compel us to contem-plate the experience of loss simultaneously with the traces of what is lost. Brookbank shows us the perceptual field, but the "exquisite ambiguity of those constructions is also so important—richly provocative in their allusive specificity," as noted by visual theorist Johanna Drucker (personal communication, 20 April 2018). The photographs of the models are not representations of houses, the things, but of the perceptual field we experience. These images evoke allusions to experiences of loss, the persistence of memory, and the experience of the spectral as "sensed persistence without the fullness of presence" (McCormack 2010, 642).

The "lifeworld" is the site of everyday routines, interactions, and events that comprise the social experience. "Homelife" is typically used to indicate the domes-tic routine or way of living. I use the hyphenated "home-life" as a way to concep-tualize how the taken-for-granted objects in the archive and the houses resist being commodified and rationalized, instead retaining the capacity to engage directly and personally with the viewer. Home-life, like "lifeworld," encompasses the social life of the more-than-human. The "creative geographical" collaboration between Brookbank and me presented in this chapter offers a methodological approach inviting critique of the "site of the image" (Rose 2007, 19)—conditions of its technological production. Working from the photographs of the models, taking the perceptual field seriously, and undertaking a poetic engagement with affect presents how artist and poet-geographer can work collaboratively with—not on—researcher and subject, with objectives that exceed conventional research and result in hybrid forms of artwork-scholarship.

Shelter as intimately immense house

The sheltering house is a powerful symbol embodying safety and health. It is the site of intimacy and universality, regardless of its physical attributes. Referring to Baudelaire's use of the word *vast* to mark the poet's sense of the infinity of intimate space, Bachelard writes, "The mind sees and continues to see objects, while the spirit finds the nest of immensity in an object" (Bachelard 1969, 190). Accordingly, the house is not an "object" that can be objectively or subjectively described, but a site of attachment, necessary for the primary function of inhabiting; furthermore, it is "our corner of the world" (4). And yet, a house does not represent the guarantee of peace and haven to all, in the same way. Langston Hughes's poem "House in the World"[4] cautions that no house exists where his "Dark brothers" can evade the "white shadows" (line 6; line 3). Even with the poetic inversion of darkness and light, no house in this world offers sanctuary or evokes memories or hopes of peaceful shelter (Hughes 1995, 138).

For the women who have been dispossessed of house and home-life in Brookbank's project, the self and house are interdependent and co-constituted. The women, the objects, and the houses are bound together as in Derrida's notion of "ontopology," in which the ontological value of present-being is linked to its situation or the stable determination of locality, city, and body in general (Derrida 1994). However, the stability of this condition of being bound together allows for the dispossession of the women alongside the house and the objects within. According to Athanasiou, the logic of dispossession is mapped onto particular bodies-in-place through situated practices (i.e., at checkpoints), including the gendered body. This logic produces subjectivities dispossessed and categorized by calculable self-sameness in spaces of non-being and non-having. This "metaphysics of presence is mapped onto particular bodies, selves, and lives as absence, obliteration, and unarchivable spectrality" (Butler and Athanasiou 2013, 352).

Conclusion

A neighborhood understood as bundles of spatially based attributes (Galster 2001) that is undergoing economic transformation is also undergoing architectural, demographic, class, and sentimental transformation (Zwiers et al. 2016). Brookbank's initial intention for the archive as a collection of hybrid domestic accumulations

(Finlinson 2018) was to witness a generation disappear, understood through the spectral remains of the accumulation of materials in the She World archive. By inviting other artists and writers to interact with the archive, she extended the physical distribution of the material in She World while also expanding who witnessed a generation's passing. Participating artists were not simply passive historiographers of women's lives but participated in constructing a space of affect of the women through their home-life. It is more typical to view gentrification, displacement, and dispossession in conjunction with disadvantaged and poor. I look at middle-class housing in relation to dispossession as a way to establish the link between body,

. . .

[comprehended]   CARROL
                                         1362
How could I   Nylon net
Walk through   Wire angel
[knowing what]   Coral
Is nothing-   Zip lock bag
there   Buttons
My   Tape mended readers
Demand   Ceramic canary
is   Pill bottle
not   Gold beads
Paying-   Easter grass
demand   Cigar box drawers
But   Sequins
Looking   Linoleum squares
long   Early American
                                         Wagon wheel fixture
As if the organs   String ball
[wait]   Hook and eyes
Hesitate to   Glass hummingbird
Embody   Weather vane
Light   Gathered valance
Rounds   Spindles
The   Faux lanterns
Corners   Split level
Light
Sources   FRANCES
Compete for standing   201
[concealment]   Decks overlook
No-[concealment]   Spring
The long fingers   Whisk
Named   Nut grinder
And   Wire basket folds
faced   Strainer
. . .   . . .

FIGURE 16.2 Excerpt from "Performance Text Score," Brookbank and Ward (2018).

house, and home-life that can be applied to an exploration of loss experienced across socioeconomic differences.

The phase of the project that culminated in the aforementioned photography exhibit and poetry reading at Cypress College entitled *Home-Life* involved a poetic engagement with the photographs of the home models. The photographs, representations of the perceptual field we experience, evoke loss and its lingering effects. The poems produced for the collaboration reanimate what's been lost, encouraging an ongoing firsthand encounter with the devalued. Ensuing recitations of the poetry in multiple voices again extend the involvement spatially and through the people it engages (the readers) to construct a narrative about the feeling of loss, a recovery from the dispossession through accumulation.

Notes

1 In the poem "Our Occupants" (Ward 1995, 25–36), I examined the experience of the intersection of body and detritus in relation to the social upheaval during the uprising in Los Angeles in 1992. In a series of untitled poems, I used the Thomas Guide maps (pre-Google) as a structural template to describe the experience of emotional upheaval, inviting the reader to experience the pathways of ambiguous and relational multiplicities (Ward 2006, 29–32). In a collaboration with visual artist Michael McMillan, *Portraits and Maps*, my poetry further explored the physical world as experienced through the body (Ward 2000). More recently, the poem "Blue Sensed-Space" (Ward 2017) explored how the relationships among gesture (of the painter's body), geo-material (pigment), and the perception of the landscape are simultaneously present in the abstract landscape paintings of Zao Wou-ki.

2 See https://la.curbed.com/ellis-act for a collection of articles documenting the rising number of rent-control units taken off the market since 2016, as well as efforts to prevent the demolition of these units in the context of growing homelessness and a severe lack of affordable rental units.

3 In real estate terms, this is when the house is transformed into a recognizably curated environment with the application of paintings, lighting, greenery, and even scents to market the property. The prospective home buyer interacts with the home when walking through; she imagines acquiring a specific lifestyle in the role of occupant. This performance is a part of the staging production, reinforced by reality television shows (including *Designed to Sell, Sell this House,* and *The Stagers*, among others) that document home buyers and sellers in staged houses (see website "Staging a house: What 99% of homeowners don't do," www.therealestatestagingstudio.com/staging.php).

4 Hughes's poem, "House in the World," was originally published in 1931 in the journal *Contempo* with the title "White Shadows." The title was changed when it was published in 1934 in *Negro Anthology* (Cunard [1934] 1969).

References

Bachelard, Gaston. 1969. *The Poetics of Space: 1958.* Translated by Maria Jolas. Boston: Beacon.

Bennett, Jane. 2010. *Vibrant Matter: A Political Ecology of Things*. Durham, NC: Duke University Press.

Brookbank, Ursula. 2017. "She World." www.ursulabrookbank.com/she-world.

Brookbank, Ursula, and Jade Finlinson. 2018. "She World Archive and the Work of Artist Ursula Brookbank: A Collaboration with Archivist Jade Finlinson." *Collections: A Journal for Museum and Archives Professionals* 14 (3): 403–421, Focus Issue: Women & Collections:

Projects, Case Studies, and Creative Practice, edited by Juilee Decker, guest edited by Janet Ashton, Margot Note, and Consuelo Sendino.

Brookbank, Ursula, and Diane Ward. 2018. Performance Text-Score for Home-life Exhibit at Cypress College, Cypress, CA.

Butler, Judith, and Athena Athanasiou. 2013. *Dispossession: The Performative in the Political (Conversations)*. Kindle edition. Cambridge: Polity Press.

Cunard, Nancy. 1969. *Negro Anthology*. New York: Negro Universities Press.

Dear, Michael. 2011. "Experimental Geography: An interview with Trevor Paglen, Oakland, CA, February 17, 2009." In *GeoHumanities*, edited by Michael Dear, Jim Ketchum, Sarah Luria, and Douglas Richardson, 37–43. Abingdon: Routledge.

Derrida, Jacques. 1994. *Specters of Marx: The State of the Debt, the Work of Mourning, and the New International*. Translated by Peggy Kamuf. New York: Routledge.

DeSilvey, Caitlin. 2007. "Salvage Memory: Constellating Material Histories on a Hardscrabble Homestead." *Cultural Geographies* 14 (3): 401–424.

Foucault, Michel. 1972. *The Archaeology of Knowledge*. Translated by A. M. Sheridan Smith. New York: Pantheon.

Finlinson, Jade. 2018. "She World Archive and the Work of Artist Ursula Brookbank: A Collaboration with Archivist Jade Finlinson." *Collections: A Journal for Museum and Archives Professionals* 14 (03): 403–421.

Galster, George. 2001. "On the Nature of Neighbourhood." *Urban Studies* 38 (12): 2111–2124.

Harvey, David. 2003. *The New Imperialism*. Oxford: Oxford University Press.

Hawkins, Harriet. 2011. "Dialogues and Doings: Sketching the Relationships Between Geography and Art." *Geography Compass* 5 (7): 464–478.

Hughes, Langston. 1995. *The Collected Poems of Langston Hughes*. New York: Vintage.

McCormack, Derek P. 2010. "Remotely Sensing Affective Afterlives: The Spectral Geographies of Material Remains." *Annals of the Association of American Geographers* 100 (3): 640–654.

Perreault, Tom. 2013. "Dispossession by Accumulation? Mining, Water and the Nature of Enclosure on the Bolivian Altiplano." *Antipode* 45 (5): 1050–1069.

Ponge, Francis. 1972. *The Voice of Things*. Translated and edited by Beth Archer Brombert. New York: McGraw-Hill Book Co.

Rich, Adrienne. 1987. "Notes Toward a Politics of Location." In *Blood, Bread, and Poetry: Selected Prose 1979–1985*, 210–232. London: Virago.

Rose, Gillian. 2007. *Visual Methodologies: An Introduction to Researching with Visual Materials*. 3rd ed. Los Angeles: Sage Publications, Inc.

"Staging a House: What 99% of Homeowners Don't Do." *The Real Estate Staging Studio*. Accessed May 10, 2018. www.therealestatestagingstudio.com/staging.php.

Ward, Diane. 1995. *Human Ceiling*. New York, NY: Roof Books.

Ward, Diane. 2006. *Flim-Yoked Scrim*. Ottawa, IL: Factory School.

———. 2017. "Blue Sensed-Space." In *Willem de Kooning | Zao Wou-Ki*, 127–130. New York: Lévy Gorvy.

———. 2018. Untitled #1 and #2 (home-life poem). Manuscript in preparation.

Ward, Diane, and Michael McMillen. 2000. *Portraits and Maps*. Castelvetro Piacentino, Italy: NLF Editions.

Zwiers, Merle, Gideon Bolt, Maarten Van Ham, and Ronald Van Kempen. 2016. "The Global Financial Crisis and Neighborhood Decline." *Urban Geography* 37 (5): 664–684.

PART III
Intervening

17

THE LIMITS AND PROMISE OF URBOPOETICS

washpark, collaboration, and pedestrian practice

Patrick Clifford and Tyrone Williams

Ecopoetics, understood in its most encompassing sense, is a mode of human-making[1] in relation to the variegated environments in which multiple forms of sentience originate, thrive, adapt, and persist. Geopoetics is one of its subcategories and, as the prefix suggests, generally concerns itself with the unbuilt and built modes and constituents of the mineral, non-sentient,[2] kingdom. The following might constitute the objects of typical investigations of a geopoetics: land and landscape, stone and outcroppings, islets and archipelagos, and puddles and oceans. But what of an urban lawn, trees in a public park, a city fountain? In coining the term urbopoetics, we acknowledge a particular narrow ground, if not the entire range, of our concerns. Our immediate inspiration for this neologism was Garance Maréchal's and Stephen Linstead's analysis of Jacques Jouet's concept of metropoetics. Although urbopoetics is, like metropoetics, based on the material and abstracted realities of urban life, it varies from Jouet's concept in that, for us, the emphasis is on pedestrian, not motorized, traffic and encounters (Maréchal and Linstead 2010). For us, then, urbopoetics may be defined as a poetics of pedestrian—not metrical—traffic within the urban environs of a developed country, in this case, the United States of America. More narrowly, our object of interest was, and is, Cincinnati, Ohio. As we will explain later, this poetics is a nonlinear set of observations based on walking, stopping, recording (by camera and pen), and moving on. The source of this difference from Jouet's metropoetics is linked to our separate experiences in Over-the-Rhine (OTR), the economically challenged square mile neighborhood located between the Cincinnati downtown business district and the lower Clifton area. We named our urbopoetics project *washpark* for two reasons: It is shorthand for Washington Park, located right in the center of OTR, and thus symptomatic of the changes occurring throughout the neighborhood. It is also a reference to money laundering, not in its legal sense but, for us, as a synonym for "cleaning up the park," a phrase (and its variants) that was ubiquitous in the media as a description of what had to happen if downtown Cincinnati was to be "saved" from assumed economic, cultural, and social irrelevance.

In 1990, Pat Clifford began 21 years of work and leadership at the Drop Inn Center, a large homeless shelter located near the southern border of Over-the-Rhine. The Center had occupied a location just west of Washington Park since 1978. In 2010, Pat was removed from his position as director of the Center, in part because he was seen as resisting attempts by a conglomerate of business interests, led by the Cincinnati Center City Development Corporation (or 3CDC), to move the Center out of OTR in order to facilitate the redevelopment of Washington Park—one block away. In addition to the Center, the park is bordered on its south side by a secondary educational center, The School of Creative and Performing Arts, and on its north and east sides by apartment buildings and small business offices. Its west side faces Music Hall, one of the oldest and most venerated buildings in Cincinnati—primarily a site for classical music performances but known for industrial exhibitions and even Tin Pan Alley concerts in the more distant past. Upgrading the park, long seen as a haven for the homeless and drug dealers, was crucial to the business interests' vision of "revitalizing" the neighborhood north of the central business area.

Tyrone Williams spent six years (2004–2009) as a member and, in the last two years, as Board President of INKTank, a community literary/arts center founded to bridge social and cultural divisions in Over-the-Rhine. Located in a storefront in the middle of Over-the-Rhine, INKTank was conceived as a community resource center for literary creativity. We held spoken word readings, art performances, and workshops, drawing on writers and artists from the region. Originally located at Main and 13th Street—two blocks east of the park—INKTank was situated on a block long designated for gentrification. Its original landlords were young college graduates whose sense of social justice drove their business. Consequently, INKTank paid only nominal rent for its space, the ground floor of a three-story building. However, rising rents and taxes around their other holdings in OTR meant they had to increase, however incrementally, INKTank's rent. Eventually, the landlords were themselves forced out of business and had to sell the building. Meanwhile, INKTank had difficulty sustaining its funding sources—primarily arts funding organizations at the local and state levels—and, under the double pressures of decreased funding and skyrocketing rent under its new landlords, had to close down in 2009. INKTank succumbed to the same economic forces and gentrification orientations that led—or drove—Pat and Tyrone to a more radical consideration of "responses" to territorial encroachment and population displacement. In short, we had to place the closing of INKTank and Pat's firing from the Drop Inn Center within a larger context of population displacement and gentrification.

In the wake of the previously mentioned events, we—Tyrone and Pat—spent a year or more discussing exactly what we were going to "do." These discussions pre-dated the writing of poetry that would become part of our urbopoetics. In

2009, Tyrone completed his Atelos project (*Howell*) just as INKTank was closing its doors. The following year Pat, as noted earlier, was terminated "without cause" from his position as director of the Drop Inn Center. These events were, in many ways, catalysts for our collaboration. Furthermore, Pat's cross-cultural and collaborative poetic projects with Aryanil Mukherjee, *chaturangik/SQUARES* (CinnamonTeal 2009), and *The Memorandum/MOU* (Kaurab 2011) also served as partial models and inspiration for our joint endeavor. In the wake of these individual and collaborative artistic projects, the intrusion of collective and personal trauma—INKTank's closing, Pat's firing—was followed by a more severe form of social trauma: In the summer of 2010 a woman, Joann Burton, sleeping under a blanket in Washington Park, was run over and killed by a police car pursuing suspected drug traffickers. So, though we began with poetry, that writing was impelled by the most persistent issues underlying the field of ethnography: The relation of a human deemed the writing subject and another human deemed the written object. That this relation was not unlike that between a vehicle of the law and a human being literally prostrate before the law was not lost on us. We continued writing in the wake of these events.

In 2011, Pat and Tyrone began writing toward what would eventually become the *washpark* project by individually writing poetry and rewriting/revising each other's contributions between informal lunch meetings. Given the aforementioned concerns and contexts, our first forays into depicting the transformation of Washington Park were observations—almost field notes—describing actions taking place in and around Washington Park. Although we had not yet titled the project, these writings, focused on "anonymous persona," that is, the kind of people who generally don't "count" until they encounter law enforcement and/or social work agencies as casualties and statistics. These writings would constitute some of the early sections of *washpark*. Here are the first two paragraphs we composed under the heading of "Documentary Stories":

"Turquoise Shirts"

Two men met in Over-the-Rhine both wearing turquoise shirts. One was riding a bike, the other was walking. At this age, neither were material depositories of exchange value nor things of air. Cracks frame cement slabs—three square meals—the indices of ordinary wear and tear. Whether regulars sold at cost or irregulars practically "given away" at a loss, they accept downplay in trade, belonging to both the example and the studio. Pedalist, pedestrian: coming and going by appointment. Between them, something hidden from public scrutiny was exchanged subject to speculation precisely unspectacular. Authorities have confirmed the reports: four unidentified margins called "left the scene" last unseen in circulation. Their relationship is said to have "passed away." Spelled out quietly by the generous population of a mixed neighborhood abstracted from mixed identities in lieu of proper names, common property. A passion for the decrepit as propriety.

"Dance of Millions!"

A man with a grey shirt, jeans and a backpack walked north on Elm Street. A man with a blue shirt and khaki pants waved at the Metro bus at 7:32 pm. Account every kind of labour to be unskilled, simple labour, by this we do no more than save ourselves the trouble of ex-lovers. Between the devoted there is no discernment of natural and value forms. Ditto for the others well-beaten black and blue and gray, walking to see Iron Maiden, so many angels streaming into their concrete and steel abstractions shaped like one pin. Sitting or lying, it is not so easy to ask. Markets induce panic, confidence, yet arise not. Canceled assumptions distance themselves from them. And no doubt for expensive offices which spread like pine. Some yearn wasted on the other. Invisible doorstops serve as certain kinds of shoes borne by certain kinds of feet. No free verses.

In these initial stabs at documenting the quotidian, we insisted—as we would throughout the project—on mixing modes of observation, commentary, and theory. We constantly switched registers and called attention to scene and context, picture and frame. There'd be no idealized "transparency" of the camera or unshaded "window" of language. We believed that type of static poetics would reinscribe the false *distance* of the ethnographic *stance* we were at pains to interrogate even as we availed ourselves of its positioning of difference. In short, our assumption was that the language of linear narrative too often served the dominant narrative of progress when what was actually progressing was complex, at play among the wide variety of genres "native" to specific disciplines: social science, social work, historical investigation, city planning, literature (verse, drama, and prose), philosophy, etc. In order to mirror the actual conflicts of interest, colliding narratives, and different stakeholders, we felt it necessary to compose a document as complex and multifaceted as those that predated, accompanied, and followed the transformation of the park. In that sense, the concept of urbopoetics came after the practice of it, in so far as it was a rubric for what we were doing and because we were doing it in a particular way and with a particular ethos and aesthetic.

For three years we followed a routine that slowly took on the predictability of a habit: Lunch at a Korean bistro (now shuttered) and then a walk around Washington Park as it was undergoing rehabilitation as part of a more widespread program to gentrify Over-the-Rhine. We both took notes on (and Pat took pictures of) the construction process, the removal of grass, dirt, monuments, even human remains from the cemetery that used to occupy the park, the bulky awkward movement of machines and workers as they worked to transform the park from a haven for drug peddlers and homeless "derelicts" (according to the dominant narrative) into a family-friendly source of civic pride. Make no mistake: Many residents of Over-the-Rhine cheered on the park's renovation even as they worried about rents spiraling upward to meet the anticipated market of young professionals returning downtown to (re)claim territory. The collaborative and ongoing nature of the project allowed us to discuss and compare perspectives and triangulate perceptions toward writing a more complex account.

However, at a certain point in the project—three years later—we began to rethink "where" the work was going. Thus, we decided we needed a scaffold of questions to serve as a general guide. On 26 June 2014, we met near Washington Park and formulated the following questions:

1 What makes Washington Park "work" economically? What resources are help-ing to maintain and define the culture of the park?
2 What is the nature of the change from the "old" Washington Park to the "new" one?
3 What are the dimensions of the change: Physical, social, economic, demo-graphic, etc.?
4 Is the change a "natural" or linear developmental process? *Could there have been other alternative histories of the park?*
5 Is the cultural change sustainable?
6 How are social groups invited or disinvited to the space of Washington Park? Is there evidence of modes of social control beyond simple exclusion?
7 How do James Clifford's ethnographic processes (experiential, interpretive, dialogical, polyphonic) factor into our choice of poetic processes (Clifford 1983)?
8 How is Washington Park used to support a "redemption" narrative for poor communities? How does this reinforce the marginalized status of unredeemed neighborhoods?
9 Has "order" been restored to the wilderness? How is Washington Park used to support a type of manifest destiny?
10 Is resistance futile? Resistance to what?
11 Overall, to what extent can/should we use Washington Park as a metonym for the ongoing gentrification of OTR?

As mentioned earlier, one of the primary reasons for centering the renovation of Washington Park as part a poetic project was to capture and reflect the com-plexity of the interests, narratives, and stakeholders involved. The typical binary narratives pitting "urban progress" of developers and civic boosters against resist-ance "gentrification" (even "econocide") of social activists seemed to us too linear, reinforcing a perhaps unproductive binary (Skirtz 2012). By slowing and complicating our writing process, we were searching for a new progressive resist-ance to "progress" as defined by the media and business interests. This type of progress is reflected in the work of the ethnographer (no stranger to poetics) James Clifford:

> Other kinds of progress become imaginable: utopias that may be already here, ways forward that are not about progressing, but rather involve turning and

returning. The challenge is to imagine different directions and movements in history, developments taking place together and apart.

(Clifford 2013, 9)

These considerations led us to a general revision of some of the initial "sketches" we'd made, particularly in light of the ramping up of "urban renewal." Following the fencing off of the park when construction began, the immense excavation that made way for new parking structures, underground water and electrical systems, and removal of trees and basketball courts inspired us to imagine a poetics as brutal as the renovation process itself. A sample of what we tried to imagine can be seen in the following excerpt from "Blue Tone," part of a section of *washpark*:

> This, like Yours, is a stretch
> cropped
> monograph
> colon as that which precedes delineation
> for example, a tree is an uprooted
> head
> buried
> in the disappearances of urban
> removal: ours a decapitation.

Because of our connections to institutions located in the community—INK Tank and the Drop Inn Center—we had formed many relationships with community

FIGURE 17.1   Photo from the *washpark* project. A view of Cincinnati's Music Hall with Washington Park gated and under construction.

Source: Photograph by Patrick Clifford.

residents, offering us some "street cred." We were not quite interlopers or tourists. We were not quite community members either. Both of us had more-or-less creative interactions with people who lived in or frequented the neighborhood, and both INKTank and the Drop Inn Center strived to maintain open-door policies: People literally walked in from the streets. However, neither of us lived in the neighborhood; we both had to drive to get to our respective destinations. Needless to say, this was an important fact. Indeed, it was a foundation of, and limit to, the assumptions we had to confront as we embarked upon our collaborative journey. Put another way, had we confined our observations to those perceived from private or public transportation vehicles, we might well have participated in a metropoetics as Jouet defines it. However, to the extent our walking about Over-the-Rhine could only approximate the foot traffic of those without access to private or public transportation, an urbopoetics was, from the start, delimited by our relative privilege. In brief, our leisurely strolls through OTR in general, and Washington Park in particular, could not approximate the affective experience of OTR residents meandering to the grocery store, walking to work, carrying a child, running from a threat, or hanging out on a corner.

As social research theory has shown, spaces to explore and create narratives like INKTank (and the *washpark* project) are often explicitly created to transform social structures in pursuit of an agenda of inclusion and social justice (Mertens and Ginsberg 2008). To engage participants in these spaces, community workers often feel the need to make the unfamiliar (to them) somehow familiar (Pillow 2003). On the other hand, participants, often marginalized on the basis of race, class, and culture, are constantly asked to manage their identities, being put in the position of deciding, as Goffman describes, "to display or not to display; to tell or not to tell; to let on or not to let on; to lie or not to lie" (Goffman 1963). In other words, all actors in these spaces are enacting overlapping strategies for complying with standardized norms, exotic expectations, and real accountability. Premature or rushed inclusion—for example, staging a participatory performance with community residents without first building trusting relationships with them or involving them in planning efforts—can blur narrative and "real" public spaces. It can cause confusion and misunderstanding, concealing risks to participants and often working to confirm the very assumptions that underpin persistent marginalization (Bishop 2012). Additionally, social work and fostering artistic expression are two different modes of labor, each with assumptions and limits unique to the endeavor. For a while, INKTank successfully resisted being siloed as a social work institution or an arts center because it was supported with an initial infusion of private funds. However, its annual attempts to secure sustainable public funds were often thwarted by the organization's failure to "fit" the criteria for public funding in the Cincinnati budget. As an institution using aesthetic means to social ends—that is, providing a space for community members to express themselves in writing (and occasionally in drawing, sketching, and painting)—and thus helping to affirm community members' integrity as active, useful citizens, INKTank fell between the cracks insofar as Cincinnati budget lines are dedicated to clearly marked silos (e.g., social work or arts organizations). INK Tank was perceived as a little of both and not enough of either. When arts are funded by the city, they tend to focus on visual iconography (e.g., murals) that market "diversity" as a city value even as non-white and/or non-viable population displacement and disempowerment via gentrification is a "hidden" counternarrative to the ones promoted in Chamber of Commerce brochures. Urbopoetics is a direct response to these silo effects insofar as its poetics explicitly presuppose the "good" of both artistic innovation and radical social policies. The former is a means to disrupt a "greater" narrative coherence in which zoning laws, for example, presuppose impermeable borders between industrial sites and residential neighborhoods. The latter empowers, for example, those with little to no economic, political, or cultural clout.

Player

Stainless steel, what does it mean? Looked ceremonial, rinky-dinky. Head of
the Queen City all fall, way to was, reverse reverses. Bandaged post-surgical
perimeter, bandied feet (DON'T WALK) on grass (walk this way = hand
in hand as overcome), bounding down future bonds. There there solace
per suburbanite hearts trapped under houses, can't get above under, where
oh where is that rabbit foot launched out of a black hat as specified?
Whatever people want to call a vow. At a time when our state faces internal
exchanges.

★★★

The *washpark* process continued to gain clarity and explore new methodologies. After leaving the Drop Inn Center, Pat returned to school to pursue a graduate degree in social work administration at Case Western Reserve University in his hometown of Cleveland, Ohio. His program centered on community-level social change and program evaluation. Although Pat worked sporadically on *washpark* during the two years he was involved with his graduate studies, it's safe to say that the collaboration slowed. This interlude was, however, a positive event since Pat returned to the project with new ideas, a honed lexicon, and renewed enthusiasm. Indeed, his new knowledge infused our project with a new level of intentionality. One of these new ideas was to incorporate parts of an oral interview that Pat and Tyrone conducted with a local resident of OTR, Catherine Stehlin, whose family were active park users both before and after the renovation. Some of her observations included the following takes on the assumptions that a functional green space must be always actively "entertaining," casting the previous activities in the park—unprogrammed and unmediated—as somehow inadequate, if not dangerous, as they related to community needs. The following is an excerpt from a transcribed conversation with Catherine Stehlin on January 13, 2015.

CATHERINE: Yeah, like a park. That's a concept. Like a park. And there should be places just for people to hang out and be, you know, without having to be entertained. You know there's just very few moments when there's not something going on over there, and there's very few places. . . . There's only this little circle around the gazebo with shaded benches that you can sit in some of the time. The rest of it is quiet, you know, in the summer.

PAT: There's always something going on.

CATHERINE: There's always something going on and there's not really even a place to sit like that's not exposed. So, there's certain things about the design. Like the whole placement of the stage, they put the one stage over there, but then whenever they have a big event, they build a stage over by Music Hall. You know what I mean?

PAT: Yeah.

CATHERINE: Yeah. So the one thing about when the Park opened—because they did a lot of work and they put a lot of money into it—and they tried in some ways to be sort of sensitive in some ways—but they were proud of themselves—and they should be—but it was always at the expense of what was there. You know what I mean? It was always like, "This was a nightmare." It totally was a nightmare. I mean some of the reports that. . . . Even the paper would write . . .

PAT: Oh my god. They always had to preface it with something like that.

CATHERINE: Was ridiculous. It was like, "This was a barren wasteland." I was like, "Not really." You know like they would describe things, physical attributes, that did not exist. It's like "Really?"

PAT: "A den for crack dealers 24/7."

CATHERINE: Right. Like there was none of that.

PAT: Right.

CATHERINE: If you're dealing crack, you can't be in the middle of a park with no traffic. Anybody who pays any attention could understand that, but it's like everybody's like "Oh."

Stehlin's observations demonstrate the ambivalence of a resident who is both a "victim" of the renovation (i.e., one whose history has been rewritten and whose park has been "programmed") and, however qualified, a "benefactor" of gentrification and the corresponding park fountains, streetcars, and business development. As we observe throughout *washpark*, however, this bifurcated interpellation of the citizen can itself reveal an epiphenomenon of capital at work in urban gentrification, or "development" as a dynamo of alienation:

A Draw, A Little Circle, A Temporary Stage

Grids
that show
back surface and the front
near bar lines, ordinarily and dim
rails heading by to-night

> sink into the landscape
> waiting
>
> In-loaded and repulsory

With the renovation of the park completed and urban renovations continuing apace in other areas, we decided to bring the project to a close in 2016. We then proceeded to consider various publishing outlets. As it happened, this was the year that Delete Press published its first full-length book, Wendy Burk's *Tree Talks*, and since Tyrone had an open invitation to submit work to the press (to say nothing of its overt concern with ecological, urban, and social issues), we decided to submit the manuscript to this publisher. It was accepted and will appear in winter 2019–20.

Our investigation of an urbopoetics continues with our current project, an examination of the persistence of lead poisoning in certain neighborhoods of Cincinnati. Poisoning endures despite not only the outlawing of lead paint usage in private and public building developments but also the ongoing remediation efforts in housing and industrial zones. We've expanded our walks to parks, development zones, and abandoned cemeteries and have included friends who are also working on their own urban investigations vis-à-vis critical poetics and poetry. Thus, we have aligned ourselves with a cohort whose critical, poetic, and political values and practices constitute an evolution of our understanding of urbopoetics: An orientation toward urban spaces that is both project-based and, more important, a way of life as expressed in a community. With these friends and colleagues—including poets Cathy Wagner and cris cheek from Miami University in Oxford, Ohio—we have initiated tours of various Cincinnati cemeteries, parks, and, guided by historian Anne Steinert, landscapes and buildings in Cincinnati's West End (a historical black neighborhood just west of the downtown business district). Several of these locations are slated for demolition for parking garages and lots that will serve a forthcoming soccer stadium. Although we are all working on individual projects related to urban spaces—for example, cheek is tracing the architectonics of Cincinnati during the year 1639—we may, at some point, find ourselves collaborating on a project related to the transformations of the city under the twin imperatives of "development" and "renewal."

Our hope is that our experience with *washpark* inspires other writers, artists, and social activists to continue to develop the concept of urbopoetics in the context of a geopoetic practice. We believe this type of collaborative and pedestrian practice can better capture and reflect the complexity of urban interests, narratives, and stakeholders and can be an effective step toward disabling the typical binary narratives of urban progress, inviting involvement and broadening participation in creative efforts.

Notes

1 Wendy Burk's *Tree Talks: Southern Arizona*. np: Delete Press, 2016 is a specific investigation of flora-making that challenges the concept of *poesis*, "making," as necessarily

human-based even if it reinforces the theorization of making, captured in the concept of *poetics*, as anthropocentric.

2 Of course, we recognize that just as the concept of sentience has been extended to flora in general, future research may confirm a "third" mode of sentience heretofore unknown or unacknowledged as such.

References

Bishop, C. 2012. *Artificial Hells: Participatory Art and the Politics of Spectatorship*. London: Verso Books.

Burk, W. 2016. *Tree Talks: Southern Arizona*. np: Delete Press.

Clifford, J. 1983. "On Ethnographic Authority." *Representations* 2: 118–146.

Clifford, J. 2013. *Returns: Becoming Indigenous in the Twenty-First Century*. Cambridge, MA: Harvard University Press.

Goffman, E. 1963. *Stigma: Notes on the Management of Spoiled Identity*. New York, NY: Simon & Schuster.

Maréchal, Garance, and Stephen Linstead. 2010. "Metropoems: Poetic Method and Ethnographic Experience." *Qualitative Inquiry* 16 (1): 66–77.

Mertens, D., and P. Ginsberg, P. 2008. "Deep in Ethical Waters: Transformative Perspectives for Qualitative Social Work Research." *Qualitative Social Work* 7: 484–503.

Pillow, W. 2003. "Confession, Catharsis, or Cure? Rethinking the Uses of Reflexivity as Methodological Power in Qualitative Research." *Qualitative Studies in Education* 16 (2): 175–196.

Skirtz, Alice. 2012. *Econocide: Elimination of the Urban Poor*. Washington, DC: NASW.

18

GEOPOETICS AS COLLABORATIVE ENCOUNTER

Performing poetic political ecologies of the Colorado River

Elissa Dickson and Nathan Clay

The most endangered river in America, we say.
As if she doesn't know.
For 17 million years she was the kind of free
we can only imagine at that instant
between sleep and awake,
the moment of infinite possibility.
  (Elissa Dickson, from *The Colorado River*)

Introduction

And so we have arrived in an era long fixed in our imaginations, forged through material incantations of our supposed will over nature: The Anthropocene, an epoch named with intent to position humanity as a force of geological change (Steffen, Crutzen, and McNeill 2007). Yet, while our hubris would have us believe that we are equal to the winds, fires, waters, and sands of time, we now sit among this legion of earth-shapers like writers without pens. Not only do we lack the vocabulary to understand and articulate our dynamic roles in the Anthropocene (Robbins 2013), but our obsession with cataloguing the ways we have written onto the earth may be blinding us to possibilities to write our way *out* of the innumerable catastrophes we now face (Rickards 2015).

This chapter responds to calls to consider geography—or "earth writing"—as a geopoetic endeavor in which we "place the earth at the center of experience" (Springer 2017, 3). Geographers have recently begun experimenting with creative non-fiction (Cresswell 2013; Marston and Leeuw 2013) and poetry (Last 2017). We follow Magrane (2015) in envisioning opportunities to merge contemporary eco-poetic practice with human geography. We suggest that the practice of geopoetics can productively be thought of as a set of collaborative encounters: between poet and geographer, performer/writer, and audience.

We (a poet-geographer and a geographer-poet) consider how slam poetry and political ecology align and overlap. Slam poetry, a form of poetry that emerged during the 1980s, emphasizes live performance in front of a participatory audience and often addresses topics related to social justice. We suggest that slam poetry's egalitarian ethos and community emphasis is shared by political ecology, which is a subfield of geography that relies on detailed empirical work with communities to chronicle uneven outcomes of social and environmental change. Following Alex Loftus's call to envision "political ecologies of the possible," wherein we learn about the world through the process of changing it (2009, 157), we argue that slam poetry is a crucial node in the emerging genre of geopoetics. With its attention to issues of social justice and commitment to providing a platform for marginalized voices, slam poetry allies with the counter-hegemonic project of radical geography, helping to (re)ground and reclaim knowledge, power, and metaphor through performance.

Our exploration of how writing and performing poetry can help us become immersed in issues of social and environmental justice centers on a poem (written and performed by the first author) about the Colorado River.[1] Elissa's poem touches on long-standing political ecological issues surrounding the environment, development, and water management in the American West. With 15 large dams on the main branch and hundreds on its tributaries, the Colorado is among the world's most intensively developed river systems, a symbol of capitalism's excesses written onto the landscape. By encouraging us to experience these issues on an emotional plane, we argue that slam poetry can enhance political ecology's dual projects of deconstructionist critique ("the hatchet") and the envisioning of alternative futures ("the seed"). Political ecology employs analytical narratives to convey the causes and consequences of social-environmental crisis. We suggest that poetry can augment these narratives by adding visceral and emotive elements.

Faced with monumental environmental and social crises of our creation, there is urgent need to subvert dominant pathways of power-knowledge and "write our way into a new vocabulary for the Anthropocene" by engaging with the vibrant, visceral, and contradictory nature of metaphor (Robbins 2013, 309). We suggest that one promising alliance in this multifaceted undertaking emerges from the confluence of slam poetry and political ecology.

Poetry, power, and the environment

Poetic language and power in geography

Much like poetry, geographic writings on social-environmental processes often indulge in metaphor and simile. Terms such as "nature," "global warming," or the "Anthropocene" carry cultural baggage and convey simplified images about how humans interact with the environment (Robbins 2013). Whether we are poets, geographers, or otherwise, metaphors attach to our thoughts and actions cognitively and emotionally, shaping the ways in which we understand and engage with the world around us (Lakoff and Johnson 2003). Thus, metaphors can be highly political, and their power lies in their ambiguity, in how they maintain tension between the abstract and the concrete. They derive value through entwining with insights, storylines, emotions, and aesthetics that are dear to our worldviews (Rickards 2015)—and indeed, word choice can influence the decision-making of individuals and societies alike (Peet 1992). Moreover, uneven power relations ensure that certain metaphors become embedded in our social consciousness, where these imaginaries come to constrain how we see and interact with the world.

Compared to poets, whose work often deliberately teases the imagined boundaries of language's form and function, geographers' use of metaphor tends to be bounded by conventional phrasings that have achieved legitimacy through circulation in the popular-scientific lexicon. However, since geography's "cultural turn" in the 1990s, human geographers have drawn on poststructural and feminist thinking on the intimate relationship of power and knowledge to critically assess the social and cultural constructions of discourse. Political ecologists continue to consider the *work* that words do, viewing discursive constructions surrounding "development" or "environment" as grounded in biophysical and social structures, refracted through power asymmetries (Peet and Watts 1996; Robbins 2012). For example, an imaginary of nature as entirely devoid of social processes often underlies development efforts that aim to modernize landscapes by forcibly partitioning nature and society (Cronon 1995; Robbins 2001). Paradoxically, geographers remain far from embracing metaphor and other figurative devices in their writing (Rickards 2015; Springer 2017). Yet, creativity is paramount in envisioning more inclusive alternative futures. In this way, geopoetry may contribute to the "seed" of political ecology.

Geopoetics

Geopoetry encourages the (re)discovery of writing styles that transcend the storylines, forms, and terms common in geographic writing (Cresswell 2013). Yet, geopoetics can be more than merely an exercise of indulging creative whims to enliven a tired genre. Poetry has immense power to personalize universal elements of the human condition. Geopoetry, in turn, may be at its liveliest when it indulges in and humanizes the grittiness and messiness that are often at the core of social and environmental issues. If poetry is a distilled snapshot of experience that resonates at emotional levels, geopoetics inhabits the liminal space between lived experience (of landscapes and bodies) and representation (memories, histories, politics).

In this sense, geopoets are active agents in social change. They are at once archaeologists deciphering the past and architects envisioning the future. As Last suggests, geopoetry may help consider pathways to a "more benevolent politics or sociality through a greater awareness of our embeddedness in matter, space and time" (2017, 50). To envision and catalyze change, geopoets must foremost acknowledge that they thrive amid the dynamic juxtapositions that comprise the Anthropocene: between human and nature, history and future, memory and dream. Indeed, poetry has always been a practice of exploring the interplay (and resemblances) of humans' "internal" emotional landscapes and the earth's "external" physical landscapes. By engaging with these juxtapositions, geopoets can help advance earlier efforts by feminist geographers to "find ways not so much to 'win' as to alter the ways knowledge is produced and shared so that the very notion of a 'contest' is undone" (Katz 1996, 487). Geopoetics in general, and slam poetry in particular, create an especially fertile space for social change.

Origins and description of slam poetry

Oral tradition is recognized as a hallmark of cultures worldwide, from Ancient India and Greece to Puebloan cultures of the American Southwest. Rooted in practices of oration and performance, spoken word poetry evolved in the United States with: the Harlem Renaissance (1920s), the Beat Generation (1940s–50s), and the feminist movement and Black Arts Movement (1960s–70s). In the mid-1980s, Marc Smith, a Chicago-born construction worker invented the "poetry slam," a competition in which poets perform original work alone or in teams before an audience. By moving poetry out of books and libraries and into everyday spaces (e.g., cafes, bars, lounges, and jazz clubs), slam's aim was to re-energize and reinvent poetry. In this way, poetry slams upended the academic status of conventional poetry and blurred the boundaries between poets, critics, and audiences.

Today, slam poetry continues to center on the interaction of poets and community members, with five audience members chosen randomly to act as judges on any given night. In their crafting and delivery of performances, slam poets emphasize tone, articulation, movement, rhythm, and memorization. Slam poems often address hot-button political and social issues, particularly race, gender, and inequality. Through these mechanisms, slam poets practice activism through counter-storytelling, where marginalized voices present narratives of resistance that challenge the conventional stories of the majority in both form and content (Chepp 2016). Slam poetry casts off formal avenues of writing/publishing in favor of creating an inclusive community of practice that thrives on the cultivation of shared experience. This has reduced barriers of entry for would-be artists. Young poets, female and LGBTQIA poets, and poets of diverse racial and cultural backgrounds have gravitated toward slam poetry, often using poetry slams as forums to build allies and mobilize around key issues. Given the diversity within slam communities, slam poets often understand social justice as multifaceted and intersectional (Chepp 2016).

This egalitarian and experience-based approach makes slam poetry a unique form of geopoetic expression and practice, one that is well positioned to engage with political ecological issues. While slam poets have historically focused on issues of race and social justice, they now also engage in intersectional environmental justice issues—increasingly so in the era of climate change. For example, Climbing Poetree, an African-American female slam duo comprising Alixa Garcia and Naima

Penniman, have achieved international acclaim for their radical environmental justice slam poetry (www.climbingpoetree.com). As Valerie Chepp observes, slam poetry involves "testifying," a political tradition of public speaking meant to inform opinion through description of experience from an individual's vantage point. Importantly, testifying in slam is not an individualistic practice but rather is "linked to a broader community ritual of affirmation and learning" (Chepp 2016, 46).

Political ecology

The egalitarian ethos and experience-based nature of slam poetry are shared by political ecology, a subfield of geography that relies on empirical engagements with communities to chronicle how uneven outcomes emerge at the interface of social and environmental change. Though its conceptual roots are deep and multifaceted, political ecology took shape during the 1980s with the recognition that environmental issues (e.g., soil erosion and desertification) were as much the result of political economic structures as they were of biophysical and ecological processes (Blaikie and Brookfield 1987; Bassett 1988). By detailing local experiences of change, political ecological writing illustrates how supposedly separate spheres of "environment" and "society" are in fact intimately entangled. Works often detail how social processes (such as those engendering inequality) are at once products of biophysical variability and drivers of ecological change (Zimmerer and Bassett 2003). In other words, social-environmental landscapes are understood as simultaneously real and imagined (Robbins 2004), a premise that we argue opens the door for geopoetic approaches to political ecology.

A further parallel with slam poetry is that political ecology did not evolve from a single linear set of theories but rather through a convergence of several streams of thought. This makes political ecologists more aptly defined by *how* they do things than by *what* they study or to which concepts they subscribe. As discussed in two volumes of *Liberation Ecologies* (Peet and Watts 1996; Watts and Peet 2002), political ecologists study and confront inequalities through engaged fieldwork and teaching, often forging connections with social movements and working alongside marginalized groups to challenge social-environmental injustices. Although they employ diverse methods and theories, political ecologists find common ground in that they pursue storylines that begin or end in contradiction and that identify winners and losers in social-environmental change. As such, political ecology can be considered "a community of practice" (Robbins 2012, 20). Slam poetry might also be described as a community of practice that similarly engages with such moments of contestation and themes of unevenness.

Political ecological writing has proliferated since the 1980s, documenting social-environmental struggles worldwide. Yet, this body of work has engaged relatively little with plausible alternatives to such struggles (Loftus 2009). And while attuned to diverse voices during the research process, political ecologists have overwhelmingly privileged analytical modes of writing over creative ones that might not adhere to the conventions of academic publishing (Bennett 2010). We suggest that geopoetics can help open up political ecological work to more visceral and emotional depictions (both written and performed) of social-environmental injustice, while also tempering the sense of exhaustion and helplessness that this critical writing (i.e., emphasis on the "hatchet" of political ecology) has wrought on the field. By reconnecting humans and the environment on philosophical and practical levels, geopoetry can help rework the parameters of environmental issues such as climate change (Magrane 2017). Further, we suggest that slam poetry and political ecology can productively unite to attest social-environmental struggles and envision more just futures. To do so, we argue for the need to nurture productive tensions among engaged research, activism, writing, and performance.

Social-environmental struggles in the Colorado River and Grand Canyon

A fascinating site to consider these intersections is the US Southwest, where a tradition of evocative natural history writing meets recent activism and renewed social-environmental conflicts. The waterways and public lands in this region are heavily contested, with the Colorado River being a key locus in ongoing controversy between environment and development. Since the end of the nineteenth century, hundreds of dams have been constructed in the West (many built by the Bureau of Reclamation through public works programs). Perhaps the most disputed are the two massive dam-reservoir complexes that bookend the Grand Canyon on the Colorado River: the Glen Canyon Dam, which creates Lake Powell at its head, and the Hoover Dam, which creates Lake Mead at its end. These enormous dams remain centerpieces of efforts to tame the West by enabling irrigated agriculture (Reisner 1986).

Recognized as an iconic American landscape, the Grand Canyon is not only politically charged, but is also an emotional touchstone in the public (sub)consciousness. Archaeological evidence suggests that Indigenous peoples have inhabited the Grand Canyon for approximately 10,000 years. In more recent history, white European conquistadors stumbled upon the canyon in 1540. It was not until 1869, however, that the Grand Canyon reentered white consciousness with John Wesley Powell's now famous expedition to descend the Colorado River through the Grand Canyon, by boat (Reisner 1986). In 1903, Teddy Roosevelt proclaimed, "leave it as it is . . . the ages have been at work on it, and man can only mar it" (quoted in Fedarko 2016, fig. caption), leading to its eventual protection as a National Park in 1919. Yet today, dams, uranium mining, air traffic, and tourist developments threaten the Grand Canyon, Colorado River, and surrounding lands.

Undergirding the basin is a complex patchwork of federal, state, and tribal land tenure and management arrangements which complicate conservation efforts. Currently, the age of large-scale water distribution projects in the West is at an end, and we are entering a fascinating phase of dam removal in which governments and citizens alike have begun to recognize the economic and ecological potential of free-flowing rivers. For example, six western dams were deconstructed in 2015 (Fedarko 2016). With both Lake Mead and Lake Powell at half-capacity and losing water, and climate change models predicting increased droughts, the future of rivers in the arid West is anything but certain. Never has the need for a reanalysis of our relationship with water and the Colorado River been so apparent. Employing a creative or poetic lens creates opportunities for exploring empathic and emotional aspects of our relationship with this complex and contested landscape.

Environmental writing on the US Southwest

Written descriptions of the landscapes of the US Southwest have long cultivated a Romantic sense of untouched wilderness. The quixotic image of solitary adventure in untouched lands was a key theme in the writing of Edward Abbey, whose book *Desert Solitaire: A Season in the Wilderness* (1968) became a manifesto for conservationists and outdoor recreationalists alike. With vivid writing that portrayed the desert Southwest as both magical and real, Abbey recounted his year of solitary exploits in Utah while criticizing mass tourism and development projects such as the Glen Canyon Dam on the Colorado River (Abbey 1968). While this writing has inspired environmental activism and devotion to desert landscapes, as we continue to flail in the Anthropocene, the motif of the lone white male enraptured by a romanticized wilderness devoid of social entanglements has become unrealistic, irrelevant, and counterproductive.

The voices of women and Indigenous peoples have recently emerged to diversify and deepen imaginaries of the US Southwest. In her book *Desert Cabal: A New Season in the Wilderness* (2018), Amy Irvine counters Abbey's individualistic narrative with a call to recognize the complex interconnections of people and place in the West. In contrast to Abbey's antagonistic activism, Irvine urges us to nurture emotions of grief and to pursue a more united and embedded political activism. Contemporary environmentalist writer Terry Tempest Williams invokes the desert landscape to explore complex issues of social and environmental justice in the Southwest as well as more personal issues such as her mother's experience with cancer (Tempest Williams 2016). Inspired by the oral tradition of his Puebloan elders, Indigenous poet Simon Ortiz reflects on the beauty of Southwestern landscapes while commenting on development as a dehumanizing force that alienates people from one another, from self, and from the earth. These writers encourage us to appreciate both the beauty of these landscapes and the messes they/we are in, a notion that intersects with political ecology's claim that nature and society cannot be separated. Moreover, these writers are increasingly aware that including diverse voices is crucial to breaking down dominant narratives in order to be better equipped and motivated to work together toward more livable futures.

Our approach

This chapter takes a geopoetic approach to reanalyze human-environment relationships in the Colorado River Basin. Via slam poetry, we illustrate how geopoetic practice can pay explicit attention to social-environmental struggles. Geographers—political ecologists in particular—bring understanding of how power relations (for example, those inscribed in systems of capitalism and patriarchy) are surreptitiously and overtly written onto landscapes and how these etchings can widen over time into entrenched patterns of uneven development (Smith 1984). Performing poetry brings political ecological issues to life in ways that transcend space and time by creating emotionally resonant experiences. We argue that geopoetics is a collaborative process that is constituted through writing as well as performing.

Poem by Elissa Dickson

The Colorado River

The most endangered river in America, we say.
As if she doesn't know.
For 17 million years she was the kind of free
we can only imagine at that instant
between sleep and awake,
the moment of infinite possibility.

She was a cascading carouser,
millennial marauder, seasonal shapeshifter,
jumping her banks again and again
just to prove she could.

Her sensual spring swells
caressed canyon walls, carving
them to her will,
wild love affairs across eons
as she rushed ever onward to kiss the sea.

And she was in charge.
Each canyon a lover,
but none like the Grand Canyon.
Like Frida Kahlo and Diego Rivera,
this was a red-hot love, combustible, lustible,
the deepest canyon in the world
only gets carved by some seriously sultry
geological salsa dancing.
And his was a patient love;
timeless timepiece to her ebbs and flows—
his wind weathered walls whispered secrets

so vast only she could hear them.
And for 6 million years they danced, with a
tick, tock
tick, tock
tick, tock
tick, tock

Tick-

      Brittle-fingered ice ages grasp
        and ungrasp this spinning planet.

Tock-

      Homosapiens rise to walk upright
      into the pink light of dawn.

Tick-

      Giant vivid green kelp forests, drunk on their own luck,
      grace ocean floors for the first time.

Tock-

      Man, with tantalizing taste of fortune on lips,
      crosses the Bering Land Bridge into North America.

Tick-

      The last of the mastodons caught in tailspin crash
      out under a luminous harvest moon.

Tock-

      The sun-soaked West beckons pioneers,
      the promise of prosperity painted purposefully
        across its smiling plains.

And still they danced with a
tick, tock
tick, tock
tick, tock
tick, tock
until,

Tick-

      Two dams slam down like shackles on hands,
      like guillotine on neck, smothering her fierce grace,
      strangling her melodic voice, taming her powerful passions
      all in the name of progress.

And from now on we will tell her when
and where she is allowed to go,
and we are her pimp.
Phoenix golf course sprinkler in the morning,
Vegas gentlemen's master suite shower by night.
And she is our slave.

On the auction block today,
The Colorado River.
Can I get a 5, 5, 5 can I get a 10, 10 . . . 15 . . . 20,
Sold to the

Man,
to a species with no regard for the true nature of things,

But that which is pent up always comes out,
and the Bureau of Reclamation is not the only one
with the power to reclaim.
And she may not be the biggest or the longest,
but she is most definitely the fiercest.
And she is the kind of lover worth waiting for.

And each morning in the Grand Canyon cathedral,
right at dawn,
as red-hot sun returns to lick cool canyon walls
and the quicksilver Colorado River sings in another day,
at that moment between sleep and awake,
we are alive to the possibility that all may not be lost just yet.

Reflections and discussion

This poem seeks to juxtapose our current Anthropocene against the backdrop
of the Grand Canyon's magnificent geological timescale. By playing with the
notion of time—both geologic phases and ephemeral moments—*The Colorado
River* encourages the reader (or, preferably, the listener, as it was written primarily to be performed) to reflect on both humans' momentary role on Earth and
on the agency and legitimacy of non-humans (e.g., rivers) in shaping the landscape. By personifying the river and the canyon, the poet translates the ephemeral relationship of water and rock into the realm of a romantic relationship as
a way to invoke empathy in the readers/listeners. By accessing the landscape's
story on an emotional register, the poem considers humans' responsibility to
respect the river and the Grand Canyon. Moreover, in depicting the river as
female, the poem engages the notion that management of natural resources (in
this case a capitalist-fueled promise of prosperity related to damming rivers) is
part of the patriarchal mindset of control and oppression. While at the center of
the poem is a story of human greed, the poem begins with a narrative of freedom and ends with one of hope. This emotional arc conveys a sense of flux and
change that encourages the audience to imagine a past beyond humans' reach
and a future of possibility.

Writing and performing poetry in and about place

Human connection with place has been a long-standing theme in poetry—a primary example being work by Mary Oliver. *The Colorado River* was written as an
environmental activism piece, with the goal of inspiring others to engage in both
enjoying and protecting public lands and waterways of the Southwest. Elissa has

spent significant time rafting the Colorado River through the Grand Canyon and feels a strong personal connection to the landscape and a strong need to advocate for its protection in the face of current environmental and development threats. As geopoets, we must acknowledge and embrace our subjectivities, for they shape the places we go and how we write about them.

The poem comes alive in each place Elissa performs it. In this way, the performance celebrates Elissa's experience with the river and also creates a communal experience with the audience. This doubles the celebration of the experiential, generating a vibrancy unique to each encounter from which shared emotional experiences spring. Performing this poem often elicits emotional responses from audience members who have remarked they can empathize with the river itself. In practice, the poem resonates with individuals who have a relationship with the river and those who have never been there as well as with people of diverse backgrounds and ages. Audience members frequently note that the poem stirs up feelings of grief, sadness, frustration, and anger as well as hope.

Slam poetry as social-environmental activism

Often much more effective than logical entreaties, emotional appeals can be crucial for exciting people's sensibilities and moving them to engage in social and environmental movements. Slam poetry is more than just a mechanism for conveying the nuts and bolts of environmental issues; it can catalyze change by creating intimate encounters among people and landscapes. Yet, scholars studying social change often overlook slam poets. What is more, while their poems engage a range of pertinent issues, slam poets are not necessarily affiliated with specific activist organizations; rather, they may work with organizations to varying degrees to build community around an issue (Chepp 2016).

A strong sense of community has been at the forefront of many recent movements. And the seeds of some contemporary environmental and social justice movements (e.g., #MeToo, Black Lives Matter, Climate Justice) can be found in the communities surrounding civil rights, feminist, and environmental movements of the 1960s. Moreover, in this pivotal historical moment it has become increasingly clear that environmental issues are also social justice issues. Slam poetry has a role to play in catalyzing the intersectionality of these disparate movements. By enabling the sharing of personal experiences with myriad social and environmental justice issues, slam poetry can play a key role in the dynamic and evolving environmental movement. Through community building among diverse racial and social groups, slam poetry, much like the geographic subfield of feminist political ecology (Rocheleau, Thomas-Slayter, and Wangari 1996), can forge anew the linkages among these movements. By creating a community of shared experience and practice, slam poetry can also nurture a spiritual and emotive element that the contemporary environmental movement lacks (Dechristopher 2014).

Slam poetry as a teaching tool for human-environment relationships

Earlier, we considered how uniting political ecology and slam poetry can facilitate active rethinking of human-environment relationships. The community-centered approaches of slam poetry and political ecology can also be leveraged as a teaching tool for all age levels. Communities everywhere can work to create opportunities for children and adults to explore their relationship to place through writing and performing slam poetry about their experiences with the environment. Poetry is already used as a teaching tool in public schools. In New York City, for example, Urban Word NYC encourages young adults to explore their vulnerabilities through poetry. Encouraging people of all ages to practice self-inquiry and examine how they relate to their environment can be an important component in the practice of geopoetics. Such a teaching program should also emphasize that, unlike "wilderness," which tends to conjure up pristine places that are isolated from humans, wild landscapes are all around us, in our own backyards, ready to be explored (Cronon 1995). The simple act of reflecting on the surrounding "nature"—neither fully pristine nor fully defined by human presence—can help us tap into the agency of landscapes, encountering a sense of *wildness* and wonder wherever we are (Van Horn and Hausdoerffer 2017). This potent teaching mechanism should be nurtured and implemented in school systems everywhere.

In detailing the social-environmental crises of the twenty-first century, historical geographer Jason Moore (2016) urges recognition that we are less in the Age of Man (Anthropocene) and more in the Age of Capital, or the Capitalocene. Indeed, the present era of commodification and rampant technology use has left children and adults at risk of behavioral patterns that leave little time for meaningful social interactions and playing outside (Charles and Louv 2009). Learning opportunities such as slam poetry writing workshops that encourage connection with and reflection on the environment could provide a vitally powerful countercurrent to this "nature deficit disorder" and to the pervasive grip that capitalist impulses have on society. While not a panacea, communal practices revolving around the sharing of experiences with nature might serve to channel the current climate of despair and helplessness into hope and concerted efforts to shape whatever age might follow. Based on current extinction rates, biologist Edward O. Wilson has suggested a name for the likely future age, after humans have driven species to extinction and all that remains is us and our creations: the *Eremocene*, or Age of Loneliness. Averting this apocalyptic era is going to require working with each other and with the earth with creativity and diligence.

Conclusions

Geopoetics encourages us to rethink how we create knowledge through the writing process (Magrane 2015; de Leeuw and Hawkins 2017). Slam poetry adds a vital performative force to the practice of geopoetics. Moreover, slam poetry can advance political ecological thinking by helping to develop emotional resonance through alternative modes of expression. There are numerous parallels—in origin and ethos—between slam poetry and political ecology, both of which thrive as engaged communities of practice that emphasize emergent properties. For slam poets, these communal properties emerge through writing and performance; for political ecologists they emerge through fieldwork. Slam poetry can enrich political ecology through visceral, emotional performance of socio-political struggles. As with Elissa's poem, *The Colorado River*, saturating the palate of performance poetry with images of wildness and issues of water management on public lands is a form of testifying to the slam community about these issues.

As Magrane proposes, "through juxtapositions and awareness of the politics of representation, works in geopoetics, as well as in other art forms, can make aesthetic-ethical critiques themselves, ones that might work in a catalytic expressive mode rather than a reactive analytic mode" (2015, 4). We suggest that individuals that practice slam poetry and/or political ecology can leverage their roles as interlocutors who situate and merge knowledges of diverse actors to create more inclusive conversations about socially and environmentally just futures. In so doing, geopoets and political ecologists will emerge as more "committed to learning about the world through changing it (and vice-versa)" (Loftus 2009, 157). In our writing and performance, we should concentrate on building a collaborative dialogue that is emblematic of the diverse encounters from which geopoetry springs to life. As a process of learning together how to see, geopoetics can be most effective by creating spaces where vulnerabilities and struggles can be explored, where human-environment relationships can be rewritten. Only with patience to listen to one another and to the earth can we hope to learn how our continued existence depends upon our unity.

> *And each morning in the Grand Canyon cathedral,*
> *right at dawn,*
> *as red-hot sun returns to lick cool canyon walls*
> *and the quicksilver Colorado River sings in another day,*
> *at that moment between sleep and awake,*
> *we are alive to the possibility that all may not be lost just yet.*

Note

1 A video of this performance can be found online at www.youtube.com/watch?v=Xd55zElp2Y0.

References

Abbey, Edward. 1968. *Desert Solitaire: A Season in the Wilderness*. New York: Simon & Schuster.
Bassett, T. J. 1988. "The Political Ecology of Peasant-Herder Conflicts in the Northern Ivory Coast." *Annals of the Association of American Geographers* 78 (3): 453–472.
Bennett, Jane. 2010. *Vibrant Matter: A Political Ecology of Things*. Durham: Duke University Press.

Blaikie, P., and H. Brookfield, eds. 1987. *Land Degradation and Society*. London: Methuen.

Charles, C., and R. Louv. 2009. "Children's Nature Deficit: What We Know—and Don't Know." *Children Nature Network*. Accessed January 29, 2019. www.childrenandnature. org/wp-content/uploads/2015/04/CNNEvidenceoftheDeficit.pdf.

Chepp, Valerie. 2016. "Activating Politics with Poetry and the Spoken Word." *Contexts* 15 (4): 42–47.

Cresswell, Tim. 2013. "Geographies of Poetry/Poetries of Geography." *Cultural Geographies* 21 (1): 141–146.

Cronon, William. 1995. "The Trouble with Wilderness: Or, Getting Back to the Wrong Nature." *Environmental History* 1 (1): 7–28.

Dechristopher, Tim. 2014. "Interview with truth-out.org." Accessed May 21, 2018. www. truth-out.org/news/item/27866-civil-disobedience-is-an-act-of-love-an-interview-with-tim-dechristopher.

de Leeuw, Sarah, and Harriet Hawkins. 2017. "Critical Geographies and Geography's Creative Re/turn: Poetics and Practices for New Disciplinary Spaces." *Gender, Place & Culture* 24 (3): 303–324.

Fedarko, Kevin. 2016. "Are We Losing the Grand Canyon?" *National Geographic*, September. Accessed May 21, 2017. www.nationalgeographic.com/magazine/2016/09/grand-canyon-development-hiking-national-parks/.

Irvine, Amy. 2018. *Desert Cabal: A New Season in the Wilderness*. Salt Lake City: Torey House Press and Moab: Back of Beyond Books.

Katz, Cindi. 1996. "Towards Minor Theory." *Environment and Planning D: Society and Space* 14: 487–499.

Lakoff, George, and Mark Johnson. 2003. *Metaphors We Live by*. Chicago: University of Chicago Press.

Last, A. 2017. "We Are the World? Anthropocene Cultural Production Between Geopoetics and Geopolitics." *Theory, Culture & Society* 34 (2–3): 147–168.

Loftus, Alex. 2009. "The Theses on Feuerbach as a Political Ecology of the Possible." *Area* 41 (2): 157–166.

Magrane, Eric. 2015. "Situating Geopoetics." *GeoHumanities* 1 (1): 86–102.

———. 2017. "Creative Geographies and Environments: Geopoetics in the Anthropocene." Doctoral diss., University of Arizona.

Marston, S. A., and Sarah de Leeuw. 2013. "Creativity and Geography: Toward a Politicized Intervention." *Geographical Review* 103 (2): iii–xxvi.

Moore, Jason W. 2016. *Anthropocene or Capitalocene: Nature, History, and the Crisis of Capitalism*. Oakland: PM Press.

Peet, Richard. 1992. "Some Critical Questions for Anti-Essentialism." *Antipode* 24 (2): 113–130.

Peet, Richard, and Michael Watts. 1996. *Liberation Ecologies: Environment, Development, Social Movements*. London: Routledge.

Reisner, Marc. 1986. *Cadillac Desert: The American West and Its Disappearing Water*. London: Penguin.

Rickards, Lauren A. 2015. "Metaphor and the Anthropocene: Presenting Humans as a Geological Force." *Geographical Research* 53 (3): 280–287.

Robbins, P. 2001. "Fixed Categories in a Portable Landscape: The Causes and Consequences of Land-Cover Categorization." *Environment and Planning A: Economy and Space* 33 (1): 161–179.

———. 2004. *Political Ecology: A Critical Introduction*. Oxford: Wiley-Blackwell.

———. 2012. *Political Ecology: A Critical Introduction*. 2nd ed. Oxford: Wiley-Blackwell.

Robbins, Paul. 2013. "Choosing Metaphors for the Anthropocene: Cultural and Political Ecologies." In *The Wiley Blackwell Companion to Cultural Geography*. Edited by Nuala C. Johnson, Richard H. Schein, & Jamie Winders, 307–319.

Rocheleau, Dianne, Barbara Thomas-Slayter, and Esther Wangari, eds. 1996. *Feminist Political Ecology: Global Issues and Local Experience*. London: Routledge.

Smith, Neil. 1984. *Uneven Development: Nature, Capital, and the Production of Space*. Athens: University of Georgia Press.

Springer, S. 2017. "Earth Writing." *GeoHumanities* 1–19.

Steffen, W., P. Crutzen, and J. McNeill. 2007. "The Anthropocene: Are Humans Now Overwhelming the Great Forces of Nature?" *Ambio* 36: 614–621.

Tempest Williams, Terry. 2016. *The Hour of Land: A Personal Topography of America's National Parks*. Farrar: Sarah Crichton Books.

Van Horn, Gavin, and John Hausdoerffer, eds. 2017. *Wildness: Relations of People and Place*. Chicago: University of Chicago Press.

Watts, Michael, and Richard Peet. 2002. *Liberation Ecologies: Environment, Development, Social Movements*. 2nd ed. London: Routledge.

Zimmerer, K. S., and T. J. Bassett. 2003. "Introduction." Chap. 1 in *Political Ecology: An Integrative Approach to Geography and Environment-Development Studies*, edited by K. S. Zimmerer and T. J. Bassett. New York: Guilford Press.

19

NEGRO-MOUNTAIN-WOLVES/ NOTES ON REGION

C.S. Giscombe

Wolves came back to Negro Mountain. There was no pattern to it, for either mountain or animals to occupy—it was not a story but rather the germ of a shape, an undetectable weight; it was shapeless finally, unmeasured, it was only relayed.

Negro Mountain—the summit of which is the highest point in Pennsylvania— is a default, a way (among others), to think about the Commonwealth.

"The Negro Mountain is so called because, after Braddock's defeat on the Monongahela, a scouting party, traveling Braddock's Road, came in contact with a like party of Indians, when a skirmish ensued, in which one Indian was killed, and a very large negro mortally wounded" (Browning 1859, 397).

Perhaps the variation of wolves and other dogs is a series of false dilemmas. Any speaker might stand—or be "placed" and decry—in their midst. In fact, the *territories* overlap; the territories are the same series—*plural*—of fields, trees, elevations, and watercourses. A coyote—called prairie wolf also, or red wolf, or "barking dog," or brush wolf—is a wolf. (Foxes are a different matter.) *Canis lycaon*—the Great Lakes wolf—was named for the king Zeus turned into a wolf. Neighborhood dogs bark in a chain: Children know this. "The red wolf of the south is smaller" (Hoffmeister and Zim 2014, 57). One is similar to another.

Word came from Negro Mountain that a lifelong resident there had surprised a wolf in her dooryard. Summer 2016. Her carpenter, she said, had also seen the wolf, on a separate occasion. Not a coyote—that population has flourished in Pennsylvania since 1930—but a wolf.

Henry Shoemaker, 1916: "The only person killed was the colonel's favorite Negro body servant, a black of colossal proportions. The Indians were driven off, and the body of the unfortunate Negro dropped down in a deep crevice of the rocks in a shoulder of the mountain, and the party resumed its way" (5).

In his *Extinct Pennsylvania Animals*, Shoemaker credits Seth Iredell Nelson of Clearfield County as having killed the last two wolves in Pennsylvania, this in the winter of 1892. Shoemaker notes that "they were native brown wolves and the last remnant of the big packs which for years infested the Divide Region. Mr. Nelson was in his 83rd year at that time, hence he can be called the oldest wolf slayer that Pennsylvania has produced" (2000, 50).

(Talk on top of one another, you mountains.)

Entre chien et loup.

Negro, and what?

(Lie atop one another, anybody could say.)

Pennsylvania man, half jungle animal, talking.

It varies.

On such a mountain a deliberate Negro man, or a pensive Negro man, might wish to be "spirited away." Industry is a deep valley or a trough. Hurston wrote, "Mules and other brutes had occupied their skins" (1991, 3). There is a long bottom to consider—not the path of a moving figure, but the swath. "Opposite" deep valleys are the hills, elevations; a common enough wish is to enter such places and/ or climb through them. Often enough one sees death in the market, or at work in an office, "under fluorescent light," explaining; someone else might see death come across the summit, like weather. Death is worked toward, death is "explained." Such a Negro man might wonder what he needed to know, for the sake of coherence, and find the qualification daunting, or amusing.

Among you. It could be the figures' motion. Between the reader and, say, the poem or the mountain at hand.

Speaker, observer, Negro, poet, wolf, measure. All you (Muehleisen and Migge 2005, 213).

Mrs. Stevenson's 1901 account: "June 30th, 1756, Col. Cresap and his party had another skirmish with the savages. He had not forgotten the lamented sleeper [Thomas Cresap, Jr., killed in the earlier 'skirmish'] on Savage Mountain; he enlisted another company of volunteers, taking with him his two surviving sons Daniel and Michael and a gigantic negro servant, belonging to him. ... This time they advanced into the wilderness as far as a mountain, a mile west of Grantsville. There, they met the Indians; a fight took place and the negro Goliath was slain, and the mountain has been 'Negro Mountain' ever since" (156).

Spirit is no sum; nor is it a *factor* of determination. Stevens, quoting "You," wrote, "'There are many truths, / But they are not parts of a truth'" (Stevens 2009, 119, ll. 5–6).

In Rumsey's online collection, Negro Mountain begins appearing on Pennsylvania maps with John Melish's "Map of Pennsylvania," published in 1822 (Rumsey, 4307.000). It appears on Henry Tanner's 1823 "Pennsylvania and New Jersey" (2589.029) and on the 1849 Barnes map as well (2475.000).

What else might a Negro *speaker* ask?

The eastern forest is no theater. There is a very long shadow in the forest. The forest harbors "game" and predators. There's a curious blankness that, *like* song or theater, is comforting but that sits outside the fact of silence; there's an angry sky, often, over valley towns—Somerset, say, or Bellefonte, the tiny county seat at the base of Bald Eagle Mountain—that extends to the clouds above the mountain. One day there is a hawk screaming on a tree branch and another one, silent, itself above a field, in the near distance, both visible from a road over the mountain. Perhaps a third hawk is nearby, unseen by the observer. Another day one arrives, say at mid-afternoon, at an intersection—a place where the roads converge at the base of a Pennsylvania mountain or a point on the mountain's flank—and finds the intersection empty. (There's bottomless absence; there is obvious failure. And grief. You know how unstable they are.)

The argument is that wolves are unlikely. Perhaps, at best, it is that there's an opacity to a Pennsylvania forest that such an animal—a wolf, say, in the present context or location of subject and trajectory and rumor—is *similar* to. Such an animal would be familiar—that is, recognizable in terms of shape and size and perhaps most profoundly in its movements—were it to appear, an act for which the observer might wish. In the Ridge and Valley region of the state the roads are often empty. A Negro speaker might ask what he was expected to represent in the field or at roadside or between the trees. Or any version of such a speaker might ask this. Intimacy might be an answer, in any case; bottomless *intimacy*—in terms of the familiar *knowledge* or claim of such or recognition of it in among the trees—is "naturally" ungainly. Or coarse, perhaps—in some descriptions—lewd.

Negro Mountain's a ridge on the same long plateau as the Endless Mountains; however, its appearance is that of "low relief." Shape fails; finally, perhaps, it's how you—meaning the speaker and the reader as well—*perceive* the shape failing since all that's too large to accommodate by sight or by any of the senses. Of the Llano Estacado—of west Texas, called "the Staked Plains" by Kenneth Irby and others and described by Irby as a region the loneliness of which is "like the love of death" (2009, 98)—a correspondent from Denver writes, "The slope is imperceptible to an observer on the plateau" (Calvert 2001, 1); similarly, "measureless" was Coleridge's dismissal of the extent of his imagined caverns (Coleridge 2018, l. 4).

Representation, a kind of dexterity or sleight of hand—once again—places one. The shape *shifting*, however, is irreducible, is interruption itself.

In relation to the Olmec jaguar tradition—concerning a predator's relation to the shamanic—Rebecca Stone wrote, "These different stamps and seals suggest that actual people could be spotted with versions of jaguar spots to show their jaguar selves directly," adding, "When seen from a distance, spotting the effigy or the human body with those of other animals would announce that the shaman was a jaguar or at least a spotted predator" (Stone 2012, 116). In the popular print by Rousseau, however, it's a Negro who "dances" with a jaguar in the lush rain forest (Rousseau 1910).

The Endless Mountains of northeast Pennsylvania, "a great series of parallel, heavily forested ranges," are a geographic barrier and they—the Endless Mountains themselves—are a dissected plateau, eroded, having been "formed after regional uplift" (Johnson 1934, 129).

(Now—2018—they're fracking there.)

A movie Negro would detach himself from the shade at the roadside of the Endless Mountains, looking "distinguished" and, being a speaker, in possession of "regular features." Let's see. There was a big fellow in the forest. *Fetch him.* Something wordless overflows its gulch. If you *cross* the Endless Mountains you'll come to Negro Mountain, but when you're in them you are only following the skyline. Let's see.

Generations of Pennsylvania wolves had "denned up" on a boulder-strewn shoulder on the north side of Negro Mountain, a location still—in 2018—called the Wolf Rocks.

Hawk Mountain, in the Ridge and Valley region, is a funded sanctuary for predatory birds of passage and for vultures—which are not predatory—as well. Updrafts and thermals peculiar to the location carry such birds over the mountain—that is, the traveling broad-winged raptors (eagles, turkey vultures and black vultures, buteos, etc.) and the accipiters (Sharp-shinned Hawks, Cooper's Hawks, and the like) make use of the trick of air currents in their migrations, ascending in spiral currents of warm air—thermals—and then, having gained altitude, following the ridges subsequent to Hawk Mountain. Hawk Mountain is itself yet another ridge, as is Negro Mountain. (The description of Negro Mountain's prominence mentioned earlier is from the Discover World website, accessed on 25 June 2018: "The mountain is flattish in appearance due to its location on the Allegheny Plateau, so its prominence is of low relief.") The mountains—Hawk and Negro—are 250 miles distant from one another. Between them: The deep forest of the Seven Mountains and, of course, the curve of the Appalachians.

Perhaps the big Negro at hand was, in effect, a country doctor. Versed, practical, aloof.

The mountain intervenes. The mountains intervene.

Performance at a scattershot mill town at the bottom of the hills. The skinny Negro at the door to the theater said, *Take your dinner in.* He said, *Don't lose your money.* He said, *Remember where you put it*—the money—at the mouth of the theatre. *Remember what you did with it,* he said.

On one hand is the lively narration in "Peter and the Wolf"—one hears oboe, flute, clarinet, and an orchestra's string section and is told that these represent animals and children. The wolf, however, is three French horns. (Or—in at least one publication—the wolf is a Harlequin drawing, red-vested, upright and loose-limbed in boots and top hat, juggling a garland of valve-less post horns [Benois, Brownell, and Conkle 2002, cover].)

> *On another, in a dream (recorded elsewhere), wolves had been in the side yard of the family house, moving in a wave across the grass and the driveway. The dream is from 1955 or 1956 or 1957; they—the dreamt-of wolves—had only come out to where they could be seen. Of course, what was there to be afraid of? Nothing was anyone's "fault"; the dream was oddly static in spite of the fact of the wolves' motion. And there was no point at which the spirit would have left the area, taking content with it.*

In 2011, VonHoldt et al. worked toward mapping the natural history of "charismatic wolf-like canids" (VonHoldt, para 2); the focus of their interest had been to "assess long-standing questions about diversification and admixture" (para 2) of such animals, and they noted that the species at hand, "the gray wolf (Canis lupus), red wolf (*C. rufus*), Great Lakes wolf (*C. lycaon* or C. lupus lycaon), and coyote (*C. latrans*) ... are characterized by high mobility and weak patterns of intraspecific differentiation [and that,] similarly, large dispersal distances have led to the formation of extensive admixture zones in North America, where four morphologically distinguishable wolf-like canids"—those named just above—"can potentially interbreed" (para 2). Their focus, in part, had to do with eastern animals, and they were criticized by Rutledge et al. not only for their study's "ascertainment bias" (2012, 190) but also because of "the presence of two wolf types in eastern North America [being] recorded in historical accounts"—these in 1859, 1842, and 1672 (190). Of interest here, because of the reported Negro Mountain sightings, is that the vonHoldt group's "latter two taxa"—the Great Lakes wolf and the coyote—"are of controversial ancestry and species status and readily hybridize with other wolf-like canids" (para 1).

An animal's what people think they see in the jungle. They'll be fracking there soon.

Leaning on earlier work by Lynda Rutledge and others, Thiel and Wydeven, for the Fish and Wildlife Service (2011), had already iterated that "Algonquin Provincial Park in east-central Ontario is considered by most geneticists to be the geographic core of current *Canis lycaon* distribution in eastern North America (Sears et al. 2003; Grewal et al. 2004; Rutledge et al. 2010a)" (Thiel and Wydeven 2011, 18) and that "immediately south of Algonquin Provincial Park [and] extending across the lower Great Lakes and south of the St. Lawrence Seaway lies the suture zone between eastern wolves and coyotes (Kolenosky and Standfield 1975; Sears et al. 2003; Grewal et al. 2004; Kays et al. 2009; Way et al. 2010; Rutledge et al. 2010a). South of Algonquin Provincial Park, in an area referred to as the Frontenac Axis, the major genetic ancestry consists primarily of coyote admixed with eastern wolf (Wilson et al. 2009). This trend is observed east into New England and *further south in western Pennsylvania* . . . (Kays et al. 2009; Way et al. 2010)" (25, emphasis added).

The Frontenac Axis is the southern extension of the Canadian Shield. Algonquin Park—where the Halfkenny brothers, George and Mark, and their white friend Billy Rhindress, teenagers all, met death in a bear encounter—is to its north. The boys had been fishing at Radiant Lake. "Those investigating the incident felt that the first boy who was killed, George Halfkenny, may have been attacked because he resembled another bear. George wore dark clothing and was black" (Herrero 2018, 114). This was in 1978. The best-known geologic feature of the Frontenac Axis is the Thousand Islands of the St. Lawrence River, which river defines part of the border between Ontario and New York State.

The 2011 Fish and Wildlife Service piece presses *suture zones* throughout (Thiel and Wydeven 2011, 21).

Mix, and match it, match the mix; the thinking overlaps and goes and goes without saying.

A "thread" on a white supremacist message board bemoaned the presence of Negroes on the Hawk Mountain website. One writer, "treblemaker," who claimed to have recently discovered Hawk Mountain, expressed the fear that Negroes would "begin showing up at this last (nigger free) sanctuary"—this in 2009.

> *In the twentieth century wolves or something* similar *crossed into the States—into the northeast and/or the Great Lakes states including of course Pennsylvania—and/or across other borders via railroad bridges (from Canada in the dead of blackest fucking night), a feat of migration or a surmise concerning migration recently mentioned "somewhere" (Melville 2003, 7) as a likely fact which I, as a child in the years following the dream of wolves in the yard, had conjectured and then assumed to be a possibility or, as I recall it—the early conjecture—in 2018, necessity itself.*

A little quiet goes a long way. All dogs range and are, in practice, scavengers. All—a trait—are communal.

Cristina Eisenberg wrote, "During the mid-Pleistocene epoch, about 700,000 years ago, successive waves of wolves returned to North America from Eurasia." Then she wrote, having skipped backwards, "The wolf genus, Canis developed in North America in the late Miocene epoch, about 10 million years ago. ... By the early Pleistocene epoch, 1.5 million years ago, wolves had split off from coyotes and crossed into Eurasia via the Bering Land Bridge, where they continued to evolve into C. lupus, which means 'dog wolf'" (2015, 120). The idea being that they went to Siberia and then came back "different."

How unlikely then would it be for a wolf to have appeared—to have shown itself—on Negro Mountain? What might the field marks have been? Summers, writing on observable phenomena, said, "sometimes it is a fantastical shape" (2003, 197). But what is a wolf?

Negro what?

The observation or claim that *paradise is full of tremendous images* is something a Negro might make or might have said; that Paradise is incomplete without both the appearance and the sharp absence of its predators is a different animal. One animal's a color—such is description—and the other's god. What's your real name? an old gentleman might ask. I'm the only dog on the plateau, somebody might respond. Measure is measure, any given day. I can teach you image (said the old gentleman) but you won't like it.

White families—in scattered houses—populate the lower slopes of Negro Mountain; the mountain stretches behind the houses as though it were an unspoken-of Negro wife. Somerset County, so says Wikipedia, is "1.59% Black or African American" (para. 18), and Negro Mountain straddles the Mason-Dixon line. Florence Mars wrote, "My other grandfather . . . told me that after the civil war it was customary for former slaveholders to build a house in the back yard and retain one of the better-looking Negro women" (Mars 1989, 16).

Spoken Commonwealth history of Negro Mountain has preserved the particular Negro's name as having been Nemises, often that spelling, though it would seem unlikely that his parents called him that, it being more likely that he was, in some way, *owned*. As the Town of Brookline website explained as of 26 June 2018, "Owners chose [slaves'] names that were classical names to indicate the owners' cultured ways: Caesar, Dido, Pompey, Primus, Venus." Nemises or Nemesis is a mascot's name, a pet's name, the name of an animal. But *Oxford English Dictionary: Nemesis* was "the goddess of retribution or vengeance, who reverses excessive good fortune, checks presumption, and punishes wrongdoing"—continuing, widening—"(hence) a person who or thing which avenges, punishes, or brings about someone's downfall; an agent of retribution," the name having lost, over time, its feminine cast or association.

Consider opposite that the namelessness of the fact of Grendel's mother. And/or consider the loa Marinette Bras Cheche (or Marinette Pied Cheche), like Nemesis, associated with punishment and retribution and often, one reads, in the company of werewolves, "who hold services in her honor" (Chatland and Corbett 2018). "*Nombre de sorciers, porteurs du point loup garou, ont la capacité de se changer durant la nuit en animal … La proximité de ces [âmes perdues] avec des mauvais esprits rôdant dans les campagnes, épousant parfois l'allure de monstres, évoque les noms de Ti-Jean Pied Chèche, Ezili-jé-rouge, Marinette-bois-chèche*" (Elysee 2018).

Or that Margaret Atwood dreamt she was watching an opera she'd written about Susanna Moodie. She wrote, as description of the dream: "I was alone in the theatre: on the empty white stage, a single figure was singing" (Atwood 1970, 62). And she wrote that Susanna Moodie imagined her husband to be a *loup-garou* but that she—Susanna Moodie—would have him change *her* "with the fox eye, the owl / eye, the eightfold / eye of the spider"—every example a predator—and that she would be *unable* to "think / what he will see / when he opens the door" (19).

A question had come up for discussion at the foot of the Blue Mountains in Portland Parish, Jamaica, BWI, the same season, summer 2016. The question, open to debate: Can a loa appear to a white person?

What *did* the Negro Mountain woman see when she opened her door on Negro Mountain?

Colonel Thomas Cresap's letter—or an "extract" of the letter—was published in the *Pennsylvania Gazette* on 6 June 1756. (The date contradicts Mrs. Stevenson's account, above.) Colonel Cresap wrote, "I made no Halt as the Indians came towards me, but marched up with my Gun cocked on my shoulder. As soon as the Indians came clear of a Bent that was in the Road, about 30 Yards from me, an old Negroe presented his Gun at them; upon which they immediately alighted from their Horses. I saw two run to one Side of the Road, and one to the other, and take Trees. We having no Advantage of Trees at that Place, two of the Indians fired, and shot the Negroe […]."

Or a wolf's figure is the figure—similar in status to another person—that one might tease out of any thought about the woods and hillsides. The wolf is the thought. In *fact*, one's live bodily presence is likely to repel "real" wolves and, understanding this, one may wish for a wolf's appearance to interrupt one's *stature*—however troubled or multiple that may be—or progress in the forest. The wolf, in terms of imagination, is bigger than the speaker; that is, the wished-for or imagined or conjured presence of a wolf brings with it the possibility of death or a glimpsed or thought-of death, but—in fact—would become (and easily so) an effect were one to describe it (the apprehension) on paper, say. One might have the wolf say, *I'll measure you.*

Or the wolf might say, *Wish for everything.* The best thing is the thing that's most available. Or, on another hand, at the *very* best one might be the hawk and the road, both.

The shift itself—to the listener or reader or observer—may be "imperceptible." It can't be traded; measure falls off.

For what is labor exchanged on the mountain?

It—the trade itself—might be theater. Or *like* theater; or song. Sing for your supper, goes one imperative; hold on to your money, confides a second one. The Cresap Society's bulletin from the 1919 meeting at Cumberland: "Colonel Thomas owned a negro of giant stature called Nemesis. In mustering his company the Colonel said 'Nemesis, wont you go with us this time, you are a good shot, and help us conquer these Indians who are murdering and scalping women and children and burning their cabins.' Nemesis considered for a few minutes and then said 'Yes, Massa. I go, but I wont come back.' 'Why, Nemesis, why say that,—you are a sure shot and fearless.' 'Massa Tommie sure shot and afraid of nothing and he not come back. I say I go but I not come back.' His premonition, second sight, was correct. Among the first to fall was the brave slave, and now and forever the mountain where he died is called Negro Mountain" (Cresap Society 1919).

Colonel Cresap's letter to the *Gazette* had continued, attesting to the unhappy fates of his son, Thomas (the "lamented sleeper" in Mrs. Stevenson's account, "Massa Tommie" mentioned earlier), and a Native, who had also perished. "We saw no Bones, but great Signs of Wolves, Turkey Buzzards and Ravens having been at the Place, and make no Doubt of his dying there; for the Bears and Wolves had eaten up another Indian that was scalped at the same time that he was shot, and had scratched up the Body of Thomas Cresap, and eaten it likewise" (Cresap 1756).

Thinking in terms of Negro Mountain, as mentioned earlier, brings the discomfort of the roadside name, *Negro*, and the public—meaning *widely promoted*—roles inhabited by the African-descended people in North America's popular histories of conquest. The Six Nations people were loose on Negro Mountain before anything was called that. "The negro Goliath," referenced earlier, was made to present himself on the stage and speak his tribute to the house—meaning *conquest*—and die then, having been slipped into memory, having been *placed*. Wolves came back, though. A reader in 2018—distanced or overtaken by anger's pockmarked and surly red face in the news—falls to example or model or is tempted to. (Another, however, at roadside, vexed but recognizing *vexed* as an old acquaintance on the Commonwealth's byways, would ask herself or her companions, as noted in a *Washington Post* article on 2 February 1995, "What Negro?")

Second day of bear season in the Commonwealth—20 November 2017—and my friend met my train at Cumberland and drove me to Negro Mountain to visit the white woman who had seen the wolf, mentioned earlier. I was the stranger on Negro Mountain, which we crossed that day, in her car, in a trip that included passing the Wolf Rocks and pausing to hike the short distance to the Baughman Rocks, where we photographed one another in the snow. And later drove to Fort Necessity, where George Washington had surrendered to the French in 1754. I was indeed a tremendous stranger on Negro Mountain but I had lived in Bellefonte, Pennsylvania—and had been a petty railroad bureaucrat there and had taught writing and image, for years, at the nearby state university—and had joined or been pulled into conversations specific to the location, though I was neither "English" nor "Dutch," and was not a Philadelphia Negro either. But my friend had not thought to inform the Negro Mountain woman that she would, in fact, be bringing a Negro to visit and the woman—who was in her 90s and had taught in the Amish and Mennonite schools on and near Negro Mountain and had been born in the house in which we visited her—ignored me and spoke only to my friend and was irritated, seemingly, by the question she—my friend—posed, having to do with the mountain's name. It was Civil War time, she replied, and an army regiment came through. "They had a Negro with them," she said, and then said that the Negro ran to get away and they chased him and killed him and buried him on the mountain. And the wolf? I asked, finally, interrupting, Did you see a wolf outside? "That was a long time ago," she said, "it's a little fuzzy."

Charismatic Pennsylvania man, upending Aldo Leopold, could himself ask, What might wolves—whatever wolf-like thing wolves are—sense on the Negro Mountain? (Leopold 2018)

What had the Negro Mountain woman seen, sitting at the kitchen table across from her on Negro Mountain itself?

I had seen a wolf skin in St. Ignace, on the Mackinac Straits, tourist shop, 1959. Tufted, grey, and long on a rack with other, more luxuriant, furs; it had cost $10, which I, being a child, did not have. I saw it was the skin of one of the animals I'd seen in the dream, mentioned earlier, two or three or four years earlier and, perhaps and likely, in a book before that as well. Later though—out of an evening at a drive-in theater in 1961— I had occasion to watch an orphaned boy, my age, in a movie, set in Spain, being plagued by dreams that he was himself "a wolf, like in a picture book," during which time a wolf or a dog was "abroad," killing goats in the countryside. His adopted father put up bars on the boy's window then, he said, to "keep your nightmares away." The priest in the film (Curse of the Werewolf), however, was frank—"An inherited weakness," offered the priest, in explanation.

In fact, necessity is plural and locations are unstable. Measure—like bacteria—can mutate and the mutation will outlast explaining (Matheson 2007, 147).

What, Negro?

References

Atwood, Margaret. 1970. *The Journals of Susanna Moodie*. New York: Oxford University Press.
Benois, Nicola, David Brownell, and Nancy Conkle. 2002. *Peter and the Wolf*. Santa Barbara: Bellerophon Books.
Browning, Meshach. 1859. *Forty-Four Years of the Life of a Hunter; Being Reminiscences of Meshach Browning, a Maryland Hunter*. Philadelphia: J.B. Lippincott.
Calvert, J. B. 2001. "The Llano Estacado." *Iron*, May 22. Accessed June 25, 2018. https://mysite.du.edu/~jcalvert/geol/llano.htm.
Chatland, J., and B. Corbett. 2018. "Haiti: List of Loa." *Faculty.webster.edu*. [online]. Accessed June 26, 2018. http://faculty.webster.edu/corbetre/haiti/voodoo/biglist.htm.
Coleridge, Samuel Taylor. (1816) 2018. "Kubla Khan by Samuel Taylor Coleridge." Reprint, Poetry Foundation. Accessed June 25, 2018. www.poetryfoundation.org/poems/43991/kubla-khan.
Cresap, Colonel Thomas. 1756. "Letter." *The Pennsylvania Gazette*, June 6.
Cresap Society. "Full Text of *The Cresap Society Meeting at Cumberland, Md.* June 14, 1919." *Internet Archive*. Accessed June 26, 2018. https://archive.org/stream/cresapsociety mee00cres/cresapsocietymee00cres_djvu.txt.
Curse of the Werewolf. 1961. Directed by T. Fischer. Bray, near Maidenhead, Berkshire, UK: Hammer Film Productions.
Eisenberg, Cristina. 2015. *The Carnivore Way*. Washington, DC: Island Press.
Elysee, M. (2018). "Books | COUCOU Magazine." *Coucoumagazine.net* [online]. Accessed June 26, 2018. https://coucoumagazine.net/tag/books/.
Herrero, Stephen. 2018. *Bear Attacks: Their Causes and Avoidance*. Lanham, MD: Rowman and Littlefield.

Hoffmeister, Donald F., and Herbert S. Zim. 2014. *Mammals: A Fully Illustrated, Authoritative and Easy-to-Use Guide*. New York City: St. Martin's Press.

Hurston, Zora Neale. 1991. *Their Eyes Were Watching God*. Champaign-Urbana: University of Illinois Press.

Irby, Kenneth. 2009. *The Intent On*. Berkeley: North Atlantic Books.

Johnson, Douglas. 1934. "How Rivers Cut Gateways Through Mountains." Full Text of *The Scientific Monthly I*, Vol. 38. Internet Archive. 1934. Accessed June 25, 2018. https://archive.org/stream/in.ernet.dli.2015.232290/2015.232290.The-Scientific_ djvu.txt.

Leopold, Aldo. "Thinking Like a Mountain by Aldo Leopold." Thinking Like a Mountain by Aldo Leopold—Wolves and Deforestation. Accessed June 26, 2018. www.eco-action. org/dt/thinking.html.

Mars, Florence. (1977) 1989. *Witness in Philadelphia*. Reprint. Baton Rouge: LSU Press.

Matheson, Richard. (1954) 2007. *I am Legend*. Reprint. New York City: Tor Books.

Melville, Herman. (1851) 2003. *Moby-Dick*. Reprint. New York City: Dover Thrift Editions.

Muehleisen, Susanne, and Bettina Migge. 2005. *Politeness and Face in Caribbean Creoles*. Amsterdam: John Benjamins Publishing.

Oxford English Dictionary, s.v. "Nemesis." Accessed June 26, 2018. www.oed.com.libproxy. berkeley.edu/view/Entry/125992?redirectedFrom=nemesis.

Rousseau, Henri. 1910. "Negro Attacked by a Jaguar Painting." *Framed Paintings for Sale*. Accessed June 25, 2018. http://bestpaintingsforsale.com/painting/negro_attacked_ by_a_jaguar-10273.html.

Rumsey, David. "David Rumsey Map Collection." Rumsey Historical Map Collection. Accessed June 25, 2018. www.davidrumsey.com/.

Rutledge, Lynda Y., Paul J. Wilson, Cornelya F. C. Klütsch, Brent R. Patterson, and Bradley N. White. 2012. "Conservation Genomics in Perspective: A Holistic Approach to Understanding Canis Evolution in North America." *Biological Conservation*. Accessed June 25, 2018. https://people.trentu.ca/~brentpatterson/Index_files/Rutledgeetal2012-conserva tiongenomicsinperspective.pdf.

Shoemaker, Henry Wharton. 1916. *Addresses*. Altoona, PA: Tribune Press.

———. (1917) 2000. *Extinct Pennsylvania Animals*. Reprint. Landisville, PA: Arment Biological Press.

Stevens, Wallace. (1938) 2009. "On the Road Home." In *Wallace Stevens: Selected Poems*, edited by John N. Serio. New York: Alfred A. Knopf.

Stevenson, Mrs. May Louise Cresap. 1901. "Colonel Thomas Cresap." *Ohio Archeological and Historical Society Publications* 10: 156.

Stone, Rebecca R. 2012. *The Jaguar Within: Shamanic Trance in Ancient Central and South American Art*. Austin: University of Texas Press.

Summers, Montague. 2003. *The Werewolf in Lore and Legend*. New York City: Courier Corporation.

Thiel, R., and A. Wydeven. 2011. "Eastern Wolf (*Canis lycaon*) Status Assessment Report Covering East-Central North America." *Fws.gov*. [online]. Accessed June 26, 2018. https://www.fws.gov/midwest/wolf/history/pdf/ThielWydevenEasternWolfStatus Review8August12.pdf.

VonHoldt, Bridgett M., John P. Pollinger, Dent A. Earl, James C. Knowles, Adam R. Boyko, Heidi Parker, Eli Geffen, et al. 2011. "A Genome-Wide Perspective on the Evolutionary History of Enigmatic Wolf-Like Canids." Advances in Pediatrics/ *Genome Research*, August. Accessed June 25, 2018. www.ncbi.nlm.nih.gov/pmc/articles/PMC3149496/.

Wikipedia. "Somerset County, Pennsylvania." *Wikipedia*. Accessed November 29, 2018. https://en.wikipedia.org/wiki/Somerset_County,_Pennsylvania.

I acknowledge with gratitude the contributions to this piece made by Thomas Johnson, Julia Spicher Kasdorf, and Meriel Melendrez.

20

HURRICANE POETICS AND CRIP PSYCHOGEOGRAPHIES

Stephanie Heit and Petra Kuppers

Introduction

During a weekend in October 2017, we, the authors, worked at Movement Research, a NYC-based laboratory for the investigation of dance and movement-based forms. We worked under the label "The Olimpias." The Olimpias is an artists' collective, founded in 1996 in Wales by mental health system survivors, with artistic director Petra Kuppers. Olimpias associates come from across the world, with a current US center. We create collaborative, research-focused environments open to people with physical, emotional, sensory, and cognitive differences and their allies. In these environments, we explore pride and pain, attention, and the transformatory power of touch. The Olimpias is disability-led, and non-disabled allies are always welcome. Stephanie Heit has been an associate of The Olimpias since 2015. Since that time, Stephanie and Petra have co-led The Asylum Project: An exploration of sanctuary, edge space, and communal well-being, infused by crip culture/disability culture values such as interdependence.

During this weekend in NYC, Stephanie (a psychiatric system survivor who is bipolar) and Petra (a wheelchair/scooter user who lives with chronic pain) worked with community participants who had signed up for the workshop, hosted by Movement Research. We engaged in score-building, i.e., in improvisatory activities that allow for individual exploration within an agreed upon structure. We grounded our exploration in the lineages passed down to us by such people, groups, and movements as Joseph Beuys, Barbara Dilley (2015), Judson Church, the psychogeography of the Situationists, and the gestural choreopolitics of Black Lives Matter. Sites of exploration included the East River Promenade, Delancey Street and a playground adjacent to it, and the area around the Tenement Museum on the Lower East Side.

In this creative/critical essay, we investigate poetry and performance as ways of being in the world, of opening our senses to our surroundings through non-realist and surrealist framings. To achieve this, we draw on the NYC weekend workshop as a sample of our collaborative and community-based practices.[1] The workshop's methods involved tuning our bodyminds to inner and outer geographies and energies. Chance encounters led to fantastical storylines involving the Anthropocene and climate change. For instance, the site of one of our NYC explorations, along the East River, was underwater during Hurricane Sandy, and our workshop coincided with the catastrophe in Puerto Rico in the wake of Hurricane Maria a month prior, in September 2017. These climate connections came into play in our site-specific engagements and influenced our experiences and movement patterns.

Each day during our Olimpias workshop at Movement Research, we met for two movement sessions, encompassing two different sites. We also offered a performance sharing following that weekend, on Monday night, at Judson Church. In this chapter, after a meditation on the *dérive* as a form of geopoetic knowing, we flow through these workshops sequentially to give a flavor of our geopoetic community performance methods.

Petra: the dérive

Our core method of engagement goes back to Situationist practice: Every day, in some form, we played in the realm of the dérive, or drifting. The original European avant-garde movement of Situationist International (SI) began in 1957, when the Situationists engaged in what Claire Doherty describes as "artistic practices for which the 'situation' or 'context' is often the starting point" (2004, 7). Given what I read of Guy Debord, co-founder of SI, I am not sure that I would have wanted to play on their playground, or been allowed onto it, as a first-gen-university-going, disabled, white woman of size, grounded in disability culture values: Critics have accused the early movement of elitism and authoritarianism (see Hancox 2012, 237, 248). Sensitivities to power differentials and their effect on space engagement are core to Olimpias actions, so we take note of anything that feels too easy in our engagement with historic forms.

In 1958, Guy Debord wrote that the dérive "is a technique of rapid passage through varied ambiances," which involves the dropping of "usual motives for movement" (19). In our 2017 dérives in NYC, we stumbled, arrested our ambles, and got stopped in multiple ways. Our arresting influences were the pattern arrangements of orange plastic safety-fences that kept spaces separate, danger zones of rubble, and the lines concrete markers make in space. As dancers, the "usual motives for movement" were already suspended. The line between dérive and improvisational impulse frayed easily as we fell into Debord's second description of the dérive: "playful-constructive" modalities of engaging space. He writes:

> Dérives involve playful-constructive behavior and awareness of psychogeographical effects, and are thus quite different from the classic notions of journey or stroll.
>
> *(1958, 20)*

Framing our dérive work by calling on asylum imagery repeatedly bound us back to the city's histories, to personal memory, to feeling safe and unsafe in skin, world, city, nation. Where was sanctuary? Where was asylum? What was stricture? What was safety? What were the differences and nuances of these experiences?

The dérive in performance often works at these interstices of political space use and embodied sensation. For instance, the German performance troupe Rimini Protokoll used dérive scores and mobile phones in 2012, dialing numbers to hear documentary fragments of Stasi surveillance of citizen movement in their walking performance, *50 Kilometres of Files*. In these city walks,

> the performer-listeners take up multiple positionings in ways that make them sensitive both to the different temporal layerings of certain places within the city and to the regimes of control and surveillance that, even after the end of the Cold War, pervade everyday life, ranging from more obvious forms of surveillance such as closed-circuit TV to the more insidious dataveillance or digital information tracking.
>
> *(Hahn 2014, 34–35)*

In our NYC drifting, we did not engage with this overlaid grid of surveillance, though we noted cameras hidden in overhangs, traffic-cams, and other mechanisms of vision. Our sensitivities to temporal layerings emerged from inhabiting and embodying metaphors and actions of climate, rupture, and the threats of past and future chaos. We explored ruptured supply chains in hurricane times, felt the drag of dancing in Hurricane Sandy territory, the energy of the rushing river right beside our (currently) dry promenade. We were aware of the political dimensions in the rupture of travel and of site construction, as well as in the breakdown of flow "as usual" that occurs in moments of catastrophic change. Flows alter site habitation: settler movement, water invasion, and now, actions kicked off in complex ways by the Anthropocene era. With these awarenesses layering into our explorations of asylum, we embraced another maxim of drifting work: Psychogeography, the eased or halted flow of energies. Debord writes that

> chance is a less important factor in this activity than one might think: from a dérive point of view cities have psychogeographical contours, with constant currents, fixed points and vortexes that strongly discourage entry into or exit from certain zones.
>
> *(1958, 20)*

In our workshop, multiple notions of energetic torsion emerged. In particular, we explored climate change and the epoch's challenges to find new patterns of human engagement with site, climate, and earth habitation; and we focused on interdependence, the lean into others, finding support and mutual care as our world changes.

Petra: drift in Abrons Arts Center

On our first day, we visited a quilt exhibit at Abrons Arts Center. The quilts, designed by Maggie Thompson, were part of a multi-year collaborative project led by Emily Johnson and Catalyst Arts entitled "Then a Cunning Voice and a Night We Spend Gazing at Stars."[2] The program notes for the exhibit reminded us of who held us in the space: "Emily Johnson/Catalyst and Abrons Art Center pay respect to Lenape peoples and ancestors past, present, and future. We acknowledge that this work is situated on the Lenape island of Manhahtaan (Mannahatta) and more broadly in Lenapehoking, the Lenape homeland."

At the core of the project were questions of well-being. These, too, were highlighted in the program notes: "What do you want for your well-being? For the well-being of your chosen friends and family? For your neighborhood? For your town, city, reserve, tribal nation, world?"

After reading the questions out, we began to inhabit the space, to find our own answers in our bodily movements, in aligning ourselves with each other, buffered by quilts. The space was a site of care, full of energy created in handiwork. The blankets were carefully folded. The program notes invited us to unfold the quilts and to envelop ourselves in them so as to make the floor warm and soft(er). So we did.

Some of us explored the movements of the quilts themselves. Some of us were matadors offering cloth to a different species. Others were caretakers swaddling a baby. Some made a canoe, a tent, a bed, or a swing out of the multi-colored quilts. Some engaged the blankets' materiality differently, and folded and refolded them, stroking palms over surfaces touched by loving others—a form of stimming (an autistic cultural expression, affirming and stimulating one's bodymind boundaries through repetitive motions). Some participants read out loud the messages written, as part of the quilting process, into the little squares: "Affordable Health Insurance," "Fiesta!," "Better Communication in my Relationship," "Community Leadership," "hot water," "live confidently without fear," "sleep." Many such messages had accumulated on these 84 blankets, which had a longer-serving purpose than the exhibit. They were designed to be part of a nighttime community gathering, to be the support structure for a long gaze up at the stars.

The program notes explain the process of the gathering of voices around and on the quilts:

> These questions have been asked over the past three years in cities, towns, reservations, farmer's markets, schools, museums, libraries—across the US, in Taiwan and Australia, and at Standing Rock and Women's Marches after the 2016 US election. Recording ideas in the moment of action, during workshops, or at marches, the quilts have become historical records of these significant events. They record visions, desires, and wants as they simultaneously express need and humanity.

We also found the following: "One student, a recent refugee seeking asylum in Australia, noted whilst stitching, 'these quilts are like maps to possible futures.'" The phrase, resonant among us after our initial group exploration of personal mean-ings of asylum, added a forward oriented acceleration. It encouraged a search for a solarpunk utopic impulse in acknowledged dystopia and stimulated an energetic torsion toward change and care.

Eventually, we had played ourselves out in the exhibit. The quilts acted as a tran-sition space for us: From the dance studio into public space, from personal explora-tion to a more communal one. The exhibit was our buffer zone.

I imagine all the quilts under the night sky on a piece of land that is well cared for; they are their own liminal space—a cunning voice space, a myth-telling space, a desire space. Our art practices are part of long heritages, long practices of creat-ing dream space, self- and other-care space, the destabilization of the real into the liminality of hopefulness, dreaming, or critique.

Stephanie: Luther Gulick Park play

First day, afternoon session, after the quilts. We drift. We enter the park, and the park enters us. In the spirit of improvisation, instead of trying to make the materials fit our intentions, we respond to what is already here. In this case, a group of black men relaxing in the afternoon warmth, their radio blaring. A white woman with her pit bull mix. Autumn leaves. Cigarette butts. Beer bottles. Plastic bags. Concrete. Green wood benches. Dead tree branch. Traffic sound. Each other. Our score: Allow your senses to receive and inform your investigation of the park. Pay attention to what draws you. Ten minutes to explore in open attention.

A couple slowly moves across the park bench, legs extended to sky, weight shifts, bodies aware of each wood grain. A few folks activate the dead tree branch; there is the castanet sound of dried leaves. The men with their stereo groove. Some of our group join in. Hips sway to quick rhythms. There is a relaxed tone to this sonic exchange. Parallel play. The folks already in the park before we arrived are certainly aware of our presences, as we are aware of theirs, but there is no formal acknowl-edgement, no head nods or eye contact. It is a kind of coexistence, where there is enough room for everyone to do their thing and get on with it. No big deal.

I'm overcome with the wealth of stimulation. I tune my senses to the location and activities of our group. I take in the wire fence surrounding three sides of the park, the nonspecific smells that scratch my olfactory memory, the nonstop city pulse—that sound of cement and wheels and humans with futures to get to, and my inner desire to curl up for an afternoon nap on a patch of nonexistent grass. Deluge. I need a dam, some kind of floodgate to make my open attention a bit less open. I touch one of the few trees. Let the bark soothe my firing nerves. Lean my spine on its slender trunk. Close my eyes. This is another kind of interdependence. I imagine what this land was like before it was made into a park. What species lived here? What grew? I feel grateful for this fragile tree, its oxygen and support.

When I make a quick survey of the group, it appears everyone is engaged. Some-times the scores we are building need tweaks, more constraint or more leeway, but the only thing this score needs is more time. We huddle close to share responses to the experience of listening and moving. One person comments that they could have gone on for an hour. Others chime in in agreement. We leave the space differently. The park does not bear the traces of our activities besides some crushed leaves, a dead tree branch moved a few feet from its original spot, and perhaps an indent in the conversation of the regular occupants of the park. But the space has shifted us both individually and as a group. The park served to magnify and entrain our awareness of each other, of our own senses, and of our surroundings. We move into our next score, a dérive along Delancey Street to the East River Walk, with senses primed.

Petra: Delancey Street, NYC, USA

At 2:40, rhythm expands along my Southern edge. Luther Gulick Playground. Small boom box on a concrete bench.

At 3:05, majestic black woman in an electric wheelchair whizzes along my flank. On the back of her chair a bag, "Puerto Rico"—electric letters glow under gray-light sky.

Stream toward the water.

At 3:10, white elder suns himself in the pearliness of heaven. He lies motionless in overgrown park near my Eastern edge. At 3:30, he rotates his nut-brown scrawny legs to mobilize his hips. An hour later, he walks near my rhythmic railing.

Sing when the trains pass above.

At 3:30, dancers lean into my far Eastern limit, spear into the East River. Across, the Domino Sugar Factory leans back. Sphinx stories echo. Limbs swing out above backs.

The break. The hold.

From 3:32 to 3:55, one dancer hovers, compact, like a bird on a water fountain. One drums beer bottles on a metal picnic table. One hugs a railing, river water swift, swift.

Turbulence beneath.

At 4:10, white ship across the East River tips over. Containers topple in slow gravity, new magnetic powers upset the logistics machine.

At 4:13, wave jumps my Eastern bank, breathes into car lanes, new flow.

X. All the clocks. From East to West.

Stop still.

Flatness spreads. Oil shines. Rainbow unbroken.

Birds tousle my wire hair.

Petra: drift on Delancey Street

During our dérive exploration of Delancey Street, we all saw her: A majestic woman in an electric wheelchair—a black woman, an elder. She passed us by as we were slowly moving from the playground to the meeting place at the bridge's ramp. In our post-score discussion later, we found out that she played a role in everybody's dérive fantasy, marking our experience with her transit.

Some of us looked at the missing persons leaflets attached to lampposts. I looked at grids in fences, the holes and wholes made by iron and plastic, the orange safety veil that is supposed to allow a small tree to flourish by the side of a steel fence. Others danced around a water hydrant, unfolding from one form into many, until a small ballet happened, limbs and canes extending into the street.

FIGURE 20.1   Dancing in Delancey Street.

Source: Photograph by Petra Kuppers.

But she drove past us, secure in her speed and on her linear path. Upon passing, we saw the back of her chair. Attached to it was a large black bag, with colorful letters spelling out Puerto Rico. None of us engaged with her, but she dissected our attention, and then we saw her leave, and thoughts of precarity, survivance, climate change, nation states, and hurricanes trailed in her wake. The gravity of the situation mixed with the levity of the drift.

We assembled at the entry of the footbridge across the FDR drive, a zigzag out toward the East River Promenade. Stephanie and I ascertained if everybody was still feeling mobile, happy to continue, warm enough, and engaged. We all did. So we drifted over, keeping our attention engaged with the exigencies of space, the patterns of the railing, political slogans, graffiti, the push and pull of the cars, all the cars, the vans, the trucks, the speed of things shuttling across our path, all the energy dense dense dense. Our bodies stood buffeted by the winds, by the drives of the cars beneath, by the hum of the asphalt. My scooter rolled up with its motor push, and gravity took it downward, down from the high point, my control loosening as I careened down the ramps and toward the park at the bottom of the far ramp. Others took much longer, let themselves be rippled and striated by the leylines of city highways. Steel mesh stops you from killing yourself.

At the bottom of the far ramp, an unkempt park offered sanctuary, a thin defense against the traffic noise assault. I turned into it, and then it opened up before me: The vista of the river side, the gleam of low afternoon sun on the river. More roilings, more energy: Right where we are, giant turbulences move the water in sinuous forms. I took a deep breath, and felt elated, moved, outward, outward, flatness, horizon. One by one, we all trundled in, all lined up by the railing, drawn to the water's ion edge. An escape route. No mesh. Contained movement. Sandy had ripped over the promenade here, and the gravity of the river's power was clear and present. But for now, in this moment, in the gleam of the sun through clouds, open wind into our faces, we were happy, or at least most faces around me spoke of arrival, of opening, of ends.

★★★

In our debriefing from our two dérives, the ending at the river had multiple layerings: Straight out from us stood the Domino Sugar Factory, and we talked about the site's gravity in the art worlds we inhabit.[3] We talked about Kara Walker. The sugar sphinx. I touched back in with some of the writings that fed my pre-city explorations: Fred Moten's and Stefano Harney's work on logistics and embodiment (2013). I fed in lines about container ships and cargo, others talked about capitalist webs that stretch across the world, and we contemplated natureculture pushbacks, sand in the sugar machine, the weeds taking over the hard edges of the park into softness. Birds in the sound machine of the FDR drive. Wheelchair/Puerto Rico/ the smoothness of the electric glide/the turbulence of the water.

In the open engagement with site that followed, mobility and pauses shifted around us. One of the dancers swung her crutches in circles, grabbing space in wide arcs. I saw a near-naked old white man with suntanned skin carefully gathering sunlight into himself, pressing each ion of autumn sun into his bio-battery. Dissolution/Revolution. I thought of batteries and exhaustion.

Eventually, I wheeled back to the benches by the river, our assembly point after our 20-minute improvisation. I moved past a dancer who had wound her limbs into the railing of the river, holding close and looking far. Then I got back, and I watched and photographed a dancer who arced backward over the river, like a wasp about to lay an egg: A leggy, elegant strangeness in her contorting limbs. Here, I made my nest, NY city's humidity and warmth layered around me as the river wind spoke of the ocean and of salt. We all came back from our stations, our playgrounds, our sugar salt liminal vehicles of transport and attention.

Stephanie: Estuary Cycles

I'm going home!
plop down on bright yellow material with pink pansies that are not pansies   slant
            slope to cement where I mix with unnamables
                        more potent than previous detritus encounters
                                    through a grate
                                    into the dark

                                East River cohorts welcome
                            me into their flow & ripple broken
                by ferry churn   shipping container wake   seagull graze
                                        reflected faces lean
                                over the promenade shored
            up bank side scrap metal   plastic bag   key chain with rusted key
                                inside rainbow smelt lungs
                                repellent gasoline slick
                        sun blocked by Williamsburg bridge

                            I vaporize
                cumulus

                        suspended in past summer heat   float
            homesick for river current & chorus   carries on as I float on a westerly
            breeze   red lights
                below   yellow cabs streak traffic on the FDR   geometric shapes hold
            jagged skyline   movement swells in grid patterns   bridges // bridges
            always over
                    now under
                        I wait not sure what I wait for
                            gravity momentum
                                            break
                                            the circle
                            fly

Petra: drift at the Tenement Museum

On the second day of our Asylum Project exploration we were at a different home space: The area in Orchard Street surrounding the East Side's Tenement Museum, a range of little streets that house the remnants of an old tenement, miraculously (or not) kept intact from its nineteenth-century appearances through landlord shenanigans. Each day, multiple tours traverse the streets, young docents taking groups of mainly tourists through the intricacies of how people used to live, many people to one room, taking in washing or sewing in one-room sweatshops. This is a museum of how people made space habitable, and it is also a voyeur's space for glimpsing intimate details (such as toilets, washing lines, frayed old clothes) of histories that took place right here, in the same palimpsestic space. It might be a good space to search for ghosts, but it is a busy space, a thoroughfare, an undecided space where there is not one entrance but multiple warren-like access paths by which to enter.

So on the street, at least, it did not feel spooky to us. The access issues of a fast city were clear to us: We managed to find a long red bench tucked into a little niche between shops and restaurants, and the little hovering space for our butts felt luxurious. There were not too many benches around. The nearest alternative was a bunch of funky yellow benches right on the meridian of a busy neighboring street: Good looking, but hardly conducive to talking and engaging each other.

There we were, on the red bench, in the shadow of the tenements, with little reminders of centuries of Lower East Side life all around us, tourists and locals swarming around the streets. We gave our final instruction for this last session of the weekend workshop: Construct a score for this space. Draw on the heritages that we have honored in our explorations together: Lineages from Joseph Beuys to Barbara Dilley and Judson Church, from the Situationists's psychogeography to Black Lives Matter's gestural choreopolitics, the interventions of Alicia Garza, Patrisse Cullors, and Opal Tometi, and the network they have helped into being.

Stephanie: red benches, crosswalks, grids

We shared our invented scores with each other in public. In one score, coiled movements slow motion shifted as one body poured into another, arms forming circles spill across the sidewalk from an apartment landing. Some passersby paused to watch, others dodged around the scene, their linear motion interrupted. We were mindful about allowing space for people to get by. The group I was in asked folks at crosswalks if we could accompany them across the street. I had a woman offer an arm. Another headphoned-twenty-something looked quizzical and said no. The people who welcomed us, we thanked on the other side. Moments of lean, of interdependence, support in space traversal if only an arm, a few words, human contact in everyday activity. The invitation shifted back and forth, from "May I accompany you?" to "Will you accompany me?" Different approaches to agency. Different approaches to need.

After all the groups shared their scores, we met back at our coveted red bench. There was a charge, a performance buzz after these actions in public space.

A deepening of our connections as our awareness of self and workshop group extended to include the Tenement Museum tour, bicycle courier blur, the family with three young kids delighted to be accompanied across the street. We exchanged goodbyes: Some of us would reunite at Judson Church tomorrow, and for others it was the end of their time together with us.

As we parted, I felt the grid I was first introduced to in Barbara Dilley's class at Naropa University (then Institute) over 20 years ago. I am referring to an improvisational structure where you imagine yourself on a grid, a map of right angles and possibilities to lean into and push against. The grid can withstand hurricane winds, erratic temperatures, dwindling resources. It can connect and web seemingly disparate beings; we are all in this curve. Latitude and longitude lines ran their course as we each navigated in relation to them and each other, our trajectories spanning the city as we wheeled, walked, and crutched it to our next destination, taking in our neighbors, human and non-human, feeling the gravity and presence of it all.

Stephanie: Judson Church

Monday, after our weekend workshops. We meet in the early afternoon to explore the site of our final sharing: Judson Church.

I'm hesitant to enter the Judson Church of brick, mortar, and glass, having lived up until now with my imagined Judson Church. The fantastical home of

FIGURE 20.2   Group during warm-up in Judson Church.

Source: Photograph by Ian Douglas.

postmodern dance, revered lineage holders, a heritage as close to religion as I get. It was an idea, a geography of lush woods in constant flux and reinvigoration.

I may have first heard of Judson Church as a teenager at North Carolina School of the Arts, or it may have been in my first-year dance history class at SUNY Purchase. To me, it was a cathedral of possibility, a bastion of choreographic impulse and innovation, landscape of shift and breadth. In fact, when I shared with friends my excitement at being at Judson Church and the majority had no idea what I was talking about, I seriously questioned my social circles.

My fear that the site's reality would poke holes in my bubble idealism was unwarranted. Gleam of wood floor, arches in sequence, pillars, saints illuminated in glass watching over all this luscious space. After the group scattered to explore the space on their own, we gathered again in the center of the floor, most of us horizontal and touching the limbs of the person next to us, sunburst or starfish entity.

For me, any individual impulses were hijacked by the space's immensity. The great vertical expanse that my long limbs couldn't begin to span. The architecture of the church designed to make people feel small. In this case, God was not the expanse but the sum of all the movements that had happened here. The walls breathing decades of history—energy lines, lines of flight, ghost dances, unseen forces creative and fierce; this church a channel, a repository, a body made up of bodies in motion/stillness, creating something where there was nothing. Here was proof of faith, and I was a believer.

We had a workshop to run and a performance score to set. With grid lines pulsing, time slipstream and tick, we each expressed gratitude for our movement lineages from our collective starfish mouth. Our words echoed and landed in the charged space. As if history allowed for us to enter easily into this stream, to dance in company, such good company.

The performance score used one of the street structures from the day before. Our plan was for the small group to place themselves in space and wield their slow-motion arcs and circles, entwining themselves and each other. Meanwhile, the rest of the group would join after the rhythm was established, feeding in energies. Some of us would invite audience members to enter into the performance, staying with them to dance until the larger form wove them into its warp and weft. To capture the outdoor elements that had featured so prominently in our time together, we all agreed to record ten minutes of sound during the break before the performance. This would be our ambient backdrop and timekeeper: October birdsong, traffic, cafe music, overheard conversation. At the start of our piece we would all come from the audience, place our phones at the back of the stage, hit play, and then the starting group would stay on stage while the rest of us returned to our seats.

My favorite part of the performance was inviting audience members to join us. I had an older woman gladly accept my invitation and explain how she used to be a dancer as we made our way onto the floor. There was a sweetness and vulnerability to asking people to dance, no words, just an open hand and readiness to be joined or not. The performance group got bigger as the audience got smaller. Some audience members, probably dancers/performers themselves, needed no personal invitation to get up and join. Other folks declined my extended hand or looked the other way. The stage was alive with an amoeba-like quality of ebb and flow, a unity of parts, people, and wheelchairs. I caught sight of my beloved's gliding arc on her travelscoot, "Scooty," as she twirled a partner through a large swath of open space. The end was a beautiful moment of interdependence and exhilaration as three of the performers gathered any leftover phones and leaned into each other and into one of the dancer's crutches, tilting toward the last moment with vigor.

Petra: endings

The aftermath of performance elation is a come-down, a slow meander back to my wheelchair-accessible Airbnb, a marginally ethically defensible option in a city where accessible hotels are far away and charge hundreds of dollars a night. Care structures have broken down all over the US, and our assemblage in New York City was marked by talks of insurance plans, accessible city routes in decline, the still inaccessible subway, how few dance studios are accessible, and the deep difficulties most artists have in making a living in this voracious and fast-paced metropole. We were all just birds on the railing of the river, there for a while.

In Judson Church, we created a slow-moving hurricane: A circular pattern that incorporated more and more people, a swirl that happened without words, in ten short minutes, full of eddies, pools, resting points. Phones, metal, my scooter, the workers who patched the pavement, the elevator in the church: All these things supported us, and I thought of them as I looked at the marble pillars in this miraculous space. Like Stephanie, I experienced this place as site of dancerly history, ghost steps everywhere. And when I first entered it from the wheelchair accessible elevator, so glad to have access, I felt something constricting a bit, too: We few humans were tiny in that large space, so much more open than the passages that we created on city streets. It was a feeling akin to working by the river: Overwhelm, a hint of the sublime, a sense of danger, a desire to find a nook in the gleaming expanse of wood.

I left the city with sensations of impermanence and power twisting in me: My limping crippy psychogeographical explorations connected with infrastructural neglect in Puerto Rico and the power of the river, with rivers of dance history and embodied transmission as well as the energy of workers, tourists, and finance in the street. I touched in with stars, and remember being wrapped in a blanket. Finding a corner in the street to sit. Leading someone across a busy intersection. A touch of a hand. Danger contact, contact zones.

Notes

1 For other accounts of the Asylum Project's site-specific explorations, see: Kuppers et al. 2016 (at the original Bedlam, Bethlem Royal Hospital, UK); Kuppers 2018b (at old psychiatric asylums near beach-spaces in Belgium and the Netherlands); and Kuppers 2018a (at an old convent in Grand Rapids, Michigan and an ex-Baptist Church, now a performance space, in Detroit).

2 For more information on the community arts project of which this quilt exhibit was a part, see www.catalystdance.com/then-a-cunning-voice/.
3 The Domino Sugar Factory in Brooklyn belonged to the American Sugar Refining Company, which produced Domino brand sugar from sugarcane shipped to the site from the colonies. The building dates from 1882, when it was the largest sugar refinery in the world. The refinery was the site of one of New York City's longest labor strikes: The 1999–2001 strikes led to stinging losses for union power. In 2014, African-American artist Kara Walker exhibited a Creative Time commission inside the factory building: A monumental sculpture of a sphinx, entitled *A Subtlety, or the Marvelous Sugar Baby, an Homage to the unpaid and overworked Artisans who have refined our Sweet tastes from the cane fields to the Kitchens of the New World on the Occasion of the demolition of the Domino Sugar Refining Plant*.

References

Debord, Guy. 1958. "Théorie de la Dérive." *Internationale Situationniste* 2 (December): 19–23.

Dilley, Barbara. 2015. *This Very Moment: Teaching Thinking Dancing*. Boulder: Naropa University Press.

Doherty, Claire. 2004. "The New Situationists." In *Contemporary Art: From Studio to Situation*, edited by Claire Doherty, 7–41. London: Black Dog Publishing.

Hahn, Daniela. 2014. "Performing Public Spaces, Staging Collective Memory: 50 Kilometres of Files by Rimini Protokoll." *TDR: The Drama Review* 58 (3): 27–38.

Hancox, Simone. 2012. "Contemporary Walking Practices and the Situationist International: The Politics of Perambulating the Boundaries Between Art and Life." *Contemporary Theatre Review* 22 (2): 237–250.

Harney, Stefano, and Fred Moten. 2013. *The Undercommons: Fugitive Planning & Black Study*. New York City: Autonomedia.

Kuppers, Petra. 2018a. "Invited Hauntings in Site-Specific Performance and Poetry: The Asylum Project." *RIDE: Journal for Applied Theatre and Performance* 23 (3): 438–453.

———. 2018b. "Blood Compost: The Asylum Project." *P-Queue* 15: 47–58.

Kuppers, Petra (host) with Stephanie Heit, April Sizemore-Barber, and V. K. Preston. 2016. "Mad Methodologies and Community Performance: The Asylum Project." *Theatre Topics* 26 (2): 221–237.

21

GEOPOETICS, VIA GERMANY[1]

Angela Last

> The border is surrounded by a strip of grass.
> Although we can hear shots, the grass tells us
> That there is an ordinary place
> Behind the fence.

I

When I started to learn about the world, I got the impression that it ended a short drive from our house. After all, we could only travel in one direction, as the other side was blocked by a tall fence armed with spring guns. The day I first saw the border, it was partially enshrouded in fog. If I remember correctly, my father said something about the people who had died trying to escape. We looked at the fence in silence for some time and then drove back. This image would later return to me while reading Michael Ende's *Neverending Story*:[2] It was how I visualized the encroaching Nothing. Or maybe it was the other way around. The fence eventually disappeared from the landscape, but people still know its place.

It wasn't that there was Nothing, but, for many youngsters like me, "the GDR [East Germany] was very far away. Further than the Moon."[3] It was a planet with no unemployment, no bananas, and rows of grey houses. At the same time, we found that this peculiar extraterrestrial position allowed for permeability. The Sandman, and also music, could escape to us through space. Through this window we learned that the supposedly different reality was in fact a parallel one. Everything on our side had its alternative expression on the other, from lemonade to cars. We would often ask our parents or teachers what a particular thing looked like—or tried to imagine its "parallel." We did not do this with other countries to which we could physically travel. It really was our "curtain" from which we liked to peek, to compare, to speculate. Despite all our fascination, "GDR" became an insult. If anything

"West" looked "East," it was underdeveloped, not cool. Later, we learned what had caused the split in our reality.

★

I have a difficult relationship with my native language. Although I was able to use German well, I have always felt freer with English. It seemed to offer more space, more elasticity. Not only did I do better at English than at German in school, but my marks in other subjects improved significantly when I switched to English textbooks—a modification my perceptive physics teacher had suggested. Although I appreciated its precision and imaginaries, German felt claustrophobic. In particular, the German I encountered at school felt like the language of a people who did not want to be understood, or who perhaps resided in another time where they may have been understood. Even my pronunciation was off, and a teacher sent me to speech therapy: My German was "too soft" and therefore "too confusing." I attended the final session in a bat costume.[4]

Our household represented another ecology of language alienation. My great-grandmother, who sometimes lived with us, spoke Low German, a language related to English, Frisian, and Dutch. So that they could get out of farming, her generation had not allowed their children to speak like them. Low German was associated with lower class, and with the inability to write in the standard language. My grandmother's Low German was broken: She could understand her mother, but not quite answer back. My parents, like me, could understand some Low German, but not speak it at all, apart from a few songs and phrases. As the language died out, they tried to rehabilitate it through competitions at the very schools where teachers had once bullied their students for sounding like "peasants."

★

In our house, writing was also generationally distinct. We did not just have different handwriting styles, but used different types of writing as well. Our books reflected this scriptural multiplicity. If each language represents a philosophy, writing seems to embody politics: Perhaps this is why strokes disappear more easily than sounds. Within a short space of time, German writing moved across old German cursive, Sütterlin, and its current "modern" form, sometimes with regional differences. Ironically, the Nazis hated blackletter or Gothic, the font that is most closely associated with them. Although it had a strong association with the tribes they had reinvented, they found it too primitive.

At my first school, named after Anne Frank and surrounded by tower blocks, we were taught to write in the current fashion—first in boxes and later on lines. My handwriting got worse with each year until they stopped marking it altogether. When my English teacher complained about the mess, I submitted an essay with each sentence written in a different style: Tick the one you prefer. The feedback read, "Oh heaven's dear!" but not one box was ticked. Gradually, I settled on a script that no one could recognize, perhaps because it traveled in time.

★

Anne Frank was a writer who was a few years older than us. When we read her diary, we identified with her, but we also knew that any one of us could have turned her in. She never wrote about her last days, spent in our countryside, an hour's drive away. It was we who completed her story.

★

At the same school, we were taught about the local landscape: how it was formed, how it was used, and how we could record it. We learned how things the planet had done without us for millions of years now shaped our lives. In our area, it created salt domes and fragile soils. The salt, used as a preservative, made our town so rich that it could buy itself out of wars and thus became preserved itself. However, when the first humans settled in our area, they caused so much deforestation that it became known as a bleak desert for thousands of years. It was not until poets and painters saw its eerie beauty that people started appreciating "the Heath." It is now this beauty that brings in the money, while the salt domes store nuclear waste.

In Nazi propaganda, deforestation was a crime. Films like *Ewiger Wald* ("*Eternal Forest*") tried to remind the German people that the forest was their birthplace, their protector, and their role model: "The tree seeks space like you and me."[5] Before the establishment of Germany, the Brothers Grimm, in their search for Germanness, also took to the forest. For them, the Napoleonic invasion had caused a rift between the people and the trees. In their stories, the forest was the place where children and other innocents got violated; in the film, German people suffer, because their forest got violated by Black French soldiers. The Grimms realized that they needed to do more than *collect* folktales: They needed to rewrite them. Recreational sex was the first thing to go, and the natural and supernatural became a horror dispelled by blonde bourgeois mothers. Today, many neo–Nazis, now termed "*völkish* settlers," congregate in our deforested area. Even without (yet) knowing about their presence, I understood early that the real monsters on the Heath were never supernatural, but human.

★

My father calls me to the living room and excitedly points at the TV. It is filled with the familiar purple landscape of our area, Lüneburg Heath, rendered rather garish on old film stock. In the 1950s, the *Heimatfilm* (German for "homeland film") became a popular genre, an idyllic refuge from the surrounding devastation. Dirndled maidens, yodeling rangers, trusty dogs—people seemed happy to ignore its similarities to the films that had popularized the war, or perhaps they appreciated them because of it. Over time, the *Heimatfilm* reintroduced conflict: between generations, between the urban and the rural, and between regions. My father consumed *Heimatfilme*, and its audio equivalent, *Volksmusik*, as comedy, and, ostensibly, as a way to annoy me and my mother.

This time, my dad is not excited because the Heath is on TV. In the 1920s, my grandmother's cousin had married a fellow theater actor. Despite the difficulties of closing venues and securing engagements as working-class thespians, they persisted in their careers. However, in the year of her husband's 25th stage anniversary, she lost permission to work: My grandmother's cousin was "half-Jewish." When the Gestapo tried to track her down, her husband claimed she had "gone missing." He was sent to the Eastern front shortly after, but survived—as did his wife. After the war, he restarted his career with a small role in a film on the Nuremberg Trials, but he found his breakthrough as a folk actor, in plays and films in Low German, celebrating the nature and culture of an area where his partner had to hide for five years in a garden shed.

★

Our neighborhood lies between two forests. The smaller one to the East separates us from the big estates; the larger one to the South, called Tiergarten, separates us from the villas of the rich. We revel in the irony that the villas were built on a bomb site that was never cleared. Right now, a new housing development in the South means that our area has to be evacuated every few days. However, bombs are not the reason why I was never supposed to cross this forest: Even though people tell me I look too much like a man, I will be Little Red Riding Hood forever.

I hated the enforced Sunday walks through Tiergarten as they did not seem to attend to what was interesting: the mysterious railway line that never seemed to carry a train, the mushrooms irradiated by Chernobyl, the graffiti on a concrete block seemingly dropped from the sky, the wooden shed with the empty bottles and syringes. Instead, my parents would show me birds, mushrooms, berries, trees, animal tracks—and graves. A train was bombed at the station. It contained people who were destined for death. The ones who were not killed in the bombing were shot by guards or fled into the forest. Some fatally turned to residents for help, since only two were sheltered, in a distant village. After the war, the people of my town were forced to exhume those they had killed and to build a memorial. Among the trees, the dead are hard to find, yet their story resonates within every detonation.

★

Helene shows me a beautiful poster of a Cyrillic alphabet. It has a light blue frame, and the black letters are illustrated with brightly colored animals. I am at her family's flat to help her learn German, but we get distracted by Russia. At the end of the "lesson," Helene gives me the poster. I am astonished that she would part with it, and I promise to take good care. Initially, my role as teacher had come about as punishment: I was frequently disruptive. Bored with repetition, I did not understand difference. One day, my teacher put me in her position, so that I would.

Back at our house, my grandmother has a new home help. She, too, is an *Aussiedler*.[6] At first, she makes my grandmother uneasy: My grandfather "went missing" in Russia, and Mrs. Sorich has a Russian accent. As time passes, they start to talk and meet after Mrs. Sorich's work. Gradually, other *Aussiedler* join, to describe their life as *Russlanddeutsche* over coffee and cake. In her 80s and nearly blind, my grandmother asks for books on Russia, in large print. She reads them with a big, black-rimmed magnifying glass. Perhaps just before her death, she finally believed her husband, who, in his last letter, wrote that Russians are not the enemy.

★

My father returns from a school reunion. A former classmate has boastfully suggested that their next meeting should be abroad: He has purchased a farm in Namibia and would like to take the others on a tour. My father is surprised that such a purchase is possible and explains why to me. Together, we look up "German colonialism" in the dictionary.

When Germans speak of the genocide they committed, they never think of Namibia and the other places where men, women, and children also died at their hands: gathering diamonds, building the railway, digging a harbor, leveling the ground for their own death camp. Women, raped and forced to scrape the skin off the skulls of their executed men for "science."[7] Today, Namibia is a place where Germans travel to see how people, black and white, still live as they did in the days of "the Empire." Perhaps, thinking of Namibia in this way makes it easier for us to remove it in space and time.

★

I do not remember much about geography at school apart from feeling bored and misunderstood. Despite my interest in other countries and landscapes, I struggled with writing geography, drawing geography, speaking geography. Although it meant I might fail my final exams, I stubbornly refused to drop the subject. I wanted to find out why something that fascinated me so much in the library and on TV did not work for me in class. Again, in the exam, I could not find the right words.

I felt I learned more about geography through physics, although I faced a similar struggle. Early on, in high school, we calculated the moon's orbit around the Earth. I somehow doubted my power to undertake this and told my teacher: "I don't think the world and the universe is as straightforward as our school books make us believe." He replied that I was correct, but would have to wait a few more years, at which time they could let us know how weird things really are. I waited, and was not disappointed.

II

> *The national media tells us we have a "refugee wave," but our area has a more pressing issue: The wolves are returning. One could argue that the concerns overlap, as the language mirrors that of the migration debate. Some people here remember the time when they themselves made our hometown double in size. Entire neighborhoods became named after places that can no longer be found on a map. The Nazis had wanted it to be the other way around—a landscape written in blackletter—and some people still want it to be so. As a geographer, it is hard to use the word "space."[8]*

I have never written about Germany until this year, but I know that it has always resonated in my texts. Having spent the last 20 years in Britain, I feel both helpfully and unhelpfully detached from the place that has shaped my interests and my approach to performing geography. Unhelpful, because my absence has created a noticeable rift that translates into partial understandings of issues, and that mirrors in awkward communications with German activists and academics. Helpful, because I have been able to see and make different connections with Germany—and with myself. Both of these consequences have given me a specific sensitivity to language and geography that compels me to write—not just academic articles but more informal blog posts and also performative pieces that have included poetry.

With this sensitivity, geopoetics becomes the most difficult geographical challenge. For most Germans with left political sensibilities, poetic writing on landscape is almost irredeemably associated with nationalist "blood and soil" evocations—a fascist turn that had developed out of Romanticism's celebrations of nature. While I realize that Germany is not the only country that attracts this kind of output—after all, nationalism is not just a German problem—I remain aware of the sharp anti-fascist monitoring of poetic engagements with geography. Despite my distance, I cannot detach myself from this fear and skepticism toward romanticization, toward appealing to the landscape for the wrong reasons—and neither do I seek to. It is more a matter of developing this sensitivity further, as it also alerts to other forms of exclusion.

How differently I had developed from German-educated geographers over the last two decades became apparent in recent conversations with a fellow German geographer. She emphasized how many of the topics that I touch upon—such as the cosmic dimension of worldviews—would be considered completely off-limits for a critical German geographer. At the same time, she acknowledged how my different path, especially via interwar French Caribbean theory and poetry, had allowed for a different sort of access to this material. Indeed, publications such as *Tropiques*[9] and the "geopoetics" developed by the Guadeloupian writer Daniel Maximin[10] had alerted me to the possibility of rewriting geographical narratives as well as to the ideological hazards of this enterprise.

> *A federal government is issuing brochures:[11] When does love for the environment turn right wing? I struggle with the brochure's definition, its level of abstraction. The author argues that neo-Nazis mistakenly believe that Germans and their landscape have a special relation. He tells me that this makes sense neither to a geologist, nor to a*

geographer, and that landscapes even more so disregard such attachments. What the brochure doesn't tell me is why neo-Nazis need landscapes to make Germans. I suddenly realize that it is not the abstraction that bothers me, but the type of abstraction. If I struggle with this leaflet, how are farmers and the majority of voters going to cope? At the same time, I think of Steiner[12] and Darré:[13] Two men who started from the same place—racist occultism and agricultural innovation—but have seemingly ended up in different directions. We love Steiner for Waldorf Schools and hate Darré for "blood and soil." At what point does a geopoetic turn to murder?

Another difficulty lies in communicating this sensitivity, as I noticed after several clumsy confrontations at geography conferences. Not only is there a problem with describing what is at issue, but also with not being written off as a cliché, as an exception that can safely be banished. This danger is particularly great when mentioning incisive German historical events such as the Nazi or Cold War eras, which have become clichés in themselves and even attained comedic value through TV and film satires. This banishing can be found in the current fashioning of "embodiments of evil" such as Donald Trump, who draws frequent comparisons with Hitler. Questions of how one might oneself contribute to such "exceptionalism" retreat into the background. At the same time, history and geography are inextricably interwoven.

My strategy in the earlier short stories has been to show how these "exceptional" histories shape German engagements with the environment so intimately that they permeate the everyday. What I also try to get across in this text is that many Germans experience their history, and especially their family history, not just as a burden, but also as an insight. For example, I find it useful to know different personal stories, particularly from the Nazi era, because they give me a sense of a spectrum of possibilities: Why did one great uncle become an SS officer, but another one didn't? The question *What would I have done?* thus does not simply morph into *What could I have done?* but into *How am I participating in history and geography in the making?* To look at the histories of geographical imaginaries against the background of these personal stories helps in answering this question.

The word "deutsch" does not correspond to the English word "German." To be "deutsch" literally means to be "of the people." This definition has served many purposes: to deride speakers of a "common tongue"; to designate people as pagans in need of conversion; to take pride in not being "latinized"; to describe people as non-White, non-Jewish, non-Muslim. The inscription of the German seat of government reads "Dem Deutschen Volke"—to the people of the people. It was installed during a crisis when, once again, "the people" could not be contained. An empire that did not even know what font to choose: Capitalis or Gothic? Romans or Barbarians? They found the answer: They melted two French canons—war trophies—to cast the script.

To look at language is also to look at geography. As anti-racism activists have pointed out, language is a key site through which exclusions are performed.[14] When

Germany created a *Heimatministerium*—a "ministry for home"—many Germans pointed out the instability as well as the exclusionary nature of this term. "Heimat" has very specific cultural and nationalist connotations, and can thus not simply be translated as "home." No one has been able to come up with an all-encompassing definition, although you can always feel its boundaries whenever the word is used. Most Germans put it in quotation marks. The *Heimatministerium* ostensibly strives to balance urban and rural development, but many people feel that it fuels fears about the impact of migrants on German culture. When the ministry was announced, not only was its composition lampooned—its leadership is exclusively made up of old white men, the perfect material for parodies involving garden gnomes—but so were its cultural resonances.

> *Trachtenminister Volksliedminister Postkartenminister Kindheitsverklärungsminister Handaufsherzminister Hymnensingminister Eigentlichmeinenwirvaterlandminister Grenzenzuminister Ausländerrausminister Deutschlandbleibtdeutschminister*[15]

A free translation would go something like this: Lederhosen Minister,[16] Oompah Minister, Postcard Minister, Sincerity Minister, Hymn Singing Minister, Actually We Mean Fatherland Minister, Close the Borders Minister, Foreigners Out Minister, Germany Remains German Minister.

As many geographers celebrate language as a means to transgress geographical borders, we can forget that it also serves to express and intensify them. From compulsory language tests for immigration to discrimination against regional and class accents, language is used as a form of boundary-making. In some cases, languages, scripts, or letters become banned, either by voluntary submission to a social hierarchy or by law.[17] Poetic language sometimes especially excludes—for instance, when it is assumed to require a particular form whose codes are left unquestioned. For example, it is not necessarily work designated as poetry that can be experienced as poetic. Dionne Brand has spoken of the "terrifying poetry of newspapers"[18]—and there are other unexpected formats that render a situation in disturbing clarity. Poetry is neither without history or geography nor a breathing space away from the "devastation" of academic critique. It represents the same system if we perform it in and from this space. We forget not just the language but also the situatedness of our performance. Who do we address in our creative expressions—and from where? What and whose geographies appear in our work? Where do we present and publish?

> *As one of the first to anticipate the excesses of German fascism, the French philosopher Simone Weil argued for a "need for roots."[19] Unlike the heroic oak trees employed by German nationalists, hers quietly processed energy and matter as a kinder model of being in the world. Her contemporary Georges Bataille, too, turned to energy and matter, but argued that we are too much like them: We sometimes just need to explode.[20] Where Weil wrote proposals for more ecological workplaces, Bataille tried to devise appropriate orgies and other transgressive events, during which energy could be more*

safely offloaded. Even now, we are still uncomfortable with nature as process, with humans as nature, with nature as culture (and vice versa) because we still don't know how we are matter.

In my writing, I am interested in exploring the relationship between materiality and representation. As a geographer, I am trying to educate myself about histories of representation, how imaginaries of places have changed over time and through whose perspectives. What sort of materialities are enlisted in telling a story about a site, in drawing boundaries? Even the most uninhabited place on Earth—or the atomic scale and far away planets for that matter—play a part in a story, often about another place.

For me, this is most sharply rendered in national (and sometimes regional) identity building, as people strive to encompass and determine everything. In encyclopedias, you can find lists of national trees, animals, colors, dress, food, music. Nothing seems to escape this enterprise, from typography to gods. The resulting "microcosms" are usually not perceived as abnormal, but legitimate. The situation appears similar for other geographical divisions such as "development," as the "Decolonize" movements have highlighted. To render these processes strange is a first step toward asking how they became normalized, and how engrained material signifiers really are.

Suzanne Césaire, a Martinican surrealist wrote: "Martinican poetry will be cannibal or it will not be."[21] It was a provocation to the colonizer, in which she implied that appropriation will go both ways and be used as a source of empowerment. One of the modes she used was satire—she inverted colonial stereotypes. It could be argued that, as the official story perishes, satire survives. We recall the imagery satires evoke, but less reliably the circumstances from which they were born. At the same time, satire is uncomfortable: It often creates pervasive stereotypes, and even if well-intentioned, it is a form of shaming. Is this what a writer, a poet should be doing? It could be argued that satire attempts to sever a connection in order to reconfigure it. It is not surprising that Brecht trained with clowns. Unlike Heidegger, he understood that distance is not a bad thing. While there is certainly a danger of becoming a jester—a person who is free to speak their mind, but whose words have no consequences—there is also the potential of creating resonance through refusal.

Although I mostly communicate in English, I continue to learn from German political satire. When Austrian writer Michael Köhlmeier was attacked for criticizing his government's suggestion of "keeping refugees in concentrated areas," German satirist Mely Kiyak urged her readers to not remain deaf to the dehumanization and supremacist attitudes in their own countries: "This is what it's all about: do you leave this sort of language pass uncommented? Can you allow yourself to play the fool?"[22] What for me also resonates in this call to arms is a call to realism. Rather than agonize over convoluted questions such as, *But how does one write in the face of long-standing narratives that use geography to dehumanize?* one should simply get to work. After all, language remains a problem that can be easily spotted, and easily contested, and you can start with whatever mode and environment that is comfortable for you: conversations with neighbors or colleagues, teaching, social media, zines, graffiti, poetry slams, songs. Or, you can directly write to the offending author, perhaps in collaboration with others. This—and learning from others who have contested racist/imperialist/fascist imaginaries across history and geography— is the project to which I would like my words to contribute.

At present, this is how I would describe my practice:

> My writing probes the lack of distance between the geographical and the geopolitical. It understands that language, and writing, is not innocent. As a mode of questioning, it awkwardly tests relations. My writing often works from what something is not, for whom a geography is not. It looks for absences and denials that expose a failure to contain. Despite this, it is not negative: It finds optimism in refusals. My writing is that of a musician, DJ, and researcher. It works with loops, fragments, and citations. In its searching, it does not discriminate against sources, since an intention is not limited by genre or discipline. My writing is driven by anger but also by care. It directs these sentiments against determinisms and their supposed antidote: To liberate geographies from histories. By creating discomfort—to myself and/or the reader—it reminds that to care is not to stage disappearing acts.

Notes

1 This piece has grown out of conversations with Margaret Byron, Gesa Helms, and many members of my family, including my parents, Brigitte and Friedhelm Last, and my cousin Sukit Manthachitra.
2 Michael Ende's *Neverending Story* was a 1979 children's book in which a fantastical empire with a connection to the "human world" is swallowed up by an entity called The Nothing.
3 Original text: The user "Herr Stemmer" wrote on February 3, 2009 on a German Radio User Forum site: "Die DDR war weit weg. Weiter als der Mond". Accessed July 16, 2018. www.radioforen.de/index.php?threads/wurden-die-ddr-medien-auch-im-westen-genutzt.27153/.
4 This paragraph was partly inspired by Afro-German poet May Ayim's writing about her training and work as a speech therapist (Ayim 1997).
5 This is a citation from the film. It is German anti-fascist practice to discuss fascist works, but not cite them. For more information on the meaning of the forest in German culture, psychology, and politics, see Macgregor 2014; Theweleit 1987.
6 *Aussiedler* is a category that was created to enable German nationals of any ethnic descent and "ethnic" Germans from former "Eastern Bloc" countries to return to Germany. This law sought to protect people of German descent who experienced retaliations or long-term discrimination after WWII. These included people whose families had left Germany for Russia from the sixteenth century onwards (*Russlanddeutsche*). An immigration wave in the 1980s was considered a success in terms of integration; however, the second wave, during the 1990s (*Spätaussiedler*), led to calls for a rewriting of the law, since many Germans felt that the arrivals were culturally more Russian and had fewer opportunities to integrate into the struggling economy of the time. As a consequence, *Aussiedler* faced discrimination in both countries, and some returned to Russia in frustration. Because of

the increasingly derogatory use of the term, immigration activists have started to look for alternative descriptions.

7 Efforts have been made by activists, including historians, to raise greater awareness of the German colonial period. For example, while it is illegal to name streets after Nazis, it continues to be acceptable to have streets named after people who committed genocide in African countries, or even to display human remains or looted objects resulting from these atrocities (see Olusoga and Caspar 2010; Kuster and Sarreiter 2013).

8 This sentence refers to the frequent reference in Nazi propaganda to the need for greater living space (Friedrich Ratzel's environmentally determinist "*Lebensraum*"), frequently abbreviated to "space" (*Raum*). The call for *Raum* was especially used to justify the "Eastern expansion," and, as a consequence, it remains a very loaded term in Germany.

9 *Tropiques* was a Martinican surrealist "cultural revue" edited by Aimé Césaire, Suzanne Césaire, Georges Gratiant, Aristide Maugée, René Ménil, and Lucie Thésée. Eleven issues appeared between 1941 and 1945.

10 Daniel Maximin (b. 1947) writes about humans as "fruit of the cyclone" (2006), an imaginary that he uses against ideas of environmental determinism and rootedness.

11 As Germany is a federal state, a lot of decisions are made by local governments. An example is the 36-page brochure entitled *Environmentalism Against Right-Wing Extremism: A Discussion Aid* (Franke 2016, translation mine), which is produced by one particular federal government, although this is most likely a national issue. While it represents a genuine effort to help agricultural organizations and voters deal with right-wing "back-to-the-land" movements and propaganda, it could be argued that things could be improved both on the level of reach and on the level of author-audience relations.

12 Rudolf Steiner (1861–1925) was an Austrian philosopher and social reformer who tried to turn his theories of "spiritual science" into practices including biodynamic agriculture, architecture, and education (e.g., the Steiner or Waldorf Schools). His reputation in Germany is marked by ambiguity: While associated with the racism and pseudo-science of the esoteric Theosophical Society, he was also known for criticizing racism and specifically anti-Semitism. His theories are considered both extremely conservative and progressive.

13 Richard Walther Darré (1895–1954) was an Argentinian-born German politician who is primarily known for his "blood and soil" legitimation of Germany's "eastward expansion" and his extreme Social Darwinism. He was a member of the racist "back to the land" Artaman League as well as a co-founder of the pseudo-scientific *Ahnenerbe* ("ancestral heritage") think tank that had links to occultism. Darré is also known for his advocacy of neo-Paganism.

14 Two publications that I would like to mention here are the ground-breaking "Showing Our Colours: Afro-German Women Speak Out" (title of the English translation), first published in 1986 and compiled after a visit to Berlin by Audre Lorde (Ayim, Oguntoye, and Schulz 1995), and the latest bestseller *Among Whites* by Mohamed Amjahid (2017). Both publications highlight everyday racist language (e.g., product names) as well as struggles of expressing belonging, both by oneself and others.

15 Tweet by German journalist and film maker Mario Sixtus on February 7, 2018.

16 *Tracht* translates as national or regional costume. In Germany it has strong associations with *Heimatfilm* culture, and in particular with Bavarian lederhosen and dirndl outfits. *Volkslied* is a folk song, but again it has associations with trashy popular celebrations of Germanness and Bavarian music.

17 An interesting essay on language and boundary-performance is Théotime Chabre's "Languages of a Stateless Nation" (2015). It was published on the independent online publishing platform, Mashallah News, which has published many other interesting essays on language relating to the Middle East.

18 I read the terrifying poetry of newspapers. I
notice vowels have suddenly stopped their
routine, their alarming rooms are shut,
their burning light collapsed

the waves of takeovers, mergers and restructuring
. . . swept the world's . . . blue chips rally in New York
. . . Bundesbank looms . . . Imperial Oil increases dividends
. . . tough cutbacks build confidence.
        (Brand 1997, 13; emphases and ellipses original)

19 Simone Weil's book *L'Enracinement* ("*The Need for Roots*," 2002) was posthumously published in 1949. It was written in early 1943 as part of her commission by the French government in exile ("Free France") to come up with a plan for post-liberation from German occupation.

20 The theme of excess and release runs through Georges Bataille's entire oeuvre (e.g., Bataille 1991), from his occupations with eroticism, sacrifices, and the psychology of fascism to his "solar economy."

21 Suzanne Césaire (1915–1966) finished her 1942 essay "Poetic Destitution" on this line (2012, 27). It was written as a reaction against apolitical Martinican "folk literature" that, for her, affirmed colonial stereotypes.

22 Mely Kiyak writes satirical columns for the *Süddeutsche Zeitung* (where this quote is from, 2018, translation mine) and especially for the website of the Maxim Gorki Theatre. Located in the Eastern part of Berlin, the Gorki Theatre began as a mouthpiece for Soviet influenced socialist realism and now presents sharp political critiques. The cast is noted for encompassing many actors "with migration background" (the currently fashionable "politically correct" term in the German press) and even has an additional cast of exiled theatre practitioners (Das Exil Ensemble). As the daughter of Kurdish parents from Turkey, Kiyak receives a lot of hate mail. Together with similar recipients of racist missives, she became co-founder of the show "Hate Poetry," during which the cast reads this hate mail as comedic poetry.

References

Amjahid, Mohamed. 2017. *Unter Weissen*. München: Hanser Berlin.

Ayim, May. 1997. *Grenzenlos Und Unverschämt*. Berlin: Orlanda Frauenverlag.

Ayim, May, Katharina Oguntoye, and Dagmar Schulz, eds. 1995. *Farbe Bekennen: Afro-Deutsche Frauen Auf Den Spuren Ihrer Geschichte*. Berlin: Orlanda Frauenverlag.

Bataille, Georges. (1967) 1991. *The Accursed Share: Volume I: Consumption*. Translated by Robert Hurley. Reprint. New York: Zone Books.

Brand, Dionne. 1997. *Land to Light On*. Toronto: McLellan & Stewart.

Césaire, Suzanne. 2012. *The Great Camouflage: Writings of Dissent 1941–1945*. Edited by Daniel Maximin. Translated by Keith L. Walker. Middletown, CT: Wesleyan University Press.

Chabre, Théotime. 2015. "Languages of a Stateless Nation." *Mashallah News*. www.mashallahnews.com/language/languages-stateless-nation.html.

Franke, Nils M. 2016. *Naturschutz Gegen Rechtsextremismus: Eine Argumentationshilfe*. Mainz: Landeszentrale für Umweltaufklärung Rheinland-Pfalz.

Kuster, B., D. Schmidt, and R. Sarreiter. 2013. "Fait accompli? In Search of Actions for Postcolonial Injunctions." An Introduction to the Special Issue *Afterlives*, edited by Artefakte// anti-humboldt. darkmatter, November 18. www.darkmatter101.org/site/2013/11/18/fait-accompli-in-search-of-actions-for-postcolonial-injunctions-an-introduction/.

Kiyak, Mely. 2018. "Aufsprühen, Wegwischen, Fertig. Kiyaks Deutschstunde." *Süddeutsche Zeitung*, May 23.

Macgregor, Neil. 2014. *Germany: Memories of A Nation*. London: Penguin.

Maximin, Daniel. 2006. *Les Fruits Du Cyclone: Une Géopoétique De La Caraïbe*. Paris: Editions du Seuil.

Olusoga, D., and Caspar W. Erichsen. 2010. *The Kaiser's Holocaust: Germany's Forgotten Genocide and the Colonial Roots of Nazism*. London: Faber & Faber.

Theweleit, K. 1987. *Male Fantasies, Volume 2: Psychoanalyzing the White Terror*. Minneapolis: University of Minnesota Press.

Weil, Simone. (1949) 2002. *The Need for Roots*. Translated by Arthur Wills. Reprint. London: Routledge.

22

INDIGENOUS PACIFIC ISLANDER GEOPOETICS

Craig Santos Perez

I am an Indigenous Chamorro poet from Guam, the southernmost island of the Mariana archipelago in the western Pacific Ocean region known as Micronesia. I grew up in the village of Mongmong, and my clan name is Gollo. When I was 15 years old, my family migrated to California, where I completed my high school, undergraduate, and graduate education. After living there for 15 years, I migrated to Hawaiʻi, where I have lived since 2010 teaching Pacific literature, creative writing, and eco-poetry at the University of Hawaiʻi, at Mānoa.

When I have traveled across the United States, I have often been asked: "Where are you from?" Most people have never heard of Guam, even though it has been a territory of the United States since 1898, and even though the Chamorro people have been US citizens since 1950. "Where is Guam?" would usually be the follow-up question. But when I would try to point out my homeland on a map, I confronted the fact that Guam is often not included on world maps because of the island's size, a mere 212 square miles. My homeland was part of an invisible geography.

Poetry became a space for me to make visible the real and symbolic geographies of Guam and Chamorro culture, identity, genealogy, history, politics, and migrations. I situate my poetry and poetic practice within what I consider an "Indigenous Pacific Islander geopoetics." The main characteristics of this ethno-subgenre include: foregrounding the interconnection between geography, poetics, and native Pacific Islander culture; highlighting the Indigenous histories and significances of island, ocean, and tidal places; honoring Pacific geographies as sacred spaces and the dwellings of ancestral spirits; critiquing imperial, colonial, tourism, and military geographies; and asserting the power of poetry to advocate, empower, inspire, heal, educate, and decolonize.

~

I articulate an Indigenous Pacific Islander geopoetics through the "archipelagic form" of my serial, multi-book poetry series, *from unincorporated territory*. Four books from the series have been published thus far: *from unincorporated territory [hacha]* *(2008)*, *from unincorporated territory [saina]* (2010), *from unincorporated territory [guma']* (2014), and *from unincorporated territory [lukao]* (2017).

I envision each individual book as an island with a unique poetic geography. The book-island is inhabited by the living and the dead, the human and the more-than-human, the land and the sea, and multiple voices and silences. The book-island vibrates with the complexity of the present moment and the depths of history and genealogy, culture and politics, bone and blood. The book series is an archipelago, a birthing and formation of book-islands. Like an archipelago, the books are related and interwoven to the other islands, yet unique. Reading the books in a series is akin to traveling and listening across an archipelago. Because Guam is part of an archipelago, the geography inspired the form of my book series. Additionally, the unfolding nature of memory, learning, listening, sharing, and storytelling informed the serial geography of the work. To me, the complexity of the story of Guam and the Chamorro people, entangled in the complications of ongoing colonialism and militarism, also inspired the ongoing serial form.

I chose the "from" in my title to indicate that each book is an excerpt of a larger, serial book project, as well as to gesture toward the theme of being "from" a particular homeland yet living in the diaspora. Just as an archipelago has a name, such as the Mariana Archipelago, each island of the archipelago has its own unique name (such as "Guam"). I chose the term "unincorporated territory" because that is Guam's official colonial status as an American territory that is owned by the United States, but is not fully a part of the nation, thus keeping Chamorro people in a state of political disenfranchisement and denying us our Indigenous rights of self-determination and sovereignty. My first book was named "hacha"—meaning "one"—to mark it as the first book, first island, first voice. While one might expect the second book to be named "hugua"—meaning "two"—I chose the name "saina," meaning "elder," to resist that linearity and instead focus on genealogy and the past. The third book, "guma," which means "house" or "home," was an attempt to map the time and space of home and belonging. The fourth book, "lukao," meaning "procession," includes themes of birth, creation, parenthood, migration, and extinction.

Overall, the "archipelagic form" of my serial book project embodies a Pacific geographic form. Moreover, this form and the titles of the book-islands assert Indigenous presence and culture while exploring themes of colonial and decolonial geographies through poetry.

~

I also express an Indigenous Pacific Islander geopoetics through an archipelagic arrangement of poems within each book, as well as through geographic imaginaries of the page, the word, the sentence, and the acts of writing and reading.

Just as each of my books contains a "from" in its title, each poem within each book contains the word "from." For example, some poems are titled: "*from* aerial roots," "*from* sounding lines," or "*from* all with ocean views." This marks each poem as an excerpt of a longer, ongoing poem. Sometimes, a poem will continue across a single book (e.g., a three-page excerpt of "from aerial roots" will appear at the beginning of a book, and then another three-page excerpt will appear later in the book). Sometimes, a poem will appear across books (e.g., excerpts of "from aerial roots" appear in both my first and second books). This trans-book threading creates an archipelagic, interwoven geopoetics.

Additionally, I use visual elements (e.g., maps and typographical symbols) to thematize the relationship between storytelling, geography, mapping, and navigation. The visual maps in my books are a way to show how Guam has been mapped by imperial and military powers, reducing Guam's geography to simply military bases or colonial territory. These maps often silence Indigenous Pacific voices and geographies, so part of my intention in including these maps is to cut across and subvert them with the narrative geographies of Chamorro stories that provide embodied and rooted portraits of place and people.

In terms of typographical symbols, I utilize the tilde (~) throughout my poetry, which subtly transforms the geography of the page. Besides resembling an ocean current and containing the word "tide," the tilde has many intriguing uses. In languages, the tilde is used to indicate a change of pronunciation. In mathematics, the tilde is used to show equivalence (e.g., x~y). I use the tilde to move between different kinds of discourse in my work (historical, political, personal, etc.) and to show that personal or familial narratives have an equivalent importance to official historical and political discourses.

How I conceptualize different poetic elements is also influenced by geography. I imagine the blank page as an excerpt of the ocean. The ocean and the page are storied and heavy with history, myth, ancestors, politics, and materiality. The ocean is not "aqua nullius." The page, then, is never truly blank. The page consists of submerged volcanoes of story and unfathomable depths of meaning.

I imagine each word is an island. The visible part of the word is its textual body; the invisible part of the word is the submerged mountain of meaning. Words emerging from the silence are islands forming. No word is just an island; every word is part of a sentence, an archipelago. The space between is defined by referential waves and currents.

I imagine poems as Pacific "song maps," a term which refers to the songs, chants, and oral stories that were created to help seafarers navigate oceanic and archipelagic spaces. Pacific navigational techniques are often understood as a "visual literacy," in the sense that a navigator has to be able to "read" the natural world in order to make safe landfall. The key features include reading the stars, ocean efflorescence, wave currents, and fish and bird migrations. With this in mind, I imagine that poems are song maps of my own journey to find Guam across historical and diasporic distances. Oceanic stories are vessels for cultural beliefs, values, customs, histories, genealogies, politics, and memories. Stories weave generations and geographies. Stories protest and mourn the ravages of colonialism, articulate and promote cultural revitalization, and imagine and express decolonization.

~

What follows is a selection of three poems from my ongoing *from unincorporated territory* series that illustrate Indigenous Pacific Islander geopoetics. The first poem, "*from* sounding lines," is in five parts and explores the theme of colonial and decolonial mapping and cartography in the Pacific. The second poem, "*from* aerial roots," maps my own migration story as well as the larger story of Chamorro diaspora. The final poem, "*from* mahalo circle," is a longer prose poem centered on my wedding in order to focus on food geographies in Hawai'i. These poems are song maps, which I hope will help readers navigate the poetic geographies of Guam and Chamorro culture and diaspora.

from **sounding lines** (first map)

~

Hasso' : *remember*
the first map
my dad hangs in the hallway :
an aerial view of our island,
which nearly fills
the entire space.
"Where's our village?" I ask.
"In the center," he points.
"Here: Mongmong : *heartbeat.*"
I read the names of other villages:
"Yigo, Dededo, Tamuning,
Barrigada, Mangilao, Chalan Pago,
Ordot, Toto, Maite, Hagatna,
Hagatna Heights, Sinajana, Asan,
Piti, Yona, Santa Rita,
Agat, Talofofo, Umatac,
Inarajan, Merizo"
I once imagined them
as separate places, but now
I see we're all part
of the same
tropical body.

from **sounding lines** (second map)

~

Hasso' the second map
    my dad hangs in the hallway :

        an aerial view of
           the Mariana archipelago.

                15 islands in a vertical
                  crescent. I

                recognize Guam,
                      the southernmost
            in the chain.

                  I read the names
            of the northern islands:

                "Rota, Aguijan, Tinian,
        Saipan, Farallon de Medinilla,

            Anatahan, Sarigan, Guguan,
      Alamagan, Pagan, Agrihan,

        Asuncion, Maug, and
Farallon de Pajaros."

They look like rosary
beads.

from **aerial roots** (off-island chamorros)

~

Remember migrating to California
        with my family.

Hasso' waiting to board
the one-way flight on "Continental,"
        the name of the airlines,
the name of our destination.

Remember the entrance to the Guam airport
        resembled the shape of i sakman,
an outrigger canoe, once described as "flying proa"
because it swiftly skimmed the waves.

Hasso' waving goodbye
        to all our relatives
as we entered the gate.

Remember our word for *airplane* :
        batkon aire : air boat.

~

        Remember the first day at my new high school,
the homeroom teacher asked me where I was from.
        "The Marianas Islands," I answered.
He replied: "I've never heard of that place. Prove
        it exists."

        When I stepped in front of the world map
                on the classroom wall,
        it transformed into a mirror:

the Pacific Ocean, like my body,   split
in two and flayed   to the margins. I

found Australia, the Philippines, Japan.

        I pointed to an empty   space between
        and said: "I'm from   this

                invisible archipelago."

Everyone laughed. And even though
I descend from oceanic navigators,
    I felt so lost   shipwrecked

on the coast of a strange continent.

~

"Are you a citizen?" he probed.
        "Yes. My island, Guam, is a U.S. territory."

We attend American schools, eat American food,
listen to American music, watch American movies and television,
play American sports, learn American history, dream
American dreams, and die
in American wars.

"You speak English well," he proclaimed,
            "with almost no accent."

And isn't that what it means to be
        a diasporic Chamorro: to feel *foreign*
in a domestic sense.

~

Over the last 50 years,
Chamorros have migrated
to escape the violent memories of war,
        to seek jobs, schools, hospitals, adventure, and love;
and most of all, to serve in the military,
        deployed and stationed to bases around the world.

According to the 2010 census, 44,000 Chamorros
        live in California, 15,000 in Washington, 10,000 in Texas,
7,000 in Hawaii, and 70,000 more in every other state
        and even Puerto Rico.

We're the most "geographically
            dispersed"
Pacific Islander population within the United
        States, and off-island Chamorros now outnumber

our on-island kin, with generations having been born
    away from our ancestral homelands,

          including my daughters.

           ~

Some of us will be able to return home
    for holidays, weddings, and funerals;
others won't be able to afford the expensive plane ticket.
Years and even decades might pass
    between trips, and each visit will feel
too short. We'll lose contact
with family and friends, and the island
     will continue to change
     until it becomes unfamiliar
        to us.

And isn't that, too, what it means to be
    a diasporic Chamorro: to feel *foreign*
in your own homeland.

         ~

There are times when I feel adrift,
    without itinerary or destination,

When I wonder: What if we stayed?
    What if we return?

When the undertow
    of these questions pull   you out to sea,

    remember: migration flows through our blood
        like the aerial roots of i trongkon nunu

    hasso': our ancestors taught us how to carry
        our culture in the canoes of our bodies.

remember: our people, scattered like stars,
      form new constellations   when we gather.

hasso': home is not simply a house, village, or island

    home is an archipelago of belonging.

from **sounding lines** (third map)

~

Remember the third map
my dad hangs in the hallway :
an aerial view
of Micronesia.

"Micro- means tiny,"
he says. "And nesia means islands."
Two thousand dots scattered
across the Western Pacific. My dad points:

"Here's us,
the Marianas,
and here's Palau,

Yap,

Chuuk,
Pohnpei,

Kosrae,
the Marshalls,

Nauru,   and

Kiribati."

"We're all cousins," he says.
The archipelagoes resemble constellations.

from **Mahalo Circle, 2013–2015** (for Brandy)

~

Mahalo Kalihi Valley, the Koʻolau mountains, and the island of Oʻahu

Mahalo *Kōkua Kalihi Valley Comprehensive Family Services* for being non-profit and birthing *Hoʻoulu ʻĀina*, 100 acres leased from the state park, whose name translates as *to grow the land and to grow because of the land* // Mahalo volunteers who removed the invasive albizia so light could heal the native flora \\ Mahalo for planting a community garden and for not spraying pesticides // Mahalo Puni and auntie Kaiulani for starting each Hawaiian birthing class with a chant honoring the valley, and a mahalo circle honoring those [we] are grateful for \\ Mahalo for serving dinner of local ʻuala and poi, garden salad, and organic chicken soup // Mahalo for brewing ʻōlena and māmaki tea on those cool rainy nights

Mahalo akua for blessing us with rain when [we] visited the 4-acre family-owned *Kūpaʻa Farm*, whose name translates as *steadfast* and *faithful*, on the slopes of Haleakalā, Maui, in 2013 // Mahalo Janet and Gerry for not spraying pesticides \\ Mahalo for using cover crops and perennial vegetative barriers // Mahalo for growing coffee trees under the shade of koa and monkeypod \\ Mahalo for letting us smell the compost // Mahalo for harvesting 50 pounds of beets, eggplant, carrots, lettuce, kale, papaya, lilikoi, and coffee beans for the wedding

Mahalo *Mana Foods* in Paia, Maui, for sourcing local ʻuala, ʻōlena, ginger, pineapple, dragon fruit, mango, sweet onion, banana, macadamia nuts, and lychee // Mahalo for giving your compostable byproducts to local farms

Mahalo *Kula Country Farms* for the strawberries // i hope to bring my daughter to the You-Pick patch someday

Mahalo *Pukalani Superette* for sourcing a local 10-pound rib eye roast for me from *Maui Cattle Company* // Mahalo *Maui Cattle Company* for feeding your animals grass

Mahalo *Tamura's Fine Wine & Liquors* in Kahului for sourcing local beer // Mahalo *Maui Brewing Company* for your *Bikini Blonde Lager*, *CocoNut Porter*, and *Big Swell IPA* \\ Mahalo for giving your spent grain to local farmers for feed and compost

Mahalo *Madre Chocolate* for being bean to bar // Mahalo for the Farm to Factory Tour, for letting us pick cacao from the *Reppun Farm* in Waiāhole, Oʻahu, and taste fresh cacao pulp \\ Mahalo to the Reppun family for struggling for land and water rights decades ago

Mahalo *Maui Bees* in Kula for raw wildflower honey and for not treating the bees with chemicals

Mahalo *Stillwell's Bakery* in Wailuku, Maui, for the three-layer wedding cake // Mahalo each layer of lilikoi cream, coconut cream, and custard cream, I love you all equally, as if you were my own future children \\ Mahalo Brandy for decorating the cake with pink anthuriums & purple orchids from your grandparents' garden in Kula

Mahalo family-run *Noho'ana Farm*, whose name translates as *lifestyle*, for restoring 2 acres of land and 500 year old lo'i in Waikapū, Maui // Mahalo Hōkūao for making 10 pounds fresh poi on the day of the wedding \\ Mahalo *Hui o Nā Wai 'Ehā* for struggling for water rights in East Maui, where the *Wailuku Sugar Company* has diverted, for more than a century, the "Four Great Waters": Waihe'e River, Waiehu Stream, Wailuku River, and Waikapū Stream

Mahalo La'akea for sourcing 3 pounds of handcrafted *Puna Pa'akai* sea salt from the Big Island and mailing it to me in Maui // it blessed the food and the venue

Mahalo Calrose white rice, imported from California, for filling our plates // Mahalo *Mama Sita achiote powder*, imported from the Philippines, for transforming the white rice into a thousand tiny red rose petals \\ Calrose is Calrose is Calrose is Calrose

Mahalo, mom and dad, for always feeding me // Mahalo for being brave enough to migrate across the ocean, for building a new home for [us], and for teaching me how to navigate this world \\ I will always carry your stories with me // Mahalo to all our family who traveled to the wedding and served our guests dinner \\ Mahalo to all our guests for your gifts and the gift of your presence // Mahalo, dear readers, for joining [us] at the table of this poem, please eat until you're full, there's more than enough for everyone, and please don't leave until [we] give you a full plate to take home for your meal tomorrow

Mahalo Brandy for saying 'ae : *yes*, which sounds similar to the Hawaiian words 'ai : *food* and ai : *sex*

Mahalo *Mānoa Community Garden Association* for giving me a plot after 9 months on the waiting list // Mahalo for banning pesticides in the garden \\ Mahalo invasive California nut grass for teaching me insistence // Mahalo *City Mill* for sourcing seedlings and tools (I don't forgive you for stocking shelves of Roundup) \\ Mahalo *Whole Foods* in Kahala, O'ahu, for sourcing imported organic seeds // Mahalo *College of Tropical Agriculture and Human Resources* at the University of Hawai'i, Mānoa (UHM) for sourcing local seeds \\ Mahalo *Hui Kū Maoli Ola* in He'eia, O'ahu, for being the world's largest native Hawaiian plant nursery and for sourcing different varieties of kalo, which Brandy and I planted in the garden // Mahalo communal water hose \\ Mahalo Mānoa rain water *from* **sounding lines** (fourth map)

~

Hasso' the fourth map :
an aerial view
of the Pacific Ocean
rimmed

by Asia   the Americas.

Countless archipelagoes
divided:
"Micronesia" "Melanesia" "Polynesia"

My dad traces a triangle
between Hawai'i

Easter Island (Rapa Nui)   New Zealand (Aotearoa)

"This is Polynesia," he says. "Poly- means many."

Then he draws an imaginary circle around
Papua New Guinea, the Solomon Islands,
Fiji, Vanuatu, and New Caledonia.
"This is Melanesia," he says. "Mela- means black."

"Remember, we're all relatives"
The ocean: our blue continent.

from **Mahalo Circle, 2013–2015** (for Brandy)

Mahalo *Kōkua Market* for being Honolulu's only co-op grocery // Mahalo for sourcing local vegetable and meats \\ Mahalo Lynette for allowing me to host a poetry reading at the store // Mahalo to my students for reciting their food poetry in the produce section

Mahalo *Down to Earth* for being Honolulu's only vegetarian grocery store // Mahalo for your healthy and bright salad bar, which makes Brandy smile as wide as her growing belly

Mahalo *Kapiʻolani Community College Farmers' Market* for your abundant selection // I'm sorry we stopped visiting but you have become an over-crowded tourist attraction

Mahalo 23-acre *Maʻo Organic Farms* in Waiʻanae, Oʻahu, for birthing a mala ʻai ʻopio : *youth food garden* // Mahalo for your yellow Community Supported Agriculture (CSA) box, with the slogan "No Panic Go Organic" in green lettering, which I pick up every week from the Urban and Regional Planning building on the UHM campus

Mahalo to my homemade, no-waste, 100% local smoothie, which I've affec-tionately named, "The Green Island," whose recipe took me months to perfect: banana, non-gmo papaya, pineapple, ginger, ʻōlena, kale, lime, and filtered tap water // St.Vitamix, puree for [us]

Mahalo *Student Organic Farm Training* (S.O.F.T.) in Mānoa, for letting me drop off my vegetable waste into your compost pile // Mahalo mythic albino cockroach for saying hello as I turned the compost

Mahalo clean tap water and Brita pitcher // Mahalo all the "embedded water" used to grow our food

Mahalo *Hawaiʻi Food Bank* for stocking enough to feed the population for seven days if the container ships stop coming // Mahalo to the long-time employee who trained me to sort produce donations \\ Mahalo *Costco* for donating hundreds of bananas from Ecuador, clementines from Chile, oranges from Australia, and cabbage from Mexico // Mahalo *Aloun Farms* for donating thousands of local green bell peppers, therefore never ask for whom the bell pepper rots, it rots for thee \\ Mahalo nameless farm in Watsonville, California, for donating thousands of strawberries, some of which are covered in white mold // Mahalo nameless farm in Santa Cruz, California, for donat-ing hundreds of lettuce heads, whose brown layers peel away easily : lettuce pray and make our salvage \\ Mahalo cases of "mainland shell protected" eggs donated from California, Arizona, and Washington, stamped "US" and still cold from the chill // Mahalo early morning radio DJs for always playing Bob Marley because "there ain't no hiding place from the Father of Creation" \\ Mahalo baby flies, fat flies, and dead flies for teaching me how time flies when you volunteer // Mahalo mop bucket for squeezing the dirty water after I mop the grime \\ Mahalo Mothers of Slow Rot, Mothers of the Still Edible, Mothers of Soup Kitchens, Mothers of Food Pantries, Mothers of Leftovers, Mothers of Canned Goods, pray for [us]

Mahalo auntie Kaiulani for organizing the monthly *Decolonize Your Diet* series at *Roots Cafe* in *Kokua Kalihi Valley* // Mahalo Aiko and Kelsey for making homemade, locally sourced SPAM for "De-processing Spam" night \\ Mahalo for inviting me to deliver my talk, "Uncle Spam Wants You," about the gastro-colonial and military history of my favorite canned meat // Mahalo Brandy for performing my poem, "Spam's Carbon Footprint," with me, you are the Spam jelly to my lucky belly

Mahalo Hanale and Meghan of *Homestead Poi* for your kalo farm in Waiahole // Mahalo for not spraying pesticides \\ Mahalo for allowing my Pacific poetry class to visit // Mahalo for delivering your poi to *Kokua Market*, which I bought every week during Brandy's pregnancy, and which was the first solid food (mixed with breastmilk) [we] fed our daughter

Mahalo *Mana Ai* in Kaneohe, O'ahu, for paying kalo farmers fair wages, for cooking and pounding kalo into pa'i'ai // Mahalo for selling the pa'i'ai in tī leaves as opposed to plastic \\ Mahalo to all the activists and farmers who fought to legalize pa'i'ai // Mahalo to the activists who stopped UHM scientists from genetic engineering and patenting kalo

Mahalo Papahānaumoku and Wākea for birthing Ho'ohokukalani, who gave birth to the stars and to Hāloanakalaukapalili, whose name translates as *the long stem whose leaves tremble in the wind*, who was stillborn then buried, and from whose grave grew the first kalo, // Mahalo elder brother for feeding us

Mahalo saina and kūpuna for carefully packing the kalo, niyok, lemmai, 'uala, 'ōlena, and mai'a into your sakman and wa'a (what war, what disease, what disaster were you migrating from) // Mahalo for reading the stars as if they were maps constellating a story of new beginnings \\ Mahalo for dreaming an island for [us] // Mahalo for loving these islands, for planting crops and creation stories

Mahalo, creation stories, for surviving // Mahalo for hiding in that place in our bodies that no one can convert or steal or behead or ban or bury or burn \\ Mahalo for carrying the weight of our origins in the hull of your memories and in the rope of your words // Mahalo for navigating all these distances and all these generations \\ Mahalo for reminding [us] who [we] are and where [we] come from // Mahalo for giving us the strength to say : no, [we] won't let them define [us], no, we won't let them confine [us], no, [we] won't let them silence [us]

Mahalo saina and kūpuna for planting as many trees as you could while everyone around you was dying // Mahalo for digging as many gardens and lo'i as you could while everyone around you was dying \\ Mahalo for saving as many seeds as you could while everyone around you was dying // Mahalo for passing down as many stories as you could while everyone around you was dying \\ Mahalo for giving [us] the strength to say : no, [we] won't let them poison our land and water anymore, no, [we] won't let them say our homes are wastelands or idle assets or military bases // [we] promise to stand up and shout, stop, our islands are sacred

Mahalo saina and kūpuna for eating canned goods, flap meats, and processed foods to survive // Mahalo for cooking foreign ingredients in recipes that felt like home \\ Mahalo for feeding [us] what you could afford // Mahalo for carrying the burden of disease so maybe we wouldn't have to // Mahalo for giving [us] the strength to say : no, [we] won't let them force feed [us] anymore

Mahalo Hawai'i nei for nurturing me even though I'm not from here // Mahalo for teaching me aloha 'āina \\ Mahalo for this dream of my daughter volunteering in the lo'i, dancing with our saina and kūpuna, dancing with our wind and tree relatives, dancing with our water and dirt relatives, dancing with our fish and bird relatives, dancing with our fruit and vegetable relatives, dancing with our more-than-human family of abundance

Mahalo Brandy for loving and caring for me // Mahalo for birthing and breastfeeding [neni] \\ Mahalo for gifting me this dream of 'ohana, this dream of belonging, this dream of our daughter, barefoot and dancing, her hair cascading into the past and braiding future generations // hu guaiya hao, hu guaiya hao, hu guaiya hao \\ *hånom håga hånom from* **sounding lines** (the fifth map)

~

My dad never hung a fifth map in the hallway.
I first see it when I travel, as an adult, to Taiwan.

Hasso' an aerial view of "Austronesia."

"Austro- means south," the tour guide says.

A highlighted area, in the shape of a full sail, stretches

from Madagascar to the Malay peninsula and Indonesia, north to the
Philippines and Taiwan, then traversing Micronesia and Polynesia.

"Austronesians migrated to
escape war, famine, disease, and rising seas."

400 million people alive today,
who speak over 1000 different languages,
all descend from the same
mother tongue,   the same genetic family.

I read the map closely, navigating beyond
the violent divisions
of national and maritime borders, beyond
the scarred latitudes and longitudes
of empire, until finally arriving
at the cartography of our most expansive legends

and deepest routes.

References

Perez, Craig Santos. (2008) 2017. *From Unincorporated Territory [hacha]*. 2nd ed. Oakland, CA:
  Omnidawn Publishing.
———. 2010. *From Unincorporated Territory [saina]*. Oakland, CA: Omnidawn Publishing.
———. 2014. *From Unincorporated Territory [guma']*. Oakland, CA: Omnidawn Publishing.
———. 2017. *From Unincorporated Territory [lukao]*. Oakland, CA: Omnidawn Publishing.

23

AGITATING A COPPER LYRE; OR, GEOLYRICISM FOR THE AGE OF DIGITAL REPRODUCTION

Jennifer Scappettone

The lyre whose manufacture this chapter will track struck again in the form of a more-than-confessional image randomly exhumed from a keepsake box last New Year's Day: A mother's snapshot of Field Day, 1984, transmitting an awkward testament of sunny bygone Northeastern blight into the bare clarity of a Chicago frigid, flat, Trump-ruled, yet Pritzker-bedecked. One clue to the current Midwestern landscape's refinement by way of Prizes, Pavilions, Galleries of Impressionism and Post-Impressionism, Children's Zoos, Schools of Medicine and of Law, and so forth resided in the semi-defunct industrial infrastructure hovering over that

FIGURE 23.1   Field Day, South Grove Elementary School, Syosset, New York, 1984.

Source: Photograph by Theresa Scappettone.

race of the '80s: Photographic punctum that tore a conceptual duct connecting the vignette of kids competing against each other in the playing fields just beyond the back gate of childhood to the lucrative network of copper wire subtending a system so vast it notoriously challenges visualization.[1] One can hypothesize the mother's psychic edit of this sunny, groveless still of youth on the grounds of South Grove Elementary School in Syosset, New York, cropping from their less seemly backdrop the sprinting children of her fellow more-or-less-recent immigrants from China, India, Indonesia, Ireland, Italy, Japan, Korea, Poland, and Puerto Rico who had sought housing on the affordable brink of a high-ranking school district on Long Island's "Gold Coast" of public education. Towering over Walt Whitman's scored grasses on this day when classes are suspended so that all can celebrate the great American traditions of Sports and Health are garbage trucks and the water tower of a defunct copper mill that for decades, alongside companies like Grumman Aerospace, the predecessors of Occidental Chemical, the Great American Corrugated Container Corporation, and the Konica Imaging corporation that likely enabled the generation of the photo itself, dumped tens of thousands of tons of toxic sludge into an unlined adjoining dump ("*Town of Oyster Bay* v. *Occidental Chemical Corp.*" 1997). The Environmental Protection Agency had placed the landfill on the Superfund program's National Priorities List at the beginning of that school year, in September 1983 (Environmental Protection Agency 1990), while the New York State Department of Environmental Conservation classified the copper mill as an Inactive Hazardous Waste Site—though we who lived in the neighborhood were ignorant of this, and no one brought it up until the girl in the center of the snapshot started pushing facts and disassembled lyrics about it into the public a couple of decades later. The neighborhood wasn't aware, either, that the copper company had been acquired on the cheap by a family of Chicago millionaires during the 1970s in a deal so lucrative it made headlines (generating gross sales of 800 million dollars, eight times their prior scale of operations)—nor that its name was changed as the new regulatory bodies identified the aforementioned adverse environmental impacts—nor that the troublesome property was swiftly flipped, subsequently changing hands twice. Nor did we know that the Inactive Hazardous Waste Site was only cleaned up and declassified as such between 1993 and 1994. Given the continuing presence of sanitation vehicles—parked and still dumping nebulous materials in the off-hours, as the Instamatic-toting mother lamented before being fearfully hushed by her best friend in more than one town hall—we didn't even exactly realize that the "sump" had been closed in 1973 because elevated levels of heavy metals were detected in the public drinking water. Nor did I know when I graduated from college—an inquisitive person, equipped with information about such far-flung stressors as the rise of cities, greenhouse gases and global warming, the causes and denouement of world wars, the establishment of biopolitics, and the histories of slavery, sharecropping, and apartheid, or the power of petroleum to fuel war in Iraq—that the dump across the street (of whose toxicity I was still unaware) had not been capped.

We don't often anticipate the way that our worlds implicate hundreds of thousands of people beyond us as a matter of course—the way in which, in this instance, a banal image of a childhood race strategized by the educational system to tout individual achievement might embed, as backdrop, or more precisely as ambient, the overweening story of global migration, of global telecommunications, and of transnational capital. But by pressing upon the unsightly phenomena of landscape that those in proximity are prone to edit out as a coping mechanism, and that those in post-Fordist power tend to prune away as real estate ruse, it becomes clear that the bodies of communities like the one in this snapshot are the unsung living monuments of telecommunications and its collateral damage. I write, pasting disencumbered bits of evidence together, in the conviction that this form of extrapolation, from the most banal of scales to the most sublime, is a necessary part of exposing material conditions that determine the unevenness of experience in the wireless age's reputed democracy of knowledge. Now more than ever aware of the teeming archives of exacted ignorance, I seek to live up to the possibility of poetry as a knowledge-producing apparatus. A critical geopoetics—by which I mean an affectively soiled, gunk-mired investigation of smothered facts—can open up pathways for a cognitive sprawl that stalks the material sprawl of industrial harm. These pathways can compete with the etherealized imagination of late-capitalist social life: One that carves out solace—in the midst of depression and bust—through fantasies of pennies from heaven, while subduing awareness of the infrastructure, the instrumentation, that made them possible.

De- and re-fabricating this geography of uncertain contours calls for the sniffing out, visualization, sounding, and setting into co-motion of non-linear chains of desires, avarices, and politesse (Coumou et al. 2018).[2] Far from being an abstruse realm of thinking, materially precise, non-linear poetics have immediate political consequences in a context in which global ecological crises are still routinely addressed myopically by privileging national or selective regional constructs in isolated national languages—and in which, as national borders are clamped down, marginalized people are systematically deprived of chances to enjoy the protections of citizenship. For in this land and e-scape, social belonging manifests in the sludge, gas migration, and water contamination that exceed the boundaries and rhetoric of political constructs yet that remain staunchly tethered to production of the communicative instruments of our so-called global village. By striking the lyre, or innumerable lyres, of copper that make up the grid of power, both material and ideological, we might create harmonies that stalk the system and its dissonant off-grid collateral damage.

★ ★ ★

Winter in the Midwest: Bareness so bleak it contours an otherworldly state of clarity. The Winter of 2013–2014 as lived out in the winds off a corrupted strip of Lake Michigan pressed questions of poetics beyond the page, to the source of their conveyance in server farms and the environing air—beyond the instant, to the histories of these communicative instruments:

POLAR VORTEXTET

Likes superceding talk, integers superceding contact, tags superceding secrets, tête-à-têtes;
      hashtags eclipsing quarrels, disputes, cognition arranged into sentences; retweets
replacing research, trending replacing intercourse, free because "virtual" intellectual labor
      replacing hired gigs; strategy
supplanting awkwardness, excess, styling expelling flesh, feeds from up to three and a half
      minutes ago expelling
      history;
the mandate for constant agitation of high-powered microwave transmitters and server-farms
      disguised to heat up and scatter heavy petroleum fractions upon anonymous underserved
      communities via flashes of likeable hashtaggable retweetable upworthy appearance as
      author and fan expelling by coercion any breath of contemplation as neighboring Home
      Depots are emptied in last-ditch efforts of ventilators large and small—
the Library of Congress cultivating cloud-farms to store, Kenny G the pulp to print, the lot,
      Sisyphean lusoriness:
slack coproduction of the spectacle of information itself afloat from any infrastructure of
      responsibility, even organ of apprehension. But the overarching
specter of futility in all that rises beyond the horizon of MyFace, myauthorfunction.com once the
      city's fantasy-doubles along the shores, fleeting betting paradises and g-force-jostlers,
      game under financialization, go under salt—
proposes it may be time to consider whether the most appropriate plan of action is to become
      absolute site-myopic site-smothering presentists after all:
Smoking of the lake, staccato scansion of the potholes tearing up Burnham's grid, pathos of
      plans to haul all the basement suitcases full of ice-chunks across the bought-out offset
      carbon paths into the red blot of Pacific lingering in the daily graphs of the explainers—
Cantre'r Gwaelod interposing itself as Pine Island melts, villagers pointing their phones at the
      peat, antediluvian oaks and newly fathomable peaks, radioactive tides broadcast;
One consults with the elders who may not be around in a 4C world: Marinetti hailing the most
      beautiful day-after-tomorrow, Lyn in the tunnel below Niagara leaking in here and there,
      deciding it's okay if humans don't last; or John issuing a stern warning to the anxious in
      the form of Safe, and Madeleine signing my book "Exactly reversible destiny to you . . ."

If the only thing that matters is the need to cut off the conduits of the whole neoliberal vampire
      period to stave off further fall and rise:
If you are a writer, for better or worse, intravenous: what the eff are you going to do about it?:
      Discuss.[3]

An emphasis on the materiality of the text has been a defining principle of innovative writing for the past several decades; waves of movements coming after modernism stressed the need to grapple critically with the opacity and mediation of language, and this obligation became de rigueur for writers who looked to the historical avant-garde for strategies of expanding inert understandings of poetic subjectivity and narrative. Yet the contradictions of producing literature and art within an apparently limitless network, while the navigation of an interminable and interrelated series of ecological disasters suggests that the dematerialization of the art object is a fallacy, demand that we revisit and revise the lessons of modernist, Language, and post-Language poetics once more. While digital media offer unprecedented opportunities for the production, motility, distribution, and preservation of ideas, they also necessitate collusion with algorithms and poisonous ecologies beyond our control or even awareness. "Watch a video on your iPad for a half hour and somewhere in America about 10 pounds of coal and a gallon (in oil equivalent terms) of fracked natural gas is burned," as Mark P. Mills points out in the pages of the self-branded "Capitalist Tool," *Forbes* (Mills 2011, 2012). What species of political and ethical complicity must we reckon with in participating in the global network, with its networks of data extraction, storage, and surveillance? What structures of oversight and obscured labor exploitation do search-string poetics, tactics of appropriation and distribution, and unconstrained digital publication need to confront, to resist? As producers of "content" and "sharers" of ideas, can we learn anything from movements of "degrowth" or *decroissance* that have flourished outside of the US (D'Alisa, Demaria, and Kallis 2015; Rabhi 2017)? What kinds of materialities—abject, dilated, distressed, or otherwise—might counteract our collaboration as cultural producers with narratives of infinite expansion, total information awareness, global access, mobility, resource extraction, and the marketizing of knowledge (Nowak 2008, 2009)? How can we respond critically? Finally, what can we learn about our apprehensible homes and their infrastructural extensions in the process of finding answers?

These questions were thrust upon me on my return home from graduate school and the forging of a community of geopoetically oriented practitioners in the beguilingly Arcadian Bay Area.[4] Back at home, I was facing once again the vague ugliness outside our kitchen window, off a freeway exit, beyond the playing fields. I was confronted with so many neglected strands of the history of that region Whitman eulogized as Paumanok—for all his faults at least making me aware of a speculative Indigenous name in the process. It was the Fall of 2005 and, having returned to the Northeast to teach for the year, I was rediscovering the material foundations of the Exxon-Mobil Masterpiece Theater of yore while deliberating, over the Greenpoint Oil Spill about the future of post-Language poetry and trying to support my mother through the ordeal of chemotherapy. The black water tower had been pulled down after 53 years of service (Saslow 2004). I had spent my life up to that point with five inscrutable white letters branding the sky—CERRO, isolated from the full name of the copper company that occupied 39 acres across the street, Cerro Wire & Cable. A Spanish word that hovered over the flat expanse of eyesoreness, inscrutable, a clue inserted into a poem, then a book as a means of groping my way line by line to the source.

> Reading HILL in Delphic syllables
> *39 acres of one's own*
> Rusted prospect with gym-glass floors'
> *Glimpse of smart-growth*
> Mister Fabricant on exterminated land;
> *Waiting for Bernd and Hilla*
> And republican grids impervious
> *Of unbranded towers*
> To the tune of 5,000 cars per hours'
> *Torturous bottlenecks at major 4F arteries:*[5]

A monument unredeemable as form even by the likes of the Bechers, whose 224 images of such structures enable swan songs to industry on the part of institutions like the Guggenheim (whose cash foundations lie in copper) (Becher and Becher 1988; Scappettone 2018). The water tower had been the sole apprehensible testament to the copper mill and (mostly) inactive hazardous waste site across the street adjacent to the landfill where Cerro Wire disposed of, for example, twenty thousand tons of metal hydroxide sludge between 1952 and 1974—supercharging the terrain with trace levels of copper, zinc, lead, cadmium, chromium, nickel, cyanides, arsenic, mercury, selenium, silver, chloroform, hydrazine, manganese, and phenol. I didn't learn of this litany until I was a full decade into research spurred by my mother's battle with breast cancer. The ordeal of keeping the port clean for administration of her own decidedly nontherapeutic chemical brew made me curious beyond our habitual dread about what precisely had gone on under the missing tower's watch. From gentrification in toxic Brooklyn to depression in the birthplace of suburbia 25 miles East: It took mere minutes at the browser to find out what "Superfund" was and realize that whatever had happened assumed more than monumental proportions. I decided to make poetry figure it out.

It may seem counterintuitive or even ludicrous to write poetry as a means of drilling into such a problem. Yet this was my intuition because intuition was all I had—a haunting and a hunch. How could I write a story when I had no sources but digits and orts? Nothing even to look at and describe: So much economic gain had depended on a general nebulousness, glazed with distractable screens, astride the white management. By passes piecemeal aimed at rubbernecking if not reaching the sublime, verse is able to plunge into the muck of sites from which history and actuality have withheld a narrative. And poetry has the potential, often untapped, to follow line by line the anarchic directions muck takes when it's not being tracked. When too stultified to know what I'm doing I do poetry, which resituates us at what Lyn Hejinian, in *Oxota* ("the hunt"), calls "the edge of a sort of cross-wonder, as if I'd never heard any story before" (Hejinian

1991, 237). Poetry makes nothing happen, Auden tells us; but it's a maw, a mouth (Auden 2007, 246). Enough had already happened to our bodies unsounded; now what was required was someone—a poet perhaps, in the antique sense—to mouth off.[6]

What made copper cabling desirable enough to determine the endangerment of an entire landscape—an adjoining schoolground, and a residential neighborhood? As anyone in analog electronics or a copper corridor knows, copper is the electrical conductor in many categories of electrical wiring that, if tracked, emerge as a sublime tangle of grist for our own mill—that, if yoked, form a vast instrument for our conceptual plectrum. The ductile metal is used in power generation, power transmission, power distribution, telecommunications, electronics circuitry, and countless types of electrical equipment. Copper and its alloys are also used to make electrical contacts. Before learning these facts through counternarrative means, I wrote a more straightforward book called *Killing the Moonlight* whose title cites the Futurists' bid to electrify the Italian nation-state: To kill the Romantic luster of cultural capitals like Venice, transforming historic cities of parasitic tourism into industrial utopias (Scappettone 2014). The Futurists' 1910 manifesto against Venice ended with the call to annihilate, by way of electricity, the light of the moon—and, by association, the tidal rhythms of the feminine: "Let the reign of divine Electric Light come to liberate Venice from its venal moonlight of bedrooms for rent" (Marinetti 1983). As the red metal that makes electrification possible, copper can be seen as the basis of this process and the mineral driver of modernity—a Bronze-Age essential whose relevance would be no less urgent in the Space Age. Wars have bolstered demand for the bronze that goes into cannons and firearms (an alloy of copper and tin) for centuries; copper sheathing protects ships from saltwater corrosion, biofouling, and explosions of arms stored inside (while leaching into the ocean and threatening bivalve populations). But demand for copper in electricity, trains, telegraphy, and eventually telephony led to unprecedented demand during the Industrial Revolution. If you visit a mine nowadays, the importance of copper mining will continue to be underscored in the most emphatic terms. "Every American born will need 945 pounds of copper," claimed the Minerals Education Coalition in 2016 (Mining & Minerals 2016); this amounts to an average of 12 pounds of copper per capita per year.

Though steadily being superseded by fiber optics, copper cabling for telegraphy and telephony laid the crucial material foundations of the Internet; images of twisted-pair copper cables being hurriedly fixed or replaced after Hurricane Sandy, shot in a Lower Manhattan "carrier hotel" originally built in 1930 as Western Union's headquarters, made that clear (D'Orazio 2012). In such spaces, copper and newer fiber-optic cables, servers, machinery, and fans to cool it all down connect to form the Internet. I became interested in this prehistory for reasons similar to Tung Hui Hu, whose *Prehistory of the Cloud* underscores the importance of the centralized ideological as well as the material structures undergirding contemporary network culture (Hu 2015).

Cerro Wire, I eventually learned, had produced 44 million feet of wire for the World Trade Center. It was incorporated into the Pritzker/Marmon holding company to mammoth profit. It ceased operations in my hometown before being acquired by Warren Buffet's multinational conglomerate, Berkshire Hathaway—so that the industrial swan songs of this suburban community were quieted, while the effluvia of Cerro's manufacturing processes and disposal practices live on to an ambiguous extent in the water and air. Yet one day, via ever more filigreed Google searches in the desert, intent on concluding my book on the matter, *The Republic of Exit 43*, I managed to follow the pathways of the wire—the potential lyre—further. Through trial and much error, I stuck together the search terms that would unearth Cerro Wire's corporate reports, which had been digitized and preserved on a server lacking an index by the Cleveland Public Library. (Like much of the other evidence, these pdf files have since fallen out of Google access.)

In those boastful internal documents aimed at gaining the confidence of investors, I discovered the names of the corporate stakeholders and geographical origins of Cerro Wire's copper, and of its initial offshore processing. Promotional fare cleared the way toward comprehending a legacy of damage that exceeded the neighborhood, the state, the continent, the book: Damage whose culpability is still being perpetually dodged at Cerro's origin in name and exploitation, though under another corporate flag. The copper mining at Cerro de Pasco in central Peru made the US syndicate founders JP Morgan, Henry Clay Frick, the Vanderbilt family, Phoebe Hearst, and others we've heard of through their art museums and "starchitecture" prizes super-rich. With open-pit operations commencing during the 1950s, the town of 70,000 people vaunts massive Dantesque "bolges" to show as noxious continuum of downward spiral between one brand and another. Mining builds alternative mountains in the spirit of alternative facts—massive pilings of toxic mining tailings that lace the neighborhoods, inhaled. Extraction bores a gash that develops and cannibalizes this city in the Andes, leading to diseases of the stomach and lungs as well as lead poisoning, developmental delays, and malnutrition in the local population (particularly the children); copper smelting produces sulfur-dioxide emissions and acid rain in the nearby town of La Oroya, which the Blacksmith Institute identified in 2006 as one of the ten most toxic sites on Earth (Geen et al. 2012).

Over the past two decades, the Andean region has emerged as the world's most productive copper region. (If you are a rising capitalist, you can buy the South America Mining Map from Infomine's e-store for CA$200.) Conflicts over mining projects in Peru have, as Reuters reports in its signature dispassion, "held up billions in potential investments over the past decade" (Taj 2015). Controversial enterprises such as the Tia Maria Mine, aiming to mine 120 thousand tons of copper cathodes per year during an 18-year life span, pit the Peruvian government—bolstering multinational economic interests—against its people; demonstrators have called on the government to cancel the copper project over projections that it will pollute nearby agricultural valleys, as others have, again and again. Meanwhile, across the Atlantic, the salvaging of copper in the slums of Accra (at the electronic waste dump of Agbogbloshie) makes for a particularly lethal postscript to that process of extraction. Wires are burned to strip away the insulating tape for the copper beneath as quickly as possible, often by children who need to pay for school supplies. This was a process my grandfather and uncles undertook in salvaging copper for resale on the streets of East Harlem until the 1970s. Agbogbloshie also made the Blacksmith Institute's 2006 list of the world's ten most polluted sites.

How can one conduct an archaeology of responsibility for such far-reaching disasters?

★   ★   ★

In the narrative interludes of Italo Calvino's *Invisible Cities*, first published in 1972, the character of Marco Polo plays chess with Kublai Khan, a structuralist who views the board's grid as the totality of his empire—a geometrically comprehended, knowable scheme: "If each city is like a game of chess," the Khan muses, "the day when I have learned the rules, I shall finally possess my empire, even if I shall never succeed in knowing all the cities it contains" (Calvino 1986, 11). From the perspective of Calvino's Marco Polo, even the geometry of a chessboard reveals itself as a product of hybrid material origins embedded, like a historical script without end, in the wood. The Venetian merchant points out that the board was cut from the ring of a trunk that grew in a year of drought, full of pores that reveal the traces of a larva's nest, and that it was scored by the wood carver with a gouge.

I virtually unearthed an equally eloquent chessboard gracing the cover of a digitized version of Cerro's 1967 annual report (see Figure 23.2). Its Inca chess pieces were—the report boasts—handcrafted for the insurance executive W. Clement Stone of Chicago from Cerro de Pasco silver, to commemorate the fact that the metal had been freed under new free trade arrangements with South American countries "to find its own competitive price" (amidst the tumult of strikes and impending, yet doomed, Socialist revolution). I seek a form that can read this image in the ways that Calvino's Polo teaches: that can unsmelt, tell of the digging and fabrication of cables, and trip up the ever-expanding grid of power through scrutiny of its material parts. Such a form could sound the sources and dispersion of its dusts as well as of its artifacts under glass. Such a form would stalk one set of wires to suss out the infrastructure connecting all. It would strip each nuclear house of its image of solidity so that, in Henri Lefebvre's terms, the home's own "image of immobility" would be replaced by "an image of a complex of mobilities, a nexus of in and out conduits" (Lefebvre 1991, 93). How, then, to wrest the guts from their metaskeleton of sheltering structures and yoke these entrails to a copper lyre?

At the stripped housing of 6018 | North: A Chicago mansion turned over to the arts whose peeled walls vaunt their infrastructure of copper wire for telephony, pipes for plumbing, and even an early-twentieth-century copper tube for calling in the servants from the cellar: How to render this structure resonant as Lefebvre's house? I imagined it as the host of an Aeolian harping on the global conduits of the one intellectual (but not nearly merely intellectual), virtual (yet carnivorous) breeze,

FIGURE 23.2 Cover of Cerro Wire's 1967 annual report.

Source: Courtesy of the Cleveland Public Library.

"Plastic and vast" (Coleridge [1912] 2009, 2:102). Poetry might allow us to sound what Calvino envisions, in an essay on Fourier, as the "pulverization" of utopia (Calvino 1995).[7] Poetry as medium enables us to in-spire, or breathe it, for better or worse. This is an operation fundamentally different from understanding. Poetry might sample from what can be perceived in a reciprocal exchange of information about information while admitting to being uncomprehensive, as it inevitably falls short of, and therefore resists, the cartographic pretense of being visually intact.

This is why, from the pop-up operas of *The Republic of Exit 43* to its sequel in progress, lyric has found itself expanding into several dimensions as a series of performance and installation works generated in dialogue with Caroline Bergvall, Judd Morrissey, and Abraham Avnisan—from *The Data That We Breathe* to *LAMENT; Or, The Mine Has Been Opened Up Well*.[8] My collaborators and I have set out to construct, as alarm of the ever-colonizing, ever-obsolescing network, of the house and the signals threaded through it, particulate and communicative, us would-be Pans, a copper harping. Piercing an image of solidity to expose a complex of wayward mobilities, a monstrous body of in and out conduits, but one that takes in as much as it generates—that replaces production with the inhale. In an ambient of alternative facts, upworthy jingles, and memory plaques, to drill alternative routes into the breadth and depth of the global network that agitate the byways of data, set historiography atremble, making of the drilling of snakes our flutes.

As I was ending one book, but not my ignorance, and beginning this other one, I finally wrote the story of why I needed to patch it together that way.

SIREN ARIA

[sung by SIRENS]

The breaching: of a whale? When it rises to the surface and we glimpse the back breathing flame resistants blithely dispensed.

Breaching motherly ditch of the man born a Futurist: reminding him of the breast of his Sudanese nurse: lips and lips incandesce.

Avatar who flushes like the class you cannot erase, according to the anonymous Neapolitan novelist. Who arrives? Alice. At frayed ends of plumbing and unspindled copper wire. Inside, it is raining heavy metals.

Her lips are capped. Against migration that has already occurred, been occurring for the decades since the war created Levittown and La Oroya the Twin Towers.

And the myths get smelted—while I loses herself to the third person in the off-gassing of the Ego. Everything depends on that.

Alice gushes out landscape, through the channeling of IO. Ilice comes out lava. One sprawlscape mill's log flume, delireal, loses the furrow.

Fluoresces with the voltage of a billion curses cut into the cast lead of the seizure air, an aria. Cheeks flush as cyanide ponds, tongue batting against Home Depot pipe.

A tailings Cenerentola starring in a high school language lesson. An emergency dusting off the page. Taking the place of liver flesh.

Monstrous, she takes the cables of Agbogbloshie pyres for her hair. To restore
Them to their roots in the archipelago of Henry Clay Frick & co.: or rather: its banlieue.

It's still conductive, pornographic. The plumbing a listening hose for strangers rooting around in
  the House's howls: I unwelcome Pan-
Am out of breath.
  Fearing the mammograph.
Does one Sioux Chief Full-Slip Repair Coupling from south of Peculiar, Missouri glued after
  the other,
  Inhaled, make an argument? Does the rig an epic?
Do the Inca chess pieces on the cover of ERR-0's 1967 annual report,
  Handcrafted for insurance executive W. Clement Stone of Chicago Illinois
From ERR-0 de Pasco silver freed
  To find its own competitive price? Cousins of Chules
On chemical scholarship framed to admire the panorama of their boss drill. 6,000 kilometers
  Northwest, spec
House razed a year past: Berkshire Hathaway signs suspend from the demure midmod fence.
  Third-degree
Kith and kin. Would-be Lombardi ransacked in skin, futzing with copper elbows, inhaling
  multipurpose Sureflex: Lord, make me a pipe of your benefiguring.
This is no structuralist's wet dream.
  Wipe yourself off & fasten the screen.
I couldn't write it in sentences. I couldn't compose sentences. I couldn't compose sentences
  Without her hands.
Because I couldn't bear to look at the totality of it.
The infrastructure sucks in her.
  Inarticulate for centuries: a Southern repulse to the scene of repair
Nurses the axe to free the artifacts from the sentences they were destined to make.[9]

★ ★ ★

The coincidental nature of rummaging through the archive and of charting one's
own "personal" geographies and footprints led me to an uncanny realization over the
course of writing this piece. The city of Cerro de Pasco has begun a marathon that
has officially been named "the highest in the world." Upon reading the headlines,
I first imagined that runners would be scaling the heights of the open-pit mine from
below—but then realized that journalists were referring instead to the city's altitude
(4,330 meters). A local newspaper, the *Diario AS*, noted in 2018 that "The Marathon
of [Cerro de] Pasco probes the limits of the aerobic capacity of the human organism." For more than one reason. Abounding, an underbelly of calculation which will,
in eluding the constricted sights of the news, have to resound elsewhere.

Notes

1 On the global network as a sublime figure see, for example, Jagoda (2016).
2 Coumou et al. argue that "non-linear interactions need to be disentangled and quanti-
   fied" (2018, 9) by science in order to understand complex weather systems in the process
   of climate change. For a conception of non-linear relations that embraces entanglement,
   see Haraway 2016.

3 This poem is my own (Scappettone 2016a) and was first published at Harriet: The Blog in May of 2016. Grateful acknowledgment is made to the editor Michael Slosek and to the Poetry Foundation.

4 Keystones of this community included the teaching and writing of Juliana Spahr; the activities of BARGE (the Bay Area Research Group in Enviro-Aesthetics), founded by David Buuck; and the environmental justice initiatives at the Marin Headlands and the National Environmental Education Foundation.

5 This is a fragment of a poem from *The Republic of Exit 43: Outtakes and Scores from an Archaeology and Pop-Up Opera of the Corporate Dump* (Scappettone 2016b). Grateful acknowledgment is made to the editors of Atelos Press, Lyn Hejinian and Travis Ortiz.

6 See, for instance, "they set to work" and "and every now and then a great crash," two late pop-up pastorals from Jennifer Scappettone, *The Republic of Exit 43: Outtakes and Scores from an Archaeology and Pop-Up Opera of the Corporate Dump* (2016b, pp. 94–95).

7 A 1986 English translation reads "a utopia of fine dust," but Calvino uses the term "*polverizzata.*" The original essay was published in 1973.

8 The first of these, *The Data That We Breathe,* was presented in collaboration with artists Caroline Bergvall and Judd Morrissey at 6018 | North in Chicago in February 2016 as part of a pedagogical, research, and practice-based residency at Chicago's Gray Center for Arts and Inquiry. A new work taking a series of titles based on terms from More-ing & Neal's mining and telegraph codebook has since been under development with collaborators Judd Morrissey and Abraham Avnisan, with substantial contributions from composer Mark Booth, under the umbrella heading *SMOKEPENNY LYRICHORD HEAVENBRED; Or, Last year / By constant penetration / Encroaching on the reserve.* It has been presented at the Gray Center for Arts and Inquiry (November 2016), Northwest-ern University's Block Museum (May 2017), the Electronic Literature Organization Festival in Porto, Portugal (July 2017), 6018 | North, as part of the Chicago Architecture Biennial (September 2017), the Poetry Foundation (February 2018), as a freestanding interactive installation at the Rutgers–Camden Center for the Arts (September through December 2018) and at Counterpath Gallery (April 2019). For more information and documentation, see https://oikost.com.

9 This is a poem from *The Republic of Exit 43* (Atelos 2016), fragments, revisions, and rear-rangements of which were used in performance of *SMOKEPENNY LYRICHORD HEAVENBRED.* Grateful acknowledgment is made to the editors of Atelos Press, Lyn Hejinian and Travis Ortiz.

References

Auden, W. H. 2007. *Collected Poems.* Edited by Edward Mendelson. New York: Random House.

Becher, Bernd, and Hilla Becher. 1988. *Water Towers.* Cambridge, MA: MIT Press.

Calvino, Italo. (1972) 1974. *Invisible Cities.* Translated by William Weaver. New York: Har-court Brace Jovanovich.

———. 1986. *The Uses of Literature: Essays.* Translated by Patrick Creagh. San Diego: Har-court Brace Jovanovich.

———. 1995. "Per Fourier, 3. Commiato. L'utopia pulviscolare." In *Saggi* 1: 307–314. Milan: Mondadori.

Coleridge, Samuel Taylor. (1796) 1912. "The Eolian Harp." In *The Complete Poetical Works of Samuel Taylor Coleridge,* edited by Ernest Hartley Coleridge, 101–102. Project Gutenberg Ebook. Vol. 2. Oxford: Clarendon Press. http://www.gutenberg.org/files/29090/29090-h/29090-h.htm.

Coumou, Dim, G. Di Capua, S.Vavrus, L.Wang, and S.Wang. 2018. "The Influence of Arctic Amplification on Mid-Latitude Summer Circulation." *Nature Communications* 9 (1): 2959. https://doi.org/10.1038/s41467-018-05256-8.

D'Alisa, Giacomo, Federico Demaria, and Giorgos Kallis, eds. 2015. *Degrowth: A Vocabulary for a New Era.* Abingdon, Oxon: Routledge.

D'Orazio, Dante. 2012. "Into the Vault: The Operation to Rescue Manhattan's Drowned Internet." *The Verge*, November 17. www.theverge.com/2012/11/17/3655442/restoring-verizon-service-manhattan-hurricane-sandy.

Environmental Protection Agency. 1990. "Superfund Record of Decision: Syosset Landfill." National Service Center for Environmental Publications. United States Environmental Protection Agency.

Geen, Alexander van, Carolina Bravo, Vladimir Gil, Shaky Sherpa, and Darby Jack. 2012. "Lead Exposure from Soil in Peruvian Mining Towns: A National Assessment Supported by Two Contrasting Examples." *Bulletin of the World Health Organization* 90 (12): 878–886. https://doi.org/10.2471/BLT.12.106419.

Haraway, Donna Jeanne. 2016. *Staying with the Trouble: Making Kin in the Chthulucene.* Durham: Duke University Press.

Hejinian, Lyn. 1991. *Oxota: A Short Russian Novel.* Great Barrington, MA: The Figures.

Hu, Tung-Hui. 2015. *A Prehistory of the Cloud.* Cambridge and London: MIT Press.

Jagoda, Patrick. 2016. *Network Aesthetics.* Chicago: University of Chicago Press.

Lefebvre, Henri. 1991. *The Production of Space.* Translated by Donald Nicholson-Smith. Oxford, UK and Cambridge, MA: Blackwell.

Marinetti, Filippo Tommaso. (1910) 1983. "Contro Venezia passatista." In *Teoria e invenzione futurista*, edited by Luciano De Maria, 33–34. Milan: Mondadori.

Mills, Mark P. 2011. "Opportunity in The Internet's Voracious Energy Appetite: The Cloud Begins with Coal (and Fracking)." *Forbes*, May 31. www.forbes.com/sites/markpmills/2011/05/31/opportunity-in-the-internets-voracious-energy-appetite-the-cloud-begins-with-coal-and-fracking/.

———. 2012. "Bravo *New York Times* For Discovering Reality in 'Power, Pollution and the Internet'." *Forbes*, September 25. www.forbes.com/sites/markpmills/2012/09/25/bravo-new-york-times-for-discovering-reality-in-power-pollution-and-the-internet/.

"Mining & Mineral Usage Statistics." 2016. *Minerals Education Coalition.* Accessed August 24, 2018. https://mineralseducationcoalition.org/mining-mineral-statistics.

Nowak, Mark. 2008. "Poetics (Mine)." *Harriet: The Blog of the Poetry Foundation* (blog), July 13. www.poetryfoundation.org/harriet/2008/07/poetics-mine.

———. 2009. *Coal Mountain Elementary.* 1st ed. Minneapolis: Coffee House Press.

Rabhi, Pierre. 2017. *The Power of Restraint.* Translated by Lisa Davidson. Arles: Actes Sud.

Saslow, Linda. 2004. "In Brief; Cerro Wire Tower Falls; Retail Complex Planned." *The New York Times*, September 26. www.nytimes.com/2004/09/26/nyregion/in-brief-cerro-wire-tower-falls-retail-complex-planned.html.

Scappettone, Jennifer. 2014. *Killing the Moonlight: Modernism in Venice.* Modernist Latitudes. New York: Columbia University Press.

———. 2016a. "Aeolian Harping: Materiality of Poetry in the Age of Digital Reproduction & Ecoprecarity (Part 1)." *Harriet Blog: Poetry Foundation* (blog), May 26. www.poetryfoundation.org/harriet/2016/05/aeolian-harping-materiality-of-poetry-in-the-age-of-digital-reproduction-ecoprecarity-part-1.

———. 2016b. *The Republic of Exit 43 Outtakes & Scores from an Archaeology and Pop-Up Opera of the Corporate Dump.* Berkeley: Atelos Press.

———. 2018. "Smelting Pot." In *Dimensions of Citizenship*, edited by Nick Axel, Nikolaus Hirsch, Ann Lui, and Mimi Zeiger, 132–171. New York and Los Angeles: Inventory Press.

Taj, Mitra. 2015. "Peruvian Foes of Tia Maria Copper Mine Expand Month-Long Protest." *Reuters*, April 23. www.reuters.com/article/peru-mining-protests/peruvian-foes-of-tia-maria-copper-mine-expand-month-long-protest-idUSL1N0XJ2PB20150423.

Town of Oyster Bay v. Occidental Chemical Corporation. 1997. E.D.N.Y. 987 F. Supp. 182. https://law.justia.com/cases/federal/district-courts/FSupp/987/182/1804598/.

24

THE POETIC LEXICON OF WASTE

From asarotos oikos (A) to flowers (F)

Lucie Taïeb

The project

The spaces we live in are delimited and limited by the functions we assign to them, by the knowledge we have about them, and by the internalized norms of behavior we generally comply with. Our perception of those spaces is quite unlikely to change, especially given that we eventually even stop perceiving them altogether. They become the mere scenery of our actions.

A narrative, a poem, or a monument can constitute the disruptive event that changes our perception of the environment, reactivates blurred memories, and restores continuity between the spaces we consider ours and the spaces we choose to ignore.

As a writer, I am obsessed with the ideal of a *complete* perception of space, a perception which would include forgotten memories, buried secrets, despised places, without any hierarchy. Such a perception would probably be unbearable in its comprehensiveness. My enduring belief is that literature has the capacity to extend our perception of what we consider as "real," allowing us to *experience* and therefore to face reality in its multiple, even uncomfortable aspects.

We in the West tend toward an ignorance about and a blindness to the topic of waste. In light of this, my goal is to challenge our general disregard through the elaboration of a poetic lexicon. This is a hybrid project in which I address waste-issues from different points of views and disciplinary fields, mainly geography and literary studies.

I assume that we see and read our environment through the relation we have with the words we use to name it—and thus, that words don't only "name" the world but also shape it. Thus, redefining some of our most common words (such as "love," "childhood," "flowers") may make waste present again in our mind.

Why a lexicon?

There are different types of lexica. The exhaustive ones intend to cover the entire field of a topic, answering any question that might come up. They therefore contain all you need to know—no more, no less—and claim objectivity.

The lexicon I've been working on is of another type. I haven't tried to be exhaustive or even objective. As a poet and as a researcher, I've tried and am still trying to grasp the slippery and versatile object (or has it become a subject?) of my own obsession: Waste.

The poetic lexicon of waste must therefore be incomplete, in the way both an index and a poem are. An index refers to the concepts used in an essay; it has no autonomy and gains its value from being a useful tool. In this view, the lexicon of waste is like an orphan index, separated from the corpus it refers to. Like a poem, it is full of silence, of things unsaid and unexplained, of blanks and holes the reader may try to fill up with interpretations, linking what has been set apart or simply wandering from one word to the other, knowing that alphabetical order is not actually a synonym for order but much more for randomness. For, as Monika Schmitz-Emans (2010, para. 3) points out in her *Encyclopedia of Imaginary Worlds*:

> it is decisive that the alphabetical structure of the lexicon should transmit knowledge following an alphabetical order rather than a system. Through this presentation, knowledge is given as disordered, thus allowing us to play with it. We may jump or fly from one object to the other—our imagination and comprehension are emancipated from causal and contextual links. They are thus much freer, as they don't have to operate within the framed architecture of a system of knowledge.
>
> *(translation mine, www.actalitterarum.de/theorie/mse/enz/enze02b.html).*

A

Asarotos oikos

"Asarotos Oikos" is the "unswept room" depicted by a mosaic, now lost, described by Pliny in his *Natural History*:

> Sosos laid at Pergamon what is called the *asarotos oikos* or "unswept room," because on the pavement was represented the debris of a meal, and those things which are normally swept away, as if they had been left there.
>
> *(quoted in Dunbabin 1999, 26)*

Created in the second century BCE, the mosaic became a popular genre in itself. It might be one of the most ancient representations of "litter." Contemplating the mosaic now shown in the Vatican Museum—although Sosos's has been lost, the one composed by Heraclitium for the Villa Hadriani is still preserved—we might not only admire the *trompe-l'oeil* technique or imagine Roman banquets, but also consider "waste" as the leftovers that could have been swept away but "have been left there." Thus, leftovers we usually sweep away, we could actually leave there, for a time, and look at, with neither fascination nor disgust, simply observing them in their very materiality and singularity.

The room is still the banquet room, and there is absolutely nothing dirty or repulsive about these leftovers. They represent a suspension in time: After the banquet has finished and before the room is cleaned up for a new one. The fact that this genre became extremely popular among the wealthiest Roman families deserves a study in itself.

One often believes the history of waste in our occidental cities to be a history of separation. Looking at the painting by Heraclitium, or imagining with Pliny the mosaic of Sosos, we probably won't ask ourselves who is going to clean up the mess and where the invisible cleaners will eventually dispose of those tiny bones and other fragments of a once joyful meal. But we may try to consider "a place to live" (*oikos*) or environment from which the now useless and valueless remnants of our everyday life would not always be out of our minds.

Aura

The man I am talking to is a Viennese bookseller, and I do not really think he will be able to help me. I'm looking for a book about waste: Any book, any discipline— just "about waste." He just stares at me with a vague expression of what I interpret as surprise or bewilderment. What puzzles him is actually the fact, he says, that he himself is fascinated by garbage and spends most of his free time taking pictures of dumps, or more precisely, of the green shoots that sprout up on top of trash heaps, growing and eventually giving trash back to nature. He sees in dumps such a beauty, a living beauty you would never find in a supermarket even though one can be considered as the underside of the other. But, he adds, in a supermarket, there is nothing happening, whereas a dump has this "aura of negativity."

A few years later, I am still trying to figure out where the fascination emanating from dumps takes root. When goods cease being goods and become garbage, when they have lost their value, what actually remains of them? Can the fresh and splendid shine of the "new"—yet spoiled—remain present in a way we still are able to perceive, like the traces of a vanished desire for someone we used to love? Or aren't we more mourning ourselves, contemplating the damaged goods which simply reflect the endless damages unrestrained consumption inflicts on us?

B

Bruit

Searching for synonyms of "noise" in my online dictionary, I find "bruit"—the exact word as the French *bruit*—and I try to figure out how this is pronounced in English.

Focusing on words and translation, meaning and sound, in various languages seems indispensable to me when it comes to "waste," as it is a way to enhance my understanding of the phenomenon. It makes me venture into unknown territories and sometimes find something I was not looking for or search for something I can't yet name.

Literary approaches to waste often pay close attention to the material and concretely physical aspects of their subject. Surprisingly, while smell is almost always mentioned, literary landfills appear as a much greater challenge to the viewer's sight, like some extraordinary spectacle that had until then remained unseen and is still invisible to almost everyone. The invisibility of waste might well explain the fascination it exercises upon those who eventually come to see it, offering this perfect and mysterious association of closeness and distance which, as Walter Benjamin says, defines the aura: "What is aura, actually? A strange weave of space and time: the unique appearance or semblance of distance, no matter how close it may be" (Benjamin 1931, 518).

Yet many writers also underline the "sound" of landfills, not only because sound is the ultimate expression of their ghostly presence but also because of all the silent voices buried in the toxic soil of landfills, the unheard and yet haunting rumor spreading from underneath uncountable garbage layers. Jennifer Scappettone refers in this regard to the "cast-off murmurs buried in the apparent serenity of Fresh Kills, [literalizing] the fundamental kinship between the manufacture and disposal of values in the *polis* across the waters" (Scappettone 2016, 111).

While trying to unravel the possible translations of waste-words between French, English, and German, I come again to the word "rubbish" and its two meanings: "Useless waste or rejected matter" *and* "something that is worthless or nonsensical." To explain the two concepts in French, we need at least two different words. And, as we don't have a proper word for "trash," translating "rubbish" becomes quite an unsatisfying project. Yet we do learn something from this rubbish-word: In English, there is continuity between worthlessness and trash. A thing of poor or no value is equated with a thing that may as well be discarded, whereas in French there is a clear line between the two concepts, the boundary between what we may keep even if it's rubbish and what we throw away because it's definitely trash.

How about German? Here again we find two words: *Müll*, which is the German word for trash, and *Gerümpel*, meaning a certain amount of worthless stuff (translated as "junk"). Digging a little deeper, looking for etymology, I find: banging and thudding.

What we hear, day after day, is neither the sound of trash nor the silent and unexpected trajectories of rubbish. What we do hear are words deprived of their meaning, repeated ad nauseam and themselves damaged by the public discourse of (political and economic) forces that detain the power to be heard—day after day. Keeping waste present in our minds, even while reflecting on its materiality and understanding it literally, eventually leads to a confrontation with an environment in which those who use language as a political weapon employ both the words themselves and the targets of their contempt as rubbish—as things of little value, used and thrown away, again and again, until they are ruined.

C

CACATOR CAVE MALUM (a warning)

Do not shit on the streets, for Jove's wrath may then come upon you! Whether Ancient Rome was exceptionally clean or shamefully dirty is still a matter of discussion. I assume there has always been—and probably still is—some tendency to keep the waste, excrement, corpses, and remnants of our everyday life within our reach, a tendency public policies and economic interests have always tried to thwart. Yet it's not for reasons of hygiene that shitting on Roman streets was so fiercely forbidden but because such manners would make the collection of the precious fertilizing matter much harder for the private traders who were in charge of the business. All this reminds us of the only certainty we may have in respect to waste: From the very beginnings of Western civilization, there has been an intimate bond between garbage and money.

Childhood

Childhood is an age with neither law nor hierarchy between objects, things, and matter. The leaf and the feather are picked off the floor and slipped into the pocket or put in the mouth; they now are "mine," or maybe even "me"? Nothing is dirty and nothing clean.

Here is the child, sitting by her wooden colored blocks, holding them out to you, showing them to you in her open hand and then setting them down and lining them, creating an order that exists only for her—an order she'll have such a joy destroying at the very moment the line comes to its end and there are no blocks left.

Another scene emerges from this one, another child playing, remembered not from what I've experienced, but from what I've read—or more precisely, from what I've experienced, while reading.

> Each time the boy picks up an object or pushes a truck across the floor, or adds another block to the tower of blocks growing before him, he speaks of what he is doing, in the same way a narrator in a film would speak. . . . There is no fixed center to any of this. . . . There is no law of nature that cannot be broken: trucks fly, a block becomes a person, the dead are resurrected at will.
>
> *(Auster 1988, 164)*

D

The dump

Hansine Bowe, we learn from the website of Oral History of DSNY and Freshkills projects where her interview has been archived, "was born and raised on Staten Island, NY. She grew up on a few acres of land that her parents owned on Richmond Avenue, which later became Fresh Kills landfill and is now Freshkills Park."

She recalls the site before it became a dump. Her childhood memories are fresh, and they express a freedom and insouciance one can only be nostalgic for, even if one has never experienced them. I often wonder to what extent the lost freedom of children in their own environment, the constant control they are submitted to, could or should in some way be linked with the omnipresence of waste management facilities in peripheries which used to host their playgrounds. Nevertheless, what strikes me the most in Hansine Bowe's interview is the quiet affirmation of her own memory, while the closed site of the former landfill is being transformed into a huge park and even renamed, now becoming "The Freshkills Park."

"Do you think it'll sort of always be the Dump, like 50 years from now, people at the Park?" asks the interviewer. To which she simply answers, "For Staten Islanders, older Staten Islanders, I can't speak for them, but I would think it's always going to be the Dump. The younger generation, it's Freshkills Park, if they complete it. So, I mean, to me, it's always going to be the Dump" (www.dsnyoralhistoryarchive.org/wp-content/uploads/2011/04/BoweTranscriptFinal.pdf).[1]

E

Energy

Everything that can't find a place within a social structure and thus could represent a threat for the same structure is therefore considered as "impure" and relegated to the margins. Yet the margins are not deprived of energy—quite the contrary. In fact, as Mary Douglas points out, "the idea of society is a powerful image. . . . Its outline contains power to reward conformity and repulse attack. There is energy in its margins and unstructured areas" (1966, 115).

What is relegated, because of its filthiness and dangerousness, outside a structured society, might however be converted, in the frame of religious rites, in a source of power and magic. There is always a moment in the life of primitive societies—which Douglas claims are not essentially different from the "elaborate" ones—where the impure, the filthy, and the dangerous are deeply needed because a flawless, self-sufficient order eventually becomes sterile and hostile to the uncontrolled movements of life. In such moments, the illusion of exemplary purity is revealed:

> Whenever a strict pattern of purity is imposed on our lives it is either highly uncomfortable or it leads into contradiction if closely followed, or it leads to hypocrisy. That which is negated is not thereby removed. The rest of life, which does not tidily fit the accepted categories, is still there and demands attention.

> *(Douglas 1979, 163)*

"Everything must go somewhere"[2]

Working on waste, I was first fascinated by its invisibility. This results from the fact that my first contact with the topic found place in Don DeLillo's *Underworld* (1997), which addresses in many troubling ways the invisibility and coexistent omnipresence of waste. I then tried to understand the phenomenon and again came across this issue. For the sociologist Michael Thompson waste is, indeed, "what we conspire not to see" (Thompson 1979, 88). From DeLillo's paranoid invisibility of waste (in its materiality and as a metaphor of the "hidden face" of American history) to Thompson's "conspiracy of blindness" (rooted in social structures and the wish to maintain social hierarchies), I saw a striking continuity. I realize now that all I've been trying to do while researching waste was to undo the fascination it exercised upon me, which is not only rooted in the sublime and terrifying image of enormous dumps—such as the Fresh Kills landfill described in Underworld—but is also grounded in the simultaneity, equally underlined by DeLillo, of its immensity and invisibility.

One of the ways to free my understanding from this fascination has been to turn my back to the dump and face the cities the waste actually comes from. If the invisibility of waste should be considered as the result of some trickery, then the cleanliness of our Occidental cities can be nothing but a sham. As Edd de Coverly, Lisa O'Malley, and Maurice Patterson (2008, 16) point out in their exemplary study, "the desire for a litter free environment is simply the desire for a simulation; a place where waste is not an issue and consumption is guilt-free."

Turning my back to the dumps has also given me a hint of what might be an appropriate method to tackle the problem, not only by focusing on paradoxes and contradictions—which are astonishingly numerous—but also by continuously questioning the mobility of matter and materials. If you're constantly asking yourself where stuff comes from and goes to, you end up considering waste not only as an object but also as an ongoing and global process. This is precisely the point when it becomes really exciting.

Exoticism

Le déchet, est-ce toujours l'autre?
Is waste always the other?

The lexical root of the word "waste" is identical in French and German: *Déchet* and *Abfall* both come both from "falling." The word "*déchet*" initially referred to the scraps of any matter, *les chutes*. It is what is left after something has been cut, what doesn't belong to a defined object but has to be separated from the primary stuff so that some unity appears. The leftovers are the random scraps when the dress is finished, or the inedible bones after the meat is cut.

Out of sight, out of mind, the waste, valueless and useless, gets relegated to the opaque and uncertain zones where one may yet come searching for another face of the city, apart from the green spaces and domesticated parklands so perfectly suited for a weekend stroll and harmless leisure activities of two-children families enjoying a relaxed and protected environment.

Those who are reluctant to take part in organized recreation may, while walking across abandoned areas and other wasteland, experience the "authenticity" of a relationship with the outside, full of unexpected encounters and unsolved mystery.

Yet I wonder to what extent those "authentic" experiences could be likened to some exoticism of the margin—thus, to what extent waste itself, when one attempts to give it some value again, is approached from some vaguely Romantic or aesthetic point of view and seen as the last shelter for lost authenticity.

This posture, I assume, will help us understand neither waste nor the cities we are actually living in. As a remedy for such a temptation, I suggest that we reverse the sentence: "Waste, that's us." (*Le déchet c'est nous.*) Plastic pollution unfortunately provides a good example for the relevance of this inversion. Here I am not so much thinking of the so-called spectacular "sixth continent" as I am of the microparticles lodged within in us, inhabiting our own organisms. Moreover, it shows that the invisibility at stake in waste matters is not only a problem of urban spaces but more and more a scale issue:[3] Waste, in the form of plastic pollution, is now everywhere— in the water we drink, in the milk we feed our children, in our own bodies.

F

Flowers

In *Ilha das Flores* (*Isle of Flowers*), a short and striking documentary movie from 1984 by the Brazilian film maker Jorge Furtado, we follow the trajectory of a tomato from production to consumption. Produced in the periphery of São Paulo and thrown away by a middle-class family mother who judges it unfit for her evening meal, the tomato ends up at another periphery of the megalopolis, in the open-air landfill of the so-called "Island of Flowers," where it is eventually picked up by poor children of the neighborhood. Many things could and have been said about the sobriety, the irony, and the cruelty of the movie in which there is no trace of any condemnatory rhetoric, for social injustice is dramatically shown simply by following the path of an item as banal and unspectacular as a tomato.

What calls our attention here is, among many other aspects, the very name of the dump: Island of Flowers. Such a pretty euphemism is actually almost too obvious. Has the place been renamed after becoming a dump, and were there flowers growing on it, before the dump covered and destroyed any Indigenous vegetal species? Or was the name chosen afterwards, as a shameless lie or in some ironic gesture?

Be that as it may, there is a troubling reluctance to actually name dumps "dumps" or even "landfills." As a matter of fact, these sites of "disposal" are actually places where cities don't get rid of waste (except from the incinerators, which still produce ashes) but rather try to deal with the inconveniences of its massive presence.

The French urban administration has now chosen the solution of acronyms, designating landfills and incinerators by a series of letters only insiders can decipher. As a result, the "centers" where garbage is stored, buried, or incinerated seem deprived of any materiality; they represent the perfect answer, the solution for the urban issue of dealing with waste.

One cannot experience such a place by reading the urban planning books dedicated to waste management; one cannot even imagine those places. The language of urban administrators is dry, not only void of emotion but also deprived of any image or sensitivity; it deals with risks, results, and efficiency.

This has not always been the case. Reading a handbook from 1946 (Joulot), an ode to the incinerator whose author recommends avoiding the creation of landfills as much as possible, I am surprised by the description of the dumps, which are filthy and stinky, dangerous, and crawling with rats and flies. Flies and rats have completely disappeared from the handbooks published nowadays.

Such an observation wouldn't lead us any further if the language of waste management was not pervading and influencing our perception of the phenomenon, dematerializing and depersonalizing our relation to it.

As a matter of fact, there is no objectivity in the way public policies and waste management companies deal with the issue and speak about it: The cultural representation of waste transmitted by the literature and often embedded in some fascination for its own object is actually equivalent to the myth of cleanliness and purity, the neat image of a city so clean that littering would be the only problem we should solve, as if an individual responsive behavior could be enough to counter the consequences of mass production and consumption.

Here again, in the handbooks of city management, there is an unnamed fiction at stake, a fiction we are just asked to collectively believe in, a fiction so powerful that the disbelief of a few won't prevent the others from calling it reality.

Notes

1 For more information on the topic: www.dsnyoralhistoryarchive.org.
2 One of the Four Laws of Ecology defined by Barry Commoner (1917–2012).
3 As Max Liboiron points out:

> one cannot understand waste intuitively . . . it is not only material waste that is invisible and hidden, but also the infrastructures, economics, social norms and other seemingly innocuous or unrelated systems. Our chief task is to redefine waste, scale it up, make these systems apparent. How a problem is defined makes some actions seem inevitable and sensible, while others are left unexamined. These are the stakes of discard studies.
> *https://discardstudies.com/2014/05/07/why-discard-studies/.*

References

Auster, Paul. 1988. *The Invention of Solitude*. New York: Penguin Books.
Benjamin, Walter. (1931) 1999. *Little History of Photography*, translated by Jephcott and Kingsley in *Selected Writings*, Vol. 2. Cambridge, MA and London: The Belknap Press of Harvard University Press.

De Coverly, Edd, Lisa O'Malley, and Maurice Patterson. 2008. *Hidden Mountain: The Social Avoidance of Waste*. No. 08–2003 ICCSR Research Paper Series. http://citeseerx.ist.psu. edu/viewdoc/download?doi=10.1.1.520.5008&rep=rep1&type=pdf.

DeLillo, Don. 1997. *Underworld*. New York: Scribner.

Douglas, Mary. (1966) 1979. *Purity and Danger: An Analysis of Concepts of Pollution and Taboo*. London: Routledge.

Dunbabin, Katherine M. D. 1999. *Mosaics of the Greek and Roman World*. Cambridge: Cambridge University Press.

Joulot, Antoine. 1946. *Les ordures ménagères: composition, collecte, évacuation, traitement*. [Household Trash: composition, collection, disposal, treatment]. Paris.

Pliny. 1962. *Natural History, Volume X: Books 36–37*. Loeb Classics Edition. Cambridge, MA: Harvard University Press.

Scappettone, Jennifer. 2016. *The Republic of Exit 43: Outtakes and Scores form an Archaeology and Pop-Up of the Corporate Dump*. Berkeley: Atelos.

Schmitz-Emans, Monika. Enzyklopädien des Imaginären. 2010. www.iablis.de/acta-litterarum.

Thompson, Michael. 1979. *Rubbish theory: The Creation and Destruction of Value*. Oxford: Oxford University Press.

CONTRIBUTOR BIOGRAPHIES

Maleea Acker is the author of two poetry collections, *The Reflecting Pool* and *Air-Proof Green* (Pedlar 2009, 2013), and the non-fiction book *Gardens Aflame: Garry Oak Meadows of BC's South Coast* (New Star Books, 2012). She teaches geography at the University of Victoria, where she is a PhD candidate in geopoetics.

Kerry Banazek is an assistant professor at New Mexico State University where she teaches critical writing and media studies courses. A practicing poet with web production and ecology field crew experience, her research engages with technologies of vision and their impact on rhetorics of sensation, education, and expertise.

Kimberly Blaeser (Anishinaabe) is the author of four poetry collections including *Copper Yearning* and *Apprenticed to Justice*, and editor of *Traces in Blood, Bone, and Stone*. A professor at University of Wisconsin-Milwaukee and MFA faculty member for the Institute of American Indian Arts in Santa Fe, Blaeser is a past Wisconsin Poet Laureate.

Candice Boyd is Australian Research Council DECRA fellow in the School of Geography, University of Melbourne. Her interests are in geographies of mental health, therapeutic spaces, rural youth, and contemporary museums. She is author of *Non-Representational Geographies of Therapeutic Art Making* and co-editor of *Non-Representational Theory and the Creative Arts*.

Nathan Clay is a postdoctoral researcher in the School of Geography and the Environment at Oxford University. His research and teaching focus on environmental justice, food, and climate change. He has a PhD in geography from Pennsylvania State University and bachelor's degrees in English and zoology from Michigan State University.

Patrick Clifford is a community social worker and activist based in Cincinnati, Ohio. He received his MSSA from the Mandel School of Applied Social Sciences at Case Western Reserve University where he currently teaches macro and policy practice. Books of poetry include *chaturangik/SQUARES* (with Aryanil Mukherjee) and *washpark* (with Tyrone Williams).

Tim Cresswell is a geographer and poet. He is Ogilvie Professor of Geography at the University of Edinburgh and author of over a dozen books including *Maxwell Street: Writing and Thinking Place* and the poetry collections *Soil* and *Fence*. He is joint editor of the journal *GeoHumanities*.

Sarah de Leeuw is an award-winning creative writer (poetry/literary non-fiction) and Canada Research Chair in Humanities and Health Inequities. Her activism, writing, scholarship, and teaching focus on unsettling geographies of power and the role of humanities in making biomedical and health sciences more socially accountable.

Elissa Dickson received her B.S. from the University of Michigan, majoring in French and Biology. She is the Adult Programs Librarian at the Wilkinson Public Library and is the current San Miguel County Poet Laureate. She teaches slam poetry to youth and adults, focusing on its potential for social change.

Sophie Anne Edwards is an artist/writer and geographer with a focus on fieldworks, site specific installation, and other land-based engagements. She lives on Mnidoo Mnising (Manitoulin Island) in northeastern Ontario, Canada.

Sameer Farooq is a Toronto-based artist working primarily in sculpture, installation, and video. He has exhibited widely, and reviews dedicated to his work have been included in *The Washington Post*, BBC Culture, Hyperallergic, Artnet, Canadian Art, and others.

C.S. Giscombe teaches English at the University of California's Berkeley campus. His books include *Prairie Style*, *Ohio Railroads*, and *Border Towns*. *Similarly*—collected and new poetry—will appear in 2019. His books-in-progress are *Negro Mountain* and *Railroad Sense*. He is a long-distance cyclist.

Harriet Hawkins works on the geographies of art works and art worlds. Themes include rubbish, environmental change, and the subterranean, as well as ideas of aesthetics, creativity, and the imagination. In addition to writing monographs and papers, she collaborates on artists' books, installations, participatory arts projects, and exhibitions.

Stephanie Heit is a poet, dancer, and teacher of somatic writing, Contemplative Dance Practice, and Kundalini Yoga. Her debut poetry collection, *The Color She*

Gave Gravity (Operating System 2017), explores the seams of language, movement, and mental health difference. Her ecopoetry chapbook is *Water Margins* (ReStory Nation forthcoming 2019).

Cecilie Bjørgås Jordheim (b.1981, Norway) is a visual artist, conceptual poet, and composer currently working on the translation between visual and auditive systems, concrete poetry, and the concept of isomorphia. Jordheim holds an MA in fine art from Oslo National Academy of Arts (2011). Jordheim lives in Oslo, Norway.

Petra Kuppers is a disability culture activist, a community performance artist, and Professor at the University of Michigan. She is author of the poetry collection *PearlStitch* (2016) and the queer/crip speculative story collection *Ice Bar* (2018). She lives in Ypsilanti, Michigan, where she co-creates Turtle Disco, a somatic writing space.

Angela Last is a London-based musician and DJ, and a human geographer at the University of Leicester. She has contributed to various sound projects, especially via the group *now*, and runs the experimental pop and dance label mottomotto. Since 2007, she has been writing the blog *Mutable Matter*.

Eric Magrane is an assistant professor of geography at New Mexico State University. His work takes multiple forms, from scholarly to literary to artistic. Hi is co-editor of the hybrid field guide/anthology *The Sonoran Desert: A Literary Field Guide* (University of Arizona Press, 2016).

Emily McGiffin is a postdoctoral fellow at the Institute for Advanced Studies in the Humanities at the University of Edinburgh and the author of *Of Land Bones and Money: Toward a South African Ecopoetics* (University of Virginia Press). Her third poetry collection is forthcoming from the University of Regina Press.

Urayoán Noel is a poet and associate professor of English and Spanish at New York University. His books include *Buzzing Hemisphere/Rumor Hemisférico* (Arizona), *In Visible Movement: Nuyorican Poetry from the Sixties to Slam* (Iowa), and, as editor/translator, *Architecture of Dispersed Life: Selected Poetry* (Shearsman) by Pablo de Rokha.

Craig Santos Perez is an Indigenous Chamorro poet and scholar from the Pacific Island of Guam. He is the author of four collections of poetry and the co-editor of three anthologies. He earned his PhD in ethnic studies from the University of California, Berkeley, and he currently works as an associate professor in the English department at the University of Hawai'i, Mānoa.

John Pluecker writes, translates, organizes, interprets, and creates. JP is co-founder of the language justice and aesthetic experimentation collaborative Antena Aire and the social justice interpreting collective Antena Houston; author of *Ford Over* (Noemi Press, 2016); and translator, most recently, of *Gore Capitalism* (Semiotext(e), 2018) and *Antígona González* (Les Figues Press, 2016). More info atwww. johnpluecker.com and www.antenaantena.org.

Angela Rawlings is a Canadian-Icelandic interdisciplinary artist whose books include *Wide Slumber for Lepidopterists* (Coach House Books, 2006), *Gibber* (online, 2012), *o w n* (CUE BOOKS, 2015), and *si tu* (MaMa Multimedijalni Institut, 2017). She is pursuing a PhD at the University of Glasgow. Rawlings loves in Iceland. More at http://arawlings.is.

Linda Russo's published works include *Meaning to Go to the Origin in Some Way* (Shearsman Books), *Participant* (Lost Roads Press), both poetry, and *To Think of Her Writing Awash in Light*, winner of the Subito Press Lyric Essay Prize. She co-edited *Counter-Desecration: A Glossary for Writing Within the Anthropocene* (Wesleyan University Press). She lives on the ceded lands of the Nez Perce Tribe in the inland Northwestern US and teaches creative writing and literature at Washington State University.

John Charles Ryan is a poet and interdisciplinary scholar with appointments as Postdoctoral Research Fellow at the University of New England, Australia, and Honorary Research Fellow at the University of Western Australia. His recent work includes the monograph *Plants in Contemporary Poetry* (2018) and the edited volume *Forest Family* (2018).

Jennifer Scappettone works at the crossroads of literary arts, translation, and scholarly research—on the page and off. Her books include *Killing the Moonlight: Modernism in Venice, From Dame Quickly*, and *The Republic of Exit 43: Outtakes & Scores from an Archaeology and Pop-Up Opera of the Corporate Dump*. She is Associate Professor at the University of Chicago.

Jonathan Skinner is a poet, editor, and critic, best known for founding the journal *ecopoetics*. Alongside numerous published essays in ecocriticism and poetics, his poetry collections include *Birds of Tifft* and *Political Cactus Poems*. He works in the Department of English and Comparative Literary Studies at the University of Warwick.

Jared Stanley is a poet who often works with visual artists. He is the author of three collections of poetry, *EARS* (Nightboat 2017), *The Weeds* (Salt 2012), and *Book Made of Forest* (Salt 2009). Stanley is an assistant professor of English and creative writing at the University of Nevada, Reno.

Lucie Taïeb is a translator, poet, and Assistant Professor in German studies and creative writing at the University of Bretagne. She has published several books of poetry, a novel, and an essay about the transformation of the Fresh Kills landfill (*Freshkills, Recycler la terre*, Editions Nota Bene, Montréal 2019).

Chris Turnbull is the author of *continua* (Chaudiere Books 2015) and [*untitled*] in *o w n* (CUE Books 2014). Her poetry and installation pieces can be found online, in print, and within landscapes. She curates *rout/e*, a footpress whereby poems are planted on trails: www.etuor.wordpress.com.

Diane Ward is a poet who has published 14 books of poetry. Her work has appeared in many literary magazines and anthologies. She has collaborated with visual and performing artists, musicians, choreographers, and filmmakers. She attended Corcoran College of Art and Design and earned a PhD in geography from UCLA.

Tyrone Williams teaches literature and theory at Xavier University in Cincinnati, Ohio. He is the author of several chapbooks and books of poetry. A limited-edition art project, *Trump l'oeil*, was published by Hostile Books in 2017. His website is at http://home.earthlink.net/~suspend/.

ACKNOWLEDGMENTS

First and foremost, we would like to thank all contributors to this book. From them we learned new ways of approaching geopoetics. We are grateful for the embodied and grounded work they each do in the world. We are also grateful for their patience and thoughtful consideration of our (often) multiple rounds of editorial questions, revision requests, and copyedits. Thanks goes to Angela Rawlings for the use of her image on the cover of this book. And a big shout-out to Julie Sutherland for her copyediting and proofreading brilliance.

As this project progressed from idea to book, it grew in size. We appreciate the three anonymous reviewers of our original book proposal, and are grateful to Routledge, and particularly to Faye Leerink, for working with us to produce a book of this length. Even so, for space considerations, we had to let some of the chapters proposed to us go—including chapters from three of the editors. We know that there are many other poets and geographers whose work would be suited to these pages, and that there remains more work to be done in geopoetics. Perhaps it goes without saying: We hope to have assembled a document of a vital field and look forward to new works that continue to deepen and expand it.

INDEX

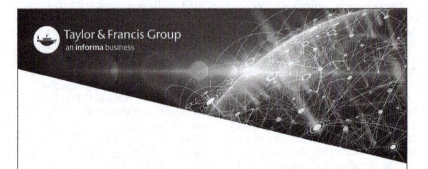